New Insights into the Genetic Regulation and Quality Improvement of Grapes

New Insights into the Genetic Regulation and Quality Improvement of Grapes

Guest Editors

Qian Zha
Meiling Tang

MDPI

Basel • Beijing • Wuhan • Barcelona • Belgrade • Novi Sad • Cluj • Manchester

Guest Editors
Qian Zha
Research Institute of Forestry and Pomology
Shanghai Academy of Agricultural Sciences
Shanghai
China

Meiling Tang
Institute of Grape
Yantai Academy of Agricultural Sciences
Yantai
China

Editorial Office
MDPI AG
Grosspeteranlage 5
4052 Basel, Switzerland

This is a reprint of the Special Issue, published open access by the journal *Horticulturae* (ISSN 2311-7524), freely accessible at: https://www.mdpi.com/journal/horticulturae/special_issues/64D25Z1RT9.

For citation purposes, cite each article independently as indicated on the article page online and as indicated below:

Lastname, A.A.; Lastname, B.B. Article Title. *Journal Name* **Year**, *Volume Number*, Page Range.

ISBN 978-3-7258-6474-4 (Hbk)
ISBN 978-3-7258-6475-1 (PDF)
https://doi.org/10.3390/books978-3-7258-6475-1

Cover image courtesy of Qian Zha

Contents

About the Editors

Qian Zha

Qian Zha is an Associate Professor at the Shanghai Academy of Agricultural Sciences, specializing in grapevine adversity physiology and molecular biology. She obtained her PhD in Pomology from China Agricultural University. Dr. Zha has published over 20 high-level papers on the high-temperature tolerance mechanism of fruit trees and cultivation technology improvement. She was awarded the "Best Poster" award at The 2nd International Electronic Conference on Horticulturae in 2025.

Meiling Tang

Meiling Tang is the director and senior agronomist of the Grape and Wine Research Institute at the Yantai Academy of Agricultural Sciences, specializing in the breeding of new grapevine varieties. She is the principal author of one book titled "Expert Q&A on High-efficiency Grape Cultivation" and has published 20 research papers. In 2013, the variety "YPG No. 1", developed through bud mutation breeding, was approved by the Shandong Provincial Crop Variety Committee; in 2014, she led the project "Research and Application of Quality Improvement and Efficiency Enhancement Cultivation Techniques for Fresh Grapes in the Yantai Production Area" as the first researcher, which won the third prize in the Yantai Municipal Scientific and Technological Progress Award.

horticulturae

Editorial

Advances in Grape Genetic Analysis, Quality Regulation, and Stress Resistance Research

Qian Zha [1,*] and Meiling Tang [2]

[1] Research Institute of Forestry and Pomology, Shanghai Academy of Agricultural Science, Shanghai 201403, China
[2] Institute of Grape, Yantai Academy of Agricultural Science, Yantai 264000, China; tmling1999@163.com
* Correspondence: zhaqian@saas.sh.cn

1. Introduction

Grapes, due to their widespread global distribution and significant commercial value, have become central research subjects in the global fruit industry. The robust development of the grape industry relies heavily on scientific research and technological support [1]. Grapevines possess diverse germplasm resources that exhibit notable variations in fruit quality, environmental adaptability, and resistance to pests and diseases. These variations provide a diverse foundation for industrial upgrading and establish higher technical demands for scientific research and exploration [2,3]. Consequently, an in-depth analysis of the genetic characteristics of various grapevine germplasms, coupled with targeted optimization of cultivation management and storage preservation techniques, is essential for establishing a high-quality grape production system. This Special Issue (New Insights into the Genetic Regulation and Quality Improvement of Grapes) invites researchers to share their research findings on the genetic regulation and quality improvement of viticulture. In order to showcase the latest scientific research achievements in the field of grapes and build an industry academic exchange platform, this Special Issue of grape research has undergone strict multiple rounds of review and finally included nine high-quality academic papers, including one review paper and eight original research papers.

2. Fundamental Research on Grapes

The breakthroughs in gene editing technology have provided precise tools for the targeted breeding of grapevines, enabling efficient improvement of key traits such as fruit quality and stress resistance, thus bypassing the labor-intensive and time-consuming processes of traditional breeding methods. In 2016, successful genome editing using CRISPR/Cas9 was first reported in grapes [4,5]; Ren et al. (contribution 1) outlined the future prospects of grapevine genome editing in model systems, precise genome editing, accelerated trait improvement, and non-transgenic genome editing, emphasizing the significant role of CRISPR/Cas in grapevine genome editing.

A systematic analysis of genetic diversity and the domestication process not only clarifies the evolutionary trajectory of grapevine germplasm resources but also provides a clear genetic map for the identification and utilization of superior genes, aiding in the innovation and selection of high-quality germplasm. Through an analysis of domestication index values, probabilities, and entropy, Fort et al. (contribution 2) found that intermediate domestication values pointed towards hybrid populations, highlighting the critical role of hybridization in the development of modern grapevine varieties. This demonstrates the significant importance of mixed populations as reservoirs of genetic diversity. Climate

https://doi.org/10.3390/horticulturae11121523

change is often regarded as the most significant challenge facing the grapevine growing industry in the 21st century. Experts are increasingly emphasizing the need to explore the biodiversity within grapevine varieties, identifying grapevine germplasm resources from the Canary Islands using Simple Sequence Repeat (SSR) markers to uncover genetic diversity (contributions 3 and 4). Additionally, reference SSR markers accepted by the global scientific community were used to identify the population structure of 'Malvasia', ensuring the authenticity of high-quality wine (contribution 5).

3. The Quality of Grape Berry

Aroma is one of the core indicators of grape fruit quality, directly determining the flavor experience of table grapes and the sensory value of wine. The related research results will lay a theoretical foundation for breeding high-aroma quality grapevine varieties and optimizing cultivation measures to enhance fruit aroma [6]. Huang et al. (contribution 6) used transcriptomic and metabolomic analysis methods to explore the berries of three grapevine varieties ("Adenauer Rose," "Mei Xiangbao," and "Kyoho") at two developmental stages, successfully identifying and quantifying various metabolites and genes related to grape aroma formation.

4. Research Directions on Stress Resistance

With global warming, high-temperature stress has become a significant factor limiting the stable development of the grapevine industry [7,8]. Analyzing the physiological and molecular regulatory mechanisms of grapes in response to high-temperature stress can provide crucial support for the development of cultivation technologies resistant to high temperatures and the breeding of high-temperature-resistant grapevine varieties. The key role of miRNAs in regulating high-temperature stress was observed by Zhang et al. (contribution 7), who found that high temperatures inhibited the expression of Vvi-miR3633a, leading to increased expression of its potential target genes Vv-Atg36 and Vv-GA3ox2. This resulted in decreased activity of the enzymes SOD and CAT, increased thermal injury, and ultimately weakened the plant's resistance to high temperatures. Biostimulants are agricultural products that contain substances capable of stimulating physiological and biochemical processes in plants, helping them adapt to various adverse environmental conditions [9]. Wu et al. (contribution 8) found that the BaZFP924 protein (patent number: ZL202110900671.3) is most effective in alleviating high-temperature stress and promoting the growth of grapes.

5. The Distribution of Diseases

The diseases in grapes are a major issue in production. Clarifying their distribution patterns and the causes of outbreaks helps to reduce the impact of diseases and pests on grape yield and quality, promoting the green and high-quality development of the industry [10]. The pathogen of downy mildew can infect all the green organs of grapevines (shoots, leaves, inflorescences, clusters) during warm and humid periods in the growing season, leading to significant losses in a short time. Nityagovsky et al. (contribution 9) identified six different amplicon sequencing variants (ASVs) of the downy mildew pathogen from metagenomic data, and through bioinformatics analysis, obtained information on potential microbial antagonists of *P. viticola*, which forms the theoretical basis for the development of biological control agents for grape downy mildew.

Overall, this Special Issue gathers cutting-edge achievements in grape research, covering the entire chain of content from basic genetic studies to industrial application technologies. The research conclusions not only hold significant academic value for basic grape research but also provide practical theoretical guidance for variety improvement, cultivation optimization, and disaster prevention in the grape industry. This has remarkable benefits for promoting the coordinated development of grape research and industry.

Author Contributions: Writing—original draft preparation, Q.Z.; writing—review and editing, M.T. All authors have read and agreed to the published version of the manuscript.

Acknowledgments: We gratefully acknowledge all authors who participated in this Special Issue.

Conflicts of Interest: The authors declare no conflicts of interest.

List of Contributions

1. Ren, C.; Mohamed, M.S.M.; Aini, N.; Kuang, Y.; Liang, Z. CRISPR/Cas in Grapevine Genome Editing: The Best Is Yet to Come. *Horticulturae* **2024**, *10*, 965. https://doi.org/10.3390/horticulturae10090965.
2. Fort, F.; Suárez-Abreu, L.R.; Lin-Yang, Q.; Asenjo, J.; Deis, L.; Canals, J.M.; Zamora, F. Origin and Possible Members of the 'Malvasia' Family: The New Fuencaliente de La Palma Hypothesis on the True 'Malvasia'. *Horticulturae* **2025**, *11*, 561. https://doi.org/10.3390/horticulturae11060561.
3. Lin-Yang, Q.; Deis, L.; Canals, J.M.; Zamora, F.; Fort, F. Uncovering the Genetic Identity and Diversity of Grapevine (*Vitis vinifera* L.) in La Palma Island (Canary Archipelago, Spain) Through SSR-Based Varietal Profiling and Population Structure Analysis. *Horticulturae* **2025**, *11*, 983. https://doi.org/10.3390/horticulturae11080983.
4. Fort, F.; Suárez-Abreu, L.R.; Lin-Yang, Q.; Deis, L.; Canals, J.M.; Zamora, F. Preliminary Clonal Characterization of Malvasia Volcanica and Listan Prieto by Simple Sequence Repeat (SSR) Markers in Free-Phylloxera Volcanic Vineyards (Lanzarote and Fuerteventura (Canary Island, Spain)). *Horticulturae* **2025**, *11*, 823. https://doi.org/10.3390/horticulturae11070823.
5. Rivera, D.; Valera, J.; Maghradze, D.; Kikvadze, M.; Nebish, A.; Ocete, R.; Ocete, C.Á.; Arnold, C.; Laguna, E.; Alcaraz, F.; et al. Heterogeneity in Seed Samples from Vineyards and Natural Habitats Along the Eurasian Vitis vinifera Range: Implications for Domestication and Hybridization. *Horticulturae* **2025**, *11*, 92. https://doi.org/10.3390/horticulturae11010092.
6. Huang, L.; Zhu, Y.; Wang, M.; Xun, Z.; Ma, X.; Zhao, Q. Integrative Analysis of Transcriptome and Metabolome Reveals the Regulatory Network Governing Aroma Formation in Grape. *Horticulturae* **2024**, *10*, 1159. https://doi.org/10.3390/horticulturae10111159.
7. Zhang, L.; Teng, Y.; Li, J.; Song, Y.; Fan, D.; Wang, L.; Zhang, Z.; Xu, Y.; Song, S.; He, J.; et al. Negative Regulatory Role of Non-Coding RNA Vvi-miR3633a in Grapevine Leaves and Callus under Heat Stress. *Horticulturae* **2024**, *10*, 983. https://doi.org/10.3390/horticulturae10090983.
8. Wu, J.; Zhong, H.; Ma, Y.; Bai, S.; Yadav, V.; Zhang, C.; Zhang, F.; Shi, W.; Abudureheman, R.; Wang, X. Effects of Different Biostimulants on Growth and Development of Grapevine Seedlings under High-Temperature Stress. *Horticulturae* **2024**, *10*, 269. https://doi.org/10.3390/horticulturae10030269.
9. Nityagovsky, N.N.; Ananev, A.A.; Suprun, A.R.; Ogneva, Z.V.; Dneprovskaya, A.A.; Tyunin, A.P.; Dubrovina, A.S.; Kiselev, K.V.; Sanina, N.M.; Aleynova, O.A. Distribution of Plasmopara viticola Causing Downy Mildew in Russian Far East grapes. *Horticulturae* **2024**, *10*, 326. https://doi.org/10.3390/horticulturae10040326.

References

1. Mostafa, S.; Wang, Y.; Zeng, W.; Jin, B. Floral scents and fruit aromas: Functions, compositions, biosynthesis, and regulation. *Front. Plant Sci.* **2022**, *13*, 860157. [CrossRef] [PubMed]
2. Khan, N.; Fahad, S.; Faisal, S.; Naushad, M. Grape Production Critical Review in the World. Available online: https://papers.ssrn.com/sol3/papers.cfm?abstract_id=3595842 (accessed on 8 February 2024).
3. Almanza-Oliveros, A.; Bautista-Hernández, I.; Castro-López, C.; Aguilar-Zárate, P.; Meza-Carranco, Z.; Rojas, R.; Michel, M.R.; Martínez-Ávila, G.C.G. Grape Pomace—Advances in Its Bioactivity, Health Benefits, and Food Applications. *Foods* **2024**, *13*, 580. [CrossRef]
4. Ren, C.; Liu, X.; Zhang, Z.; Wang, Y.; Duan, W.; Li, S.; Liang, Z. CRISPR/Cas9-mediated efficient targeted mutagenesis in Chardonnay (*Vitis vinifera* L.). *Sci. Rep.* **2016**, *6*, 32289. [CrossRef] [PubMed]

5. Malnoy, M.; Viola, R.; Jung, M.H.; Koo, O.J.; Kim, S.; Kim, J.S.; Velasco, R.; Nagamangala Kanchiswamy, C. DNA-Free Genetically Edited Grapevine and Apple Protoplast Using CRISPR/Cas9 Ribonucleoproteins. *Front. Plant Sci.* **2016**, *7*, 1904. [CrossRef]
6. Lin, J.; Massonnet, M.; Cantu, D. The genetic basis of grape and wine aroma. *Hortic. Res.* **2019**, *6*, 81. [CrossRef] [PubMed]
7. Zha, Q.; Xi, X.; He, Y.; Yin, X.; Jiang, A. Effect of Short-Time High-Temperature Treatment on the Photosynthetic Performance of Different Heat-Tolerant Grapevine Cultivars. *Photochem. Photobiol.* **2021**, *97*, 763–769. [CrossRef]
8. Zha, Q.; Xi, X.; He, Y.; Jiang, A. Transcriptomic analysis of the leaves of two grapevine cultivars under high-temperature stress. *Sci. Hortic.* **2020**, *265*, 109265. [CrossRef]
9. Du Jardin, P. Plant biostimulants: Definition, concept, main categories and regulation. *Sci. Hortic.* **2015**, *196*, 3–14. [CrossRef]
10. Koledenkova, K.; Esmaeel, Q.; Jacquard, C.; Nowak, J.; Clément, C.; Ait Barka, E. Plasmopara viticola the Causal Agent of Downy Mildew of Grapevine: From Its Taxonomy to Disease Management. *Front. Microbiol.* **2022**, *13*, 889472. [CrossRef]

horticulturae

MDPI

Review

CRISPR/Cas in Grapevine Genome Editing: The Best Is Yet to Come

Chong Ren [1,2,3], Mohamed Salaheldin Mokhtar Mohamed [2,3,4,5], Nuremanguli Aini [2,3,4], Yangfu Kuang [1,2,3] and Zhenchang Liang [1,2,3,*]

1 State Key Laboratory of Plant Diversity and Specialty Crops, Beijing 100093, China; chongr@ibcas.ac.cn (C.R.); kuangyf@ibcas.ac.cn (Y.K.)
2 Beijing Key Laboratory of Grape Sciences and Enology, Beijing 100093, China; mohamed14@ibcas.ac.cn (M.S.M.M.); nuremanguliaini22@mails.ucas.ac.cn (N.A.)
3 China National Botanical Garden, Beijing 100093, China
4 University of Chinese Academy of Sciences, Beijing 100049, China
5 Maryout Research Station, Genetic Resources Department, Desert Research Center, 1 Mathaf El-Matarya St., El-Matareya, Cairo 11753, Egypt
* Correspondence: zl249@ibcas.ac.cn

Abstract: The advent of Clustered Regularly Interspaced Palindromic Repeat (CRISPR)/CRISPR-associated (Cas) proteins as a revolutionary innovation in genome editing has greatly promoted targeted modification and trait improvement in most plant species. For grapevine (*Vitis vinifera* L.), a perennial woody plant species, CRISPR/Cas genome editing is an extremely promising technique for genetic improvement in a short period. Advances in grapevine genome editing have been achieved by using CRISPR technology in recent years, which promises to accelerate trait improvement in grapevine. In this review, we describe the development and advances in CRISPR/Cas9 and its orthologs and variants. We summarize the applications of genome editing in grapevine and discuss the challenges facing grapevine genome editing as well as the possible strategies that could be used to improve genome editing in grapevine. In addition, we outline future perspectives for grapevine genome editing in a model system, precise genome editing, accelerated trait improvement, and transgene-free genome editing. We believe that CRISPR/Cas will play a more important role in grapevine genome editing, and an exciting and bright future is expected in this economically significant species.

Keywords: CRISPR/Cas; genome editing; grapevine; trait improvement; transgene-free

1. Introduction

Though the time of the first discovery of Clustered Regularly Interspaced Palindromic Repeats (CRISPRs) dates back to 1987 [1], not until twenty years later had CRISPR been found to provide an antiviral defense in prokaryotes [2–4]. In 2012, two studies revealed that the RNA-guided CRISPR-associated protein 9 (Cas9) system could be used to cut DNA sequences [5,6]. The application of CRISPR/Cas9 gene editing was first achieved in mammalian cells in 2013 [7]. Since then, the CRISPR technology has revolutionized genome editing and has been widely used in various research areas.

Prior to CRISPR/Cas9, zinc-finger nucleases (ZFNs) and transcription activator-like effector nucleases (TALENs) had been developed to target genomic sites of interest for gene editing [8–10]. In the two systems, DNA-binding domains are fused with the FokI nuclease domain, which should form a dimer to activate the nuclease activity, thereby cutting the target DNA sequence to create a double-stranded break (DSB) [11]. However, both of the two systems rely on protein–DNA interactions, and new proteins should be engineered for targeting different sites, which is usually labor-intensive, precluding the applications of ZFNs and TALENs. In contrast, Cas9 protein is an RNA-guided site-specific nuclease and recognizes target DNA sequence by forming Watson–Crick base pairings with its guide RNA [11]. In theory, Cas9 could be programmed to target different sites just by changing the guide RNA. To date, CRISPR/Cas9 has been the primary CRISPR tool for genome editing in both animals and plants.

The applications of CRISPR/Cas9 in plants were reported in 2013 in rice [12,13], tobacco, and Arabidopsis [13]. In horticultural plants, CRISPR/Cas9-mediated genome editing was first documented in tomato and citrus in 2014 [14,15]. Grapevines are economically important perennial fruit crops widely cultivated in the world. Global viticulture is constantly challenged by climate changes and the prevalence of diseases and pests, which result in a reduction in grapevine production and berry quality. Thus, it is imperative to develop elite cultivars of grapevine with superior traits. As is known, conventional breeding, relying on hybridization across species or cultivars, is tedious and time-consuming. The emergence of CRISPR/Cas technology enables precision breeding to genetically improve grapevine traits of interest in a designed fashion, bypassing the labor-intensive and time-consuming process of conventional breeding method. Successful genome editing in grapevine by CRISPR/Cas9 was first reported in 2016 [16,17]; since then, this CRISPR system has been used as an important tool for gene functional research and trait improvement in this species. As in the other plant species, CRISPR/Cas9 is the most commonly used genome editor in grapevine, and gene knockout is still the major editing type. In addition, CRISPR activation systems based on nuclease-dead Cas9 (dCas9) had also been developed for endogenous gene activation in grapevine [18]. Recently, the base editing of grapevine genes had been successfully documented [19]. However, the great potential of CRISPR/Cas technology for grapevine genome editing has not been fully exploited so far. Here, we reviewed the currently used CRISPR/Cas systems and their derived CRISPR tools, applications of CRISPR/Cas technology in grapevine, and the challenges and future prospects in grapevine genome editing, expecting to provide an overview of recent advances and promote the applications of CRISPR/Cas in grapevine genome editing in the future.

2. CRISPR/Cas Nucleases

As mentioned above, CRISPR/Cas system functions as an adaptive immune system in prokaryotes against invading DNA or RNA molecules [2–4]. Based on the Cas proteins required for the immune response, CRISPR systems have been classified into two major classes: class 1 systems (type I, III, and IV) use multiple effector proteins to form a large complex for target cleavage, while class 2 systems (type II, V, and VI) only require one effector protein for target cleavage [20] (Figure 1). Notably, most of the CRISPR/Cas systems from both class 1 and class 2 systems are found to target DNA molecules, while several CRISPR/Cas systems are directed to target RNA sequences [20]. Therefore, according to the types of target sequences, CRISPR systems can be generally divided into DNA-targeting and RNA-targeting systems.

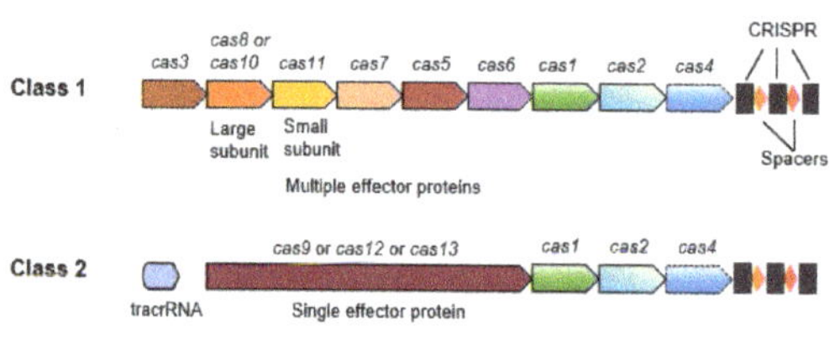

Figure 1. Illustration of the generic organizations of the class 1 and class 2 CRISPR/Cas loci. Class 1 systems have effector modules composed of multiple Cas proteins that function in protein complex during the editing. Class 2 systems have a single, multidomain effector protein that is functionally analogous to the effector protein complex of class 1. Some of the class 2 systems like Cas9 proteins require trans-acting CRISPR RNA (tracrRNA).

2.1. DNA-Targeting Cas Proteins

2.1.1. Cas9 Nucleases

The most widely used *Streptococcus pyogenes* Cas9 (SpCas9) protein is from class 2 type II CRISPR system and requires the CRISPR RNA (crRNA) and trans-activating crRNA (tracrRNA) duplex to recognize target DNA through a 20 base-pair (bp) sequence called spacer in crRNA [5,7]. In an engineered CRISPR/Cas9 system, the crRNA and tracrRNA were fused into a single guide RNA (sgRNA) [7]. The Cas9 protein is guided by sgRNA to cut the target DNA in the presence of a protospacer-adjacent motif (PAM) located immediately 3′ of the protospacer, resulting in blunt-end DSB [7]. The introduced DSBs are the most deleterious DNA damages and could be repaired by several conserved mechanisms such as non-homologous end-joining (NHEJ), microhomology-mediated end joining (MMEJ, also knowns as alternative EJ), and homology-directed repair (HDR) [21,22]. The predominant NHEJ process, for example, is to re-ligate the broken DNA ends, usually leading to nucleotide deletions or insertions (indels) at the DSB sites (Figure 2). However, NHEJ repair could also be precise, but the restored sequence could be targeted again and again by Cas9 until mutations are produced [22]. MMEJ depends on short microhomology sequences surrounding the broken sites, and increasing evidence shows that MMEJ is an active repair way in human cells [23,24]. HDR is a high-fidelity repair pathway and triggered by the presence of a donor DNA template (Figure 2). However, the efficiency of HDR is less than NHEJ and MMEJ [22].

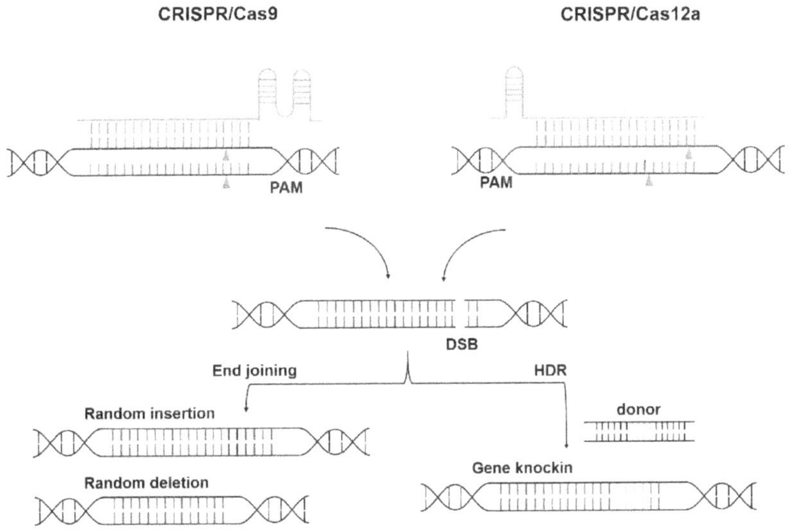

Figure 2. Genome editing induced by CRISPR/Cas9 and CRISPR/Cas12a. Both CRISPR/Cas9 and CRISPR/Cas12a generate double-stranded break (DSB), which can be repaired via end joining or the homology-directed repair (HDR) pathway, resulting in indel (random insertion or deletion) or knock-in mutations. The inserted nucleotides via end joining and desired mutations carried by donor template are indicated in red and yellow, respectively.

The native Cas9 protein contains two distinct nuclease domains, namely, a HNH nuclease domain and RuvC-like nuclease domain. The mutation of either of the two nuclease domains generates a Cas9 nickase (Cas9n), which could be used to improve editing specificity when combined with a pair of sgRNAs [25,26]. The catalytically inactivation of both nuclease domains produces dCas9 that can be repurposed for different applications, such as transcriptional regulation and epigenetic modifications [27–29].

The widely used SpCas9, consisting of 1368 amino acids, recognizes a common NGG (where N is an A, T, G, or C) or a weaker NAG PAM [30]. PAM availability limits the applications of CRISPR/Cas9 during genome editing. To expand the targeting scope of SpCas9, researchers have developed a number of Cas9 variants harboring mutations of different amino acids. For instance, the SpCas9 mutants, SpCas9-EQR, SpCas9-VQR, and SpCas9-VRER, recognize NGAG, NGA, and NGCG PAMs, respectively [31]. xCas9-3.7 exhibits higher activities on NGT and NGA PAMs than that of SpCas9 [32,33]. In addition, SpCas9-NG (targets NG PAMs), SpG (targets NGN PAMs), and SpRY (targets NRN/NYN PAMs, where Y is C or T) have successively been developed with less restrictive PAM compatibilities [34,35]. In addition to SpCas9 variants, various Cas9 orthologs have been isolated and characterized from *Staphylococcus aureus* [36], *Streptococcus thermophiles* [37], and many other organisms [38–41]. These Cas9 proteins have different overall size, guide RNA structure, and PAM requirement. For example, the Cas9 protein isolated from *Staphylococcus aureus* (SaCas9) contains 1053 aa and recognizes NNGRRT (where R is an A or G) PAMs [36]. All these Cas9 orthologs have been studied and developed as tools for genome editing in bacteria or mammalian cells. Some of them have been tested and used in plant genome editing.

Another major research focus has been the development of Cas9 variants with higher DNA specificity. As described above, paired Cas9n combined with two sgRNAs targeting opposing strands of the DNA target can reduce off-target effects [25,26]. Engineered SpCas9 variants, eSpCas9(1.1) and SpCas9-HF1, were developed with reduced off-target editing by decreasing the binding affinity between Cas9/sgRNA complex and DNA targets [42,43]. The combination of mutations used in eSpCas9(1.1) and SpCas9-HF1 makes the HeFSpCas9 variant [44]. An enhanced fidelity variant named HypaCas9 was later developed using rational design approaches [45]. The selection of a library of SpCas9 variants with different mutations in yeast and *Escherichia coli* identified the evoCas9 and Sniper-Cas9 with greater fidelity, respectively [46,47].

2.1.2. Cas12 Nucleases

Unlike Cas9, Cas12 nucleases contain a single RuvC-like nuclease domain for the DNA cleavage of both strands. Moreover, many Cas12 effectors only require a crRNA for efficient DNA cleavage. Cas12a (also named Cpf1) is the first characterized and widely used Cas12 protein for genome editing [48]. Cas12a has RNase activity and could process its pre-mature crRNAs to generate mature crRNAs. Cas12a recognizes T-rich PAMs and cuts the target regions distal to the PAM sequences [48,49] (Figure 2). Cas12a orthologs from *Acidaminococcus* sp. *BV3L6* (AsCas12a), *Francisella novicida* (FnCas12a), *Lachnospiraceae bacterium ND2006* (LbCas12a), *Eubacterium rectale* (ErCas12a), and many other organisms have been studied and applied for genome editing in plants [50–55].

Notably, Cas12a is more sensitive to temperatures than Cas9, and much effort had been made to develop Cas12a variants with lower temperature sensitivity [56–59]. Temperature tolerant LbCas12a (ttLbCas12a and LbCas12a-D156R) was first developed to display enhanced editing efficiency at a low temperature [58]. LbCas12a-RRV and LbCas12a-RVQ were also constructed by conducting saturation mutagenesis in *E. coli* [56]. Based on structure-guided protein engineering, AsCas12a-Plus was developed with increased activity and specificity [57]. Additionally, to expand the targeting scope of Cas12a, nine Cas12a orthologs were tested in plants, and an engineered Mb2Cas12a-RVRR variant was finally identified with more relaxed PAM requirements [55]. Moreover, Cas12a has also been repurposed for transcriptional regulation and base editing [51,60,61].

Recently, phylogenetic analysis revealed that TnpB and Fanzor of the OMEGA (obligate mobile element guided activity) system might be the evolutionary ancestor of Cas12, and experimental evidence demonstrated that TnpB of Deinococcus radiodurans ISDra2 and Fanzor are RNA-guided nucleases; both of them could be reprogrammed for human genome engineering applications [62,63], suggesting that TnpB and Fanzor could be used as novel systems for genome editing.

2.1.3. Base Editors

Base editors could introduce targeted point mutations without generating DSBs. There are three main types of base editors that have been developed and are currently in use: cytosine base editors (CBEs), adenine base editors (ABEs), and glycosylase base editors (GBEs) [64–66]. CBE was initially developed by fusing a rat cytidine deaminase rAPOBEC1 to the N-terminus of dCas9 [64]. The rAPOBEC1 can deaminate C into U (uracil) in the non-target DNA strand, and C-to-T substitution could be achieved through subsequent DNA repair and replication [64]. However, the G:U base pair could be detected as a mismatch by cellular base excision repair mechanism, and the resulting U is likely to be removed by uracil *N*-glycosylase (UNG), which results in a low editing efficiency of the CBE1 system [64]. Thus, an uracil DNA glycosylase inhibitor (UGI) was fused to the C-terminus of dCas9 in the CBE1 to develop the CBE2, which was tested with improved editing efficiency [64]. In the CBE3, a Cas9n (D10A) was used instead of dCas9 to generate a nick in the target strand to increase editing efficiency by promoting the cellular repair process [64]. The fourth-generation CBE, termed CBE4, was constructed by fusing two UGIs to the C-terminus of Cas9n in the CBE3, and the new CBE4 could enhance base editing efficiency and decrease the frequency of undesired C-to-A or C-to-G transversions [67].

The applications of CBEs are limited by its narrow editing window and strict PAM requirement for SpCas9. To expand the editing window, different deaminases were tested for base editing. The human AID mutant (hAID*Δ), Petromyzon marinus cytidine deaminase (PmCDA1), and human deaminase APOBEC3A (hA3A) were all successfully used to develop CBEs [68–70]. The phage-assisted continuous evolution of current deaminases is a promising way to obtain superior deaminase variants. For example, the test of a series of evolved TadA8e mutants led to the development of several TdCBEs, which enable higher editing accuracy and efficiency [71,72]. Moreover, the early versions of CBEs were developed with SpCas9, which generally recognizes NGG PAMs. To expand the targeting scope of base editors, Cas9 orthologs and variants, such as ScCas9, SpCas9-NG, xCas9, and SpRY, were used instead of SpCas9 to develop CBE variants recognizing alternative PAMs [73–75].

Similar to CBEs, ABEs consist of a Cas9n and an artificially evolved adenosine deaminase, which catalyzes the conversion of A to I (inosine). A:T-to-G:C substitutions are created through subsequent DNA repair and replication [65]. The first developed ABE7.10 was further improved by codon optimization and the addition of nuclear localization sequence (NLS) [76]. Additional NLS was added to both ends of ABE7.10 to develop the ABEmax, which could install A:T-to-G:C conversions in rice with the efficiencies ranging from 17.6% to 62.3% [77,78]. To achieve high editing efficiency, a more efficient adenine deaminase mutant named TadA8e had been evolved from TadA7.10 [79,80]. The developed ABE8e with TadA8e displayed significantly enhanced efficiency of A-to-G conversions [80]. When applied in rice, ABE8e was optimized by combining codon-optimized TadA8e with bis-bpNLS [81]. The resulting rice ABE8e (rABE8e) displayed nearly 100% editing efficiency at most tested targets [81]. Single-stranded DNA binding domain (DBD) was fused with TadA8e to develop high-efficiency ABE (PhieABE) toolbox based on the SpCas9 nickase variants SpCas9n, SpGn, and SpRYn [82]. Among these PhieABEs, hyper ABE8e-DBD-SpRYn (hyABE8e-SpRY) was tested with extremely high editing efficiency, and a high proportion of homozygous base substitutions were also detected [82]. Two mutations (V82S and Q154R) introduced into TadA8e resulted in the generation of TadA9, which expands the editing window in rice when compared to the previous version [83].

Both CBEs and ABEs catalyze only base transitions (C-to-T and A-to-G). GBEs were developed to produce base transversions. GBEs are composed of a Cas9n, a cytidine deaminase, and a UNG [66,84]. As mentioned before, the U base created by cytidine deaminase could be excised by UNG, resulting in the formation of an apurinic/apyrimidinic (AP) site that induces the DNA repair process [84]. GBEs containing AID-Cas9n-UNG and rAPOBEC1-Cas9n-UNG had been successfully used in *E. coli* and mammalian cells, respectively [84]. An optimized GBE was also generated to achieve C-to-G editing in rice [85]. The combination of an ABE with hypoxanthine excision protein N-methylpurine DNA glycosylase makes the adenine transversion base editor, AYBE, for A-to-Y base editing in mammalian cells [86]. Furthermore, base editors for G-to-Y conversion had also been developed recently [87]. However, the feasibility of these novel base editors in plants remains to be tested in the future.

It is worth noting that multiplex base editing could be accomplished by using different deaminases simultaneously. For instance, a dual adenine and cytosine base editor (A&C-BEmax) was developed by fusing cytosine and adenine deaminases with a Cas9n to achieve C-to-T and A-to-G conversions [88]. Simultaneous and wide editing induced by a single system (SWISS) based on CRISPR/Cas9 was also developed to induce C-to-T and A-to-G substitutions [89]. Likewise, a new dual deaminase-mediated base editor, AGBE, was developed by fusing CGBE (a GBE) with ABE for inducing four types of base conversions (C-to-G, C-to-T, C-to-A, and A-to-G) simultaneously in mammalian cells [90].

2.1.4. Prime Editors

Prime editors (PEs), comprising a Cas9n (H840A) and an engineered reverse transcriptase, can install all 12 possible types of base substitutions, small deletions, and insertions in a precise manner [91]. PEs are guided by an engineered prime editing guide RNA (pegRNA), which contains a spacer for targeting the specific site and an extension carrying the desired edit and a primer binding sequence (PBS) [91]. Once binding to the target, the Cas9n generates a nick at the non-target strand, and the PBS in the pegRNA could hybridize with the 3′ end of the nicked DNA strand to prime reverse transcription to install the carried mutation into the genomic DNA through DNA repair and replication [91]. Several generations of PEs have been developed and characterized. The first developed PE1 is a fusion of Cas9n and Moloney murine leukemia virus (M-MLV) reverse transcriptase (RT). Then the wild-type M-MLV RT in PE1 was replaced by an engineered pentamutant M-MLV RT to generate PE2. Compared to PE2, PE3 harbors an additional sgRNA for nicking the non-edited strand to increase the editing efficiency [91]. However, a high frequency of undesired indel products were detected with PE3, and a PE3b system was therefore developed to reduce unwanted indel mutations [91]. Later on, PE4 and PE5 were developed by fusing an endonuclease-impaired MLH1 protein to the C-terminus of PE2 and PE3, respectively, with an enhanced editing efficiency by an average of 7.7- and 2.0-fold in mammalian cells [92].

The design of pegRNA is a predominant determinant of prime editing efficiency. pegRNA parameters such as the length of PBS and RT template, GC content, secondary structures within pegRNA, and pegRNA stability should be considered [21,93]. In general, efficient PBSs are 8–15 nt in length, while the length of RT templates falls in a range from 10 to 20 nt [21]. A study in rice showed that the optimal melting temperature of PBS was 30 °C [94]. Low GC content generally requires longer PBS to ensure efficient annealing with the nicked non-target strand [91]. Moreover, to avoid undesired base pairing, the last base of RT template should not be a cytidine, which could pair with G81 in the pegRNA scaffold [91]. Modifying the secondary structures of pegRNA is an efficient way to increase prime editing efficiency [95,96]. The instability of pegRNA is thought to be a factor affecting prime editing efficiency because the 3′ end extension of pegRNA is susceptible to exonucleolytic degradation in cells [97]. To enhance the stability of pegRNAs, RNA motifs like evopreQ1 and mpknot were added to the 3′ end of pegRNAs to protect them from degradation, thereby improving the editing efficiency of by 3- to 4-fold in mammalian cells and plants [97]. Recently, several robust methods based on PEs, such as PRIME-Del [98], twinPE [99], PEDAR [100], and GRAND editing [101], were reported to achieve precise large DNA fragment deletions, insertions, or replacement.

2.2. RNA-Targeting Cas Proteins

2.2.1. Cas13 Nucleases

Cas13 nucleases belong to Class 2 type VI CRISPR systems and are known to exclusively cleave RNA molecules [102]. All Cas13 nucleases have RNase activity for pre-crRNA processing and two HEPN (Higher Eukaryotes and Prokaryotes Nucleotide-binding) domains required for RNA degradation [102]. Upon binding to the targets, Cas13/crRNA complexes exhibit conformational changes and cleave both the target RNA molecules (*cis*-cleavage) and bystander RNAs (*trans*-cleavage) [103]. Unlike DNA-targeting Cas nucleases, Cas13 effectors appear not to rely on consensus PAMs; some of Cas13 members prefer a specific sequence termed the protospacer flanking sequence (PFS), which is located just 3′ of the crRNA complementary sequence of the target RNA [103–105]. Cas13 systems have been employed as powerful tools for RNA degradation (gene knockdown) [103,104], RNA editing [106], nucleic acid detection [107], and so on.

Cas13a (formerly C2c2) is the most widely studied Cas13 effector. In addition, Cas13b, Cas13c, Cas13d, and the other Cas13 proteins have also been successively identified and characterized [108]. As for applications of Cas13 systems in plants, CRISPR/Cas13-based RNA interference could protect plants against RNA virus infections. The CRISPR/Cas13-based RNA editing in plants has been reviewed by Kavuri et al. recently [109]. However, the application of CRISPR/Cas13 in plants is still limited.

2.2.2. RNA-Targeting Cas9

Previous studies showed that SpCas9 can also bind and cleave single-stranded RNA [110,111], even though the exact mechanism underlying this action of SpCas9/sgRNA remains unknown. In addition, various Cas9 proteins from other bacteria have been characterized for RNA targeting (RCas9) [108]. For example, the *Francisella novicia* Cas9 (FnCas9) was capable of inhibiting the infection by the hepatitis C RNA virus in human cells by using a small CRISPR-associated RNA [112]. Nevertheless, most of the found RCas9, such as *Staphylococcus aureus* Cas9 and *Campylobacter jejuni* Cas9, were tested *in vitro* [113,114], and their activities for RNA editing *in vivo* have not been reported yet.

3. Genome Editing in Grapevine

Since the efficacy of CRISPR/Cas9 system was demonstrated in 2016, more and more reports on grapevine genome editing by using this CRISPR system have been released in recent years. The optimization of CRISPR/Cas9 was performed to improve the editing efficiency in grapevine. Additionally, several other CRISPR/Cas systems were also tested and applied in grapevine (Figure 3). In summary, the CRISPR/Cas9 and CRISPR/LbCas12a systems are available now for gene knockout in grapevine, while RNA targeting effector like Cas13a has also been developed for RNA editing. Elevated gene expression could be achieved by using CRISPR/dCas9-mediated gene activation. Precise point mutations are expected to be accomplished by using CBE or PE.

3.1. Proof-of-Concept Studies on CRISPR/Cas9

In early 2016, the availability of suitable target sites for CRISPR/Cas9 was thoroughly analyzed in grapevine (*V. vinifera*) genome, and over 7 million highly specific potential targets were found to distribute uniformly in grape genome [115]. Interestingly, the coding regions of grape genes have the highest abundance of predicted target sites [115]. This study showed that CRISPR/Cas9 could be theoretically applicable to grapevine, and gene coding regions might be suitable regions for genome editing. Later on, the efficacy of CRISPR/Cas9 system was experimentally demonstrated by targeting the *L-idonate dehydrogenase* (*IdnDH*) gene in Chardonnay [16]. Successful editing was achieved in both grape calli and regenerated plants with an efficiency as high as 100%, and the targeted mutagenesis of *IdnDH* resulted in a lower tartaric acid content in grape calli [16]. This was the first report on CRISPR/Cas9-mediated genome editing in grapevine, demonstrating the effectiveness of CRISPR/Cas9 in this species. In late 2016, DNA-free genome editing by using purified CRISPR/Cas9 ribonucleoproteins (RNPs) was reported in Chardonnay protoplasts, and the susceptible gene *MLO7* involved in powdery mildew resistance was successfully targeted, even though the editing efficiency was as low as 0.1% [17]. Based on the experimental methods from the two studies [16,17], a detailed protocol, including plasmid-mediated genome editing and CRISPR/Cas9 RNP-mediated genome editing, was established for grapevine [116]. The development of the protocol, which can be referred to during grapevine genome editing, is expected to prompt applications of CRISPR/Cas9. Subsequently, a research group from Japan reported the editing of *V. vinifera phytoene desaturase* (*VvPDS*) gene in Neo Muscat by using CRISPR/Cas9 system, with an efficiency ranging from 2.7% to 86% [117].

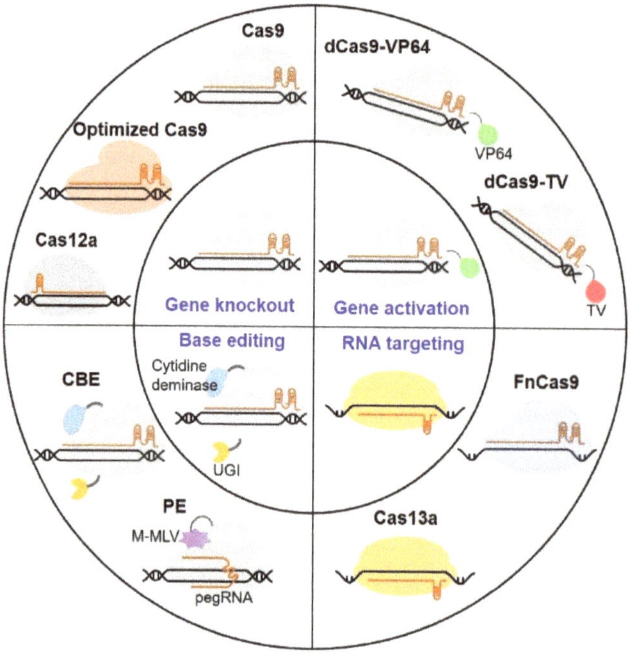

Figure 3. Current CRISPR editing in grapevine. CRISPR/Cas9 (including optimized Cas9 system) and CRISPR/LbCas12a have been employed for gene knockout. CRISPR activation systems based on nuclease-dead Cas9 (dCas9), namely, dCas9-VP64 and dCas9-TV, have been developed for gene activation. Cytosine base editor (CBE) and prime editor (PE) have been used for base editing. RNA-targeting effectors, FnCas9 and LshCas13a, have also been reported in grapevine.

3.2. Applications of CRISPR/Cas9 in Grapevine

CRISPR/Cas9 is efficient in inducing the targeted mutagenesis of genes of interest, resulting in improved traits in grapevine. Most of the reported applications in grapevine mainly focus on the improvement of disease resistance (Table 1). A transcription factor gene *VvWRKY52* was edited by CRISPR/Cas9 in Thompson Seedless, and 4 out of 22 independent *wrky52* mutants were tested with enhanced resistance to *Botrytis cinerea* [118]. Two grape *MLO* genes, *VvMLO3* and *VvMLO4*, were targeted in Thompson Seedless by CRISPR/Cas9, with the editing efficiency ranging from 0% to 38.5% [119]. Four *VvMLO3*-edited grapevine plants showed enhanced resistance to powdery mildew [119]. In a study reported in 2021, four putative grape susceptibility genes, namely, *auxin induced in root culture 12* (*AIR12*), *sugars will eventually be exported transporter 4* (*SWEET4*), *lesion initiation 2* (*LIN2*), and *dimerization partner-e2f-like 1* (*DEL1*), were successfully edited in Thompson Seedless, respectively [120]. These edited grapevine plants were subjected to *Erysiphe necator* and *Botrytis cinerea*, and a *DEL1*-edited plant displayed over 90% reduction in symptoms induced by powdery mildew infection [120]. To investigate the function of a *pathogenesis-related protein 4b* (*VvPR4b*) gene in grapevine resistance to downy mildew, the *VvPR4b* gene was knockout in Thompson Seedless by CRISPR/Cas9 system, and the loss of function of *VvPR4b* increased the susceptibility of edited plants to the oomycete pathogen *Plasmopara viticola* [121]. In contrast, the knockout of the *downy mildew resistant 6* (*DMR6*) gene, which encodes a negative regulator of plant immunity, could reduce the susceptibility of grapevine to downy mildew [122,123]. Grapevine contains two copies of *DMR6*: *VvDMR6-1* and *VvDMR6-2*. *VvDMR6-1*-edited biallelic and chimeric plants of 41B rootstock (*V. vinifera* × *V. berlandieri*) exhibited reduced growth as well as susceptibility to *P. viticola* [122]. In another study, Giacomelli et al. developed *dmr6-1*, *dmr6-2*, and *dmr6-1_2* mutants of two different grapevine cultivars and tested their resistance to *P. viticola* by inoculating *P. viticola* to the detached leaf discs of young plants. The expected reduction in susceptibility was only detected in Crimson Seedless but not in Sugraone [123]. Further evaluation with the support of a robust statistical method showed that double mutant *dmr6-1_2* always displayed a significant reduction in susceptibility to *P. viticola* in both cultivars as compared to wild-type plants [123]. Elevated levels of salicylic acid were detected in *dmr6* mutants in both of the two studies [122,123], suggesting that the enhanced resistance to downy mildew may be associated with increased salicylic acid content.

Table 1. CRISPR/Cas genome editing in grapevine.

Effector	Target	Explant	Delivery Method	Editing Type	Trait	Phenotype	Reference
SpCas9	*IdnDH*	EC of Chardonnay	*Agrobacterium tumefaciens*	KO	Tartaric acid synthesis	Decreased TA content	[16]
	MLO7	Protoplasts from Chardonnay EC	PEG	KO	Powdery mildew resistance	ND	[17]
	VvPDS	EC of Neo Muscat	*Agrobacterium tumefaciens*	KO	Carotenoid biosynthesis	Albino phenotype	[117]
	VvPDS	EC of Chardonnay and 41B	*Agrobacterium tumefaciens*	KO	Carotenoid biosynthesis	ND	[124]
	VvPDS	Protoplasts from EC of 101-14, Cabernet Sauvignon, Chardonnay, Merlot, Thompson Seedless, Colombard, GRN1, *V. arizonica*, and Pixie	PEG	KO	Carotenoid biosynthesis	ND	[125]
	VvPDS	EC of Nebbiolo	PEG and lipofectamines	KO	Carotenoid biosynthesis	Albino phenotype	[126]
	VvWRKY52	EC of Thompson Seedless	*Agrobacterium tumefaciens*	KO	*Botrytis cinerea* resistance	Enhanced resistance	[118]
	CCD8	EC of 41B	*Agrobacterium tumefaciens*	KO	Shoot branching	Increased branches	[127]
	VvPR4b	EC of Thompson Seedless	*Agrobacterium tumefaciens*	KO	Downy mildew resistance	Decreased resistance	[121]
	VvMLO3, *VvMLO4*	EC of Thompson Seedless	*Agrobacterium tumefaciens*	KO	Powdery mildew resistance	Enhanced resistance	[119]
	TAS4, *MYBA7*	EC of rootstock 101-14	*Agrobacterium tumefaciens*	KO	Anthocyanin accumulation	No accumulation	[128]

Table 1. *Cont.*

Effector	Target	Explant	Delivery Method	Editing Type	Trait	Phenotype	Reference
SpCas9	*PDS* *TMT1, TMT2*	EC of 41B	*Agrobacterium tumefaciens*	KO	Carotenoid biosynthesis sugar accumulation	Albino phenotype Decreased sugar content	[129]
	VvAIR12, VvSWEET4, VvLIN2, VvDEL1	EC of Thompson Seedless	*Agrobacterium tumefaciens*	KO	*Botrytis cinerea* resistance	Enhanced resistance	[120]
	VvPLATZ1	EC of microvine, genotype 04C023V0006 (H/H)	*Agrobacterium tumefaciens*	KO	Female flower morphology	Reflex stamens	[130]
	VvbZIP36	EC of Thompson Seedless	*Agrobacterium tumefaciens*	KO	Anthocyanin accumulation	Increased anthocyanin content	[131]
	VvDMR6	EC of Crimson seedless and Sugraone	PEG	KO	Downy mildew resistance	ND	[132]
	VvMLO6				Powdery mildew resistance	ND	
	VvMYBA1	EC of Chardonnay EC of Shine Muscat	*Agrobacterium tumefaciens*	KO	Anthocyanin accumulation	ND ND	[133] [134]
	VvEPFL9-1	EC of Sugraone	*Agrobacterium tumefaciens*	KO	Stomata formation	Reduced stomatal density	[135]
	VvDMR6-1	EC of 41B	*Agrobacterium tumefaciens*	KO	Downy mildew resistance	Enhanced resistance	[122]
	GFP	Protoplasts from Thompson Seedless EC	PEG	KO	GFP fluorescence	Loss of GFP fluorescence	[136]
	VvDMR6-1, VvDMR6-2	EC of Crimson seedless and Sugraone	*Agrobacterium tumefaciens*	KO	Downy mildew resistance	Enhanced resistance	[123]
zCas9i	*LysM receptor-like kinase gene (Vitvi05g00623)*	EC of Chardonnay	*Agrobacterium tumefaciens*	KO	Immune response	ND	[137]
LbCas12a	*TMT1, TMT2, DFR1*	EC of 41B	*Agrobacterium tumefaciens*	KO	Sugar accumulation flavonoid accumulation	Altered sugar and flavonoid contents	[138]
PE	*VvDXS1*	EC of Scarlet Royal	*Agrobacterium tumefaciens*	Base editing	Monoterpenes biosynthesis	Increased monoterpenes content	[19]

EC, embryogenic calli; KO, gene knockout; ND, not detected.

In addition to the studies on pathogen resistance, the targeted mutagenesis of grape genes involved in plant development and secondary metabolite accumulation was also reported. The biosynthesis of strigolactones, which play a key role in controlling axillary bud outgrowth, is affected by two enzymes: carotenoid cleavage dioxygenase 7 (CCD7) and CCD8. Grapevine *CCD7* and *CCD8* were knocked out in 41B by using CRISPR/Cas9, and the obtained *ccd8* grapevine mutants exhibited increased shoot branches [127]. Epidermal patterning factor like 9 (EPFL9) induces stomata formation in vascular plants, and the targeted mutation of *VvEPFL9-1*, one of the two grape *EPFL9* genes, resulted in reduced stomatal density in the edited Sugraone plants [135]. Intrinsic water-use efficiency was also improved in *epfl9-1* mutants as compared to the wild-type control [135]. The CRISPR/Cas9-mediated editing of *plant AT-rich sequence- and zinc-binding protein1* (*VvPLATZ1*) in a hermaphrodite genotype showed that homozygous, edited lines produced flowers with reflex stamens, suggesting a role for VvPLATZ1 in controlling female flower morphology in grapevine [130]. A grapevine bZIP family gene named *VvbZIP36* was edited in Thompson Seedless, and corresponding mutants with monoallelic mutations were created with the increased accumulation of anthocyanins in leaves [131]. Moreover, the *trans-acting small-interfering locus4* (*TAS4*) and *MYBA7* genes are thought to be involved in pathogen-induced anthocyanin accumulation, and grapevine mutants of the two genes were developed in the rootstock 101-14, but no visible anthocyanin accumulation was observed in these mutants [128].

3.3. Optimization of CRISPR/Cas9 for Grapevine Genome Editing

The function of CRISPR/Cas9 system relies on two components: sgRNA and Cas9. The efficient cleavage of the target is associated with the expression of sgRNA and *Cas9* in cells. During the editing of *IdnDH* gene in grape calli and plants, the edited lines exhibited a higher expression level of sgRNA as compared to the control [16]. Previous studies showed that increasing sgRNA expression could improve CRISPR/Cas9 genome editing [139,140]. To increase the expression of sgRNA and *Cas9*, four VvU3 and VvU6 promoters and two ubiquitin (UBQ) promoters were identified and amplified from grape genome; the use of grape promoters significantly promoted the expression of sgRNA and *Cas9* in grape, thereby resulting in a higher editing efficiency [129]. Furthermore, CRISPR/Cas9 editing efficiency was surveyed with different parameters: sgRNA GC content, different varieties, and the expression level of *Cas9* [124]. The results showed that genome editing in 41B was more efficient than in Chardonnay, and sgRNA with a high GC content (50–65%) yielded higher editing efficiency independent of the grape varieties [124]. The *Cas9* expression level also had an effect on editing efficiency [124,137]. In a recent study, a maize-codon-optimized Cas9 containing 13 introns (zCas9i) was used to achieve up to 100% biallelic mutations in Chardonnay [137]. Another effective method is using geminivirus-derived vectors for the expression of sgRNA and *Cas9*. After transformation, a large number of geminivirus replicons could be produced through rolling-circle replication in plant cells, so as to greatly improve the expression level of sgRNA and *Cas9* [141]. Luckily, the commonly used bean yellow dwarf virus (BeYDV) had been successfully modified and used for grape genome editing [120], which provides an alternative to common vectors during grapevine genome editing.

3.4. Multiplex Genome Editing in Grapevine

An advantage of CRISPR/Cas9 system over ZFNs and TALENs is that it is easier to conduct multiplex genome editing. In general, multiple sgRNAs driven by different U3 or U6 promoters could be simply stacked in one CRISPR vector for achieving multiplex genome editing [142]. In addition, self-cleaving ribozymes (RZ), tRNA, and Csy4 have also been employed to produce different sgRNAs from a single transcript array [143,144]. In fact, in several studies, researchers used over two sgRNAs to target the same gene to guarantee efficient cleavage during the editing in grapevine [118,119,137]. Importantly, the use of multiple sgRNAs could likely generate large DNA fragment deletion, which is exemplified recently by the removal of a 10-kb *Gret1* transposon from *VvMYBA1* promoter in grapevine [133,134]. Moreover, in addition to the multiplex genome editing system based on multiple sgRNA expression cassettes, the polycistronic tRNA-sgRNA cassette (PTG) was also reported to target the grape *tonoplastic monosaccharide transporter 1* (*TMT1*) and *TMT2* genes simultaneously, resulting in reduction in the contents of maltose, glucose, and fructose [129].

3.5. CRISPR/dCas9-Mediated Gene Activation

Both loss-of-function and gain-of-function mutations are necessary for gene functional study. The native Cas9 protein is efficient in generating loss-of-function mutations, whereas dCas9 can be used to develop CRISPR activation (CRISPRa) systems for gene activation. Two major strategies are commonly used to develop CRISPRa systems. One is the fusion of transcriptional activation domains (TADs) to dCas9, and the other is modifying sgRNA scaffold to recruit TADs [145]. The grape CRISPRa systems were developed by using dCas9-VP64 (four tandem repeats of the Herpes simplex viral protein 16) and dCas9-TV (six copies of TALEs and two copies of VP64) fusion proteins, and their efficiency in gene activation was tested by targeting *UDP-glucose f lavonoid glycosyltransferases* (*UFGT*) and *C-repeat binding factor 4* (*CBF4*) genes in grapevine [18]. The effectiveness of the dCas9-VP64 system in the transcriptional activation of *UFGT* gene was limited to 1.6- to 5.6-fold in grape cells; in grapevine plants, the expression of *CBF4* was activated by 19.3- to 42.3-fold [18]. Notably, the genetic manipulation of gene promoters has emerged as a robust approach for altering gene expression in crops [146]. Promoter editing can be performed in a designed manner or be exploited to introduce random mutations to generate novel genetic variations with altered expression patterns [146]. Additionally, the genome editing of upstream open reading frames (uORFs) has also been demonstrated to be an effective way to regulate mRNA translation [147]. All these strategies could be tested and applied in grapevine for the regulation of gene expression in the near future.

3.6. Base Editing in Grapevine

Increasing evidence suggests a relationship between single nucleotide polymorphisms (SNPs) and grapevine traits [148,149]. The modification of single nucleotide has great potential in grapevine genome editing. An initial attempt using CBE in grapevine was made by targeting *gibberellin insensitive1* (*GAI1*) in 41B [150]. Though successful C-to-T substitutions were achieved, the editing efficiencies were only 2.4–15% [150]. Moreover, PE was adopted for the creation of a single-point substitution of a lysine with an asparagine at position 284 in *VvDXS1* recently, and most of the edited plants were identified with one allele being successfully edited [19]. These edited grapevine plants had higher contents of monoterpenes in their leaves than the control [19]. The applications of base editors and PEs in grapevine remain to be reinforced in the future.

3.7. Other CRISPR/Cas Systems Used in Grapevine

To expand the editing scope in grapevine, the CRISPR/LbCas12a was developed and applied for the targeted mutagenesis of *TMT1* and *dihydrofavonol-4-reductase 1* (*DFR1*) genes in 41B [151]. Short-term heat treatment could improve the performance of CRISPR/LbCas12a in genome editing, and truncated crRNAs with 20 nt guide sequences were efficient enough to induce targeted mutagenesis as original crRNAs with 24 nt guides [151]. The knockout of *DFR1* gene by CRISPR/LbCas12a led to the alteration in flavonoid accumulation in *dfr1* mutant cells [151]. Moreover, a method based on CRISPR/Cas12a had been developed for the detection of grapevine red-blotch virus (GRBV) [138], which could be deployed for the rapid detection of viral infections in the vineyard. The CRISPR/xCas9 system was also tested by targeting 12 designed targets with different PAMs within *GAI1*, *TMT1*, and *CCD8*. The results showed that no desired mutations were detected except for the first target of *GAI1* (GAI1-g1) with TGG PAM, but the editing efficiencies were less than 1.89% [150]. The optimization of the CRISPR/xCas9 system should be conducted prior to further applications in grapevine.

RNA-targeting Cas effectors, CRISPR/FnCas9 and CRISPR/LshCas13a, have been used to interfere with grapevine leafroll-associated virus 3 (GLRaV-3) in grapevine plantlets [152]. The transient expression of CRISPR/FnCas9 and CRISPR/LshCas13a reagents in *in vitro* Cabernet Sauvignon plantlets inhibited viral accumulation, and LshCas13a outperformed FnCas9 in virus inhibition [152]. This study provides novel antiviral strategies and serves as a successful example for the applications of RNA-targeting Cas effectors in grapevine.

3.8. Online Tools for Guide RNA Design

Early efforts to identify potential targets for Cas9 in grape genome gave birth to the Grape-CRISPR, which is a database containing all the predicted protospacers and detailed information about the target sites [115]. An upgraded version named Plant-CRISPR was developed later by incorporating 138 plant genomes and potential protospacers for Cas9 and Cas12a editing [153]. The target design and test of the effectiveness of designed guide RNA are supported by the web tool of Plant-CRISPR [153]. In addition, other CRISPR websites such as CRISPR-P v2.0 [154] and CRISPR-GE are also good online tools for target design [155].

4. Challenges for Grapevine Genome Editing

Though a number of encouraging studies have been reported in recent years, grapevine genome editing is still facing challenges that restrict the extensive applications of CRISPR technology in this species.

4.1. Lack of Efficient System for Testing Cas Effectors

For a novel CRISPR/Cas system, its editing efficacy should be tested before further applications. However, CRISPR/Cas9 genome editing conducted with grape embryogenic calli requires ~9 months to obtain transgenic calli, and the experimental period would be prolonged to at least 12 months to get regenerated plants [118] (Figure 4). Thus, a rapid and efficient system for testing Cas nucleases in grapevine is important. In rice, for example, protoplasts have been used as an efficient system to verify the functions of CRISPR/Cas systems, with the transfection rates reaching as high as 90% [156,157]. In contrast, the transfection rates of grape protoplasts with CRISPR/Cas9 RNPs were ~20% [136,158], and successful editing was tightly related to the amount of RNPs [136]. In the first report on CRISPR/Cas9 editing in grape protoplasts, the editing efficiency was only 0.1% [17]. These results suggest that the current transfection method of grape protoplasts is not suitable for the evaluation of the editing efficiencies of CRISPR/Cas systems, given that the editing mediated by CRISPR/Cas systems with low activities may not be detected in grape protoplasts.

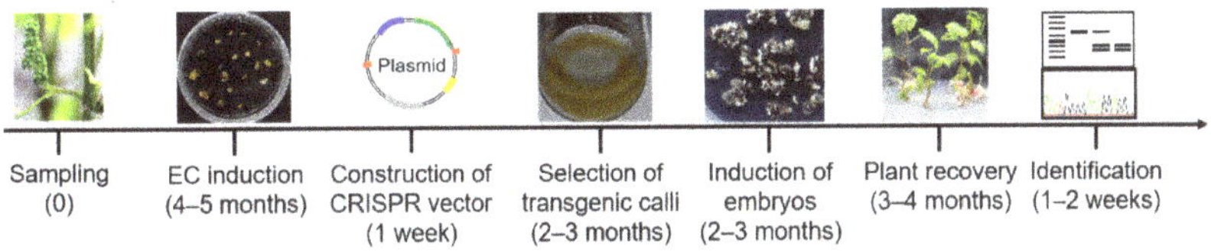

Figure 4. Time schedule of Agrobacterium-mediated CRISPR editing in grapevine.

As with grapevine, the plant transformation of soybean is also laborious and time-consuming, and soybean hairy root system is therefore developed as an efficient and rapid research tool for assessing genome editing efficiency within 14 days [132,159]. Importantly, the editing efficiency detected in soybean hairy roots is totally comparable to that detected in stable transformed plants [159]. For grapevine, hairy root cultures had been used to produce secondary metabolites, such as resveratrol and cyanogenic glucoside [160–162]. However, the use of hairy roots for grapevine genome editing has not been reported yet. According to the previous results, the transformation efficiency in grape hairy roots could be further improved [162,163] if this system is applied to genome editing in the future.

4.2. Limited Available Explants

Almost all the studies on grapevine genome editing are conducted using embryogenic calli (EC) or protoplasts isolated from EC as the explants, which are expected to generate stable grapevine plants via somatic embryogenesis. EC could be induced from anthers, ovaries, filaments, or whole flowers [117,164,165]. Nevertheless, EC are only successfully induced in limited varieties, such as Chardonnay [16,165], Thompson Seedless [118,119], Crimson Seedless [123], and Sugraone [123,135]. Furthermore, the induction efficiency was usually low and varied in different cultivars [166,167]. Meanwhile, the frequency of EC induction changed according to the adopted induction media [165]. In theory, it is better to establish an optimal system for EC induction for a given grape variety to get plenty of EC more easily.

4.3. Delivery of CRISPR/Cas Reagents and Plant Regeneration

A prerequisite for successful editing is to deliver CRISPR/Cas reagents into plant cells. The lack of efficient transformation system is a bottleneck that restricts the applications of CRISPR systems in grapevine. Some proven genetic transformation methods, including *Agrobacterium* infection, particle bombardment, and polyethylene glycol (PEG) or lipofectamine-mediated delivery, have been successfully applied in grapevine. The *Agrobacterium tumefaciens*-mediated transformation of EC is the commonly used method for grape transformation. This approach is cheap and user-friendly, but the transformation rate, however, is generally not high. Efficient transformation requires a relatively long antibiotic-dependent selection process to enrich stably transformed cells from a pool of untransformed cells [118,127,129]. To facilitate the selection of transformed plants, fluorescent markers like EGFP and DeRed2 were used to help to enrich transformed cells during the selection process [137,151]. The biolistic method can deliver biomolecules to a wide range of plant materials, but the transformation rate is low, and plant tissue is often damaged under high bombardment pressures [126,168]. The high cost and requirement of specialized equipment also restrict its application. Though the biolistic transformation of grapevine had previously been reported [169,170], this method is now rarely used during grapevine transformation. As discussed above, CRISPR/Cas reagents can also be delivered into plant cells by protoplast transfection [17,136,158]. PEG is commonly used to mediate the delivery of RNPs [17,136,158]. Very recently, lipofectamine-mediated protoplast transformation was reported in grapevine, and edited grapevine plants were obtained 5 months after the transfection [167]. The biggest advantage of PEG/lipofectamine-mediated delivery is its support for DNA-free genome editing. However, this delivery method is high skilled, and the stability of protoplasts after transfection is important for efficient transformation. Unfortunately, the transformation rate of protoplasts is low (~20%) [17,136], and the editing efficiency reported so far in grapevine protoplasts is about 0.1–42.3% [17,136,158,167], much lower than that obtained with *Agrobacterium*-mediated transformation. Additionally, it is difficult to get calli from protoplasts, and plant regeneration efficiency from protoplasts is less than 30% [158]. Other transformation methods such as electroporation and microinjection have not been used in grapevine, and the presence of the thick cell wall of plant cells limits the applications of the two delivery methods [126,168].

The bottleneck of grapevine transformation and regeneration is species and genotype dependence. To promote genotype-independent transformation, several strategies have been developed in recent years. One strategy is using morphogenic regulators, which are typically *baby boom* (*BBM*) and *Wuschel2* (*Wus2*) isolated from maize [171,172]. The overexpression of the two genes could produce high transformation frequencies in recalcitrant maize inbred lines [171,172]. Similarly, the overexpression of *BBM1* gene from apple could significantly improve transgenic plant production in apple [173]. These morphogenic regulators might be experimentally tested in grapevine, expecting to increase the transformation as well as regeneration frequencies. However, the constitutive overexpression of morphogenes causes negative phenotypic and reproductive outcomes [172], making it necessary to remove these morphogenes after inducing embryogenesis. Alternatively, the use of tissue-specific promoters or inducible expression systems would be good choices when using morphogenic regulators. The growth regulating factor 4 (GRF4) and GRF interacting factor 1 (GIF1) chimeric protein was reported to enhance transformation and regeneration efficiency in wheat, and grape GRF4-GIF1 chimeric protein was also revealed to improve the transformation and regeneration in citrus [174], which suggests that the grape chimeric protein might be used to enhance the transformation and regeneration efficiency in grape, too. To sidestep the need for tissue culture, Maher et al. [175] reported a method to directly induce transgenic or edited shoots from *in vitro* plants by the de novo induction of meristems using developmental regulators (DRs). A proof-of-concept study had been conducted in grapevine, and transgenic plants with luciferase luminescence were successfully produced [175]. This approach promises to overcome the bottleneck in grape varieties that are recalcitrant to *Agrobacterium* transformation. However, the occurrence of genetic chimeras is a big concern when using this approach. Alternatively, nanoparticles could penetrate plant cell wall and can be employed as genotype-independent transformation method for exogenous biomolecule delivery [176,177].

5. Regulation of Gene-Edited Grapevine Plants

Public concern is often triggered by the cultivation and release of genetically modified (GM) organisms. The regulation of GM crops and foods is positive and necessary to protect human safety, the environment, and the economy [178]. However, one issue is that the current legislation might no longer be fit for the regulation of GM crops due to the rapid advances in molecular biology. Should the gene-edited (GE) plants or products produced by using CRISPR/Cas9 be regulated under the same regulatory framework as conventional GM organisms? In general, GM regulations can be categorized as process- and product-oriented [125,179]. According to process-oriented regulations, new breeding technologies could be regarded as novel techniques when compared with conventional methods, thus requiring specific legislation for the regulation. In contrast, the novel characteristics of produced products were emphasized under the product-oriented regulations [180]. In fact, the mutations introduced by editing generally rely on the formation of DSBs, which also occur naturally. The role of Cas nucleases is to generate DSBs at specific sites, and the DSBs are repaired by cellular repair mechanisms. In theory, the specific changes in GE plants cannot be discerned from the same mutations occurring naturally in unedited plants [181,182] if the editing components are not integrated or removed from plant genome after the editing.

Some countries have been reevaluating their legislation and considering how to regulate GE plants or products. In USA, the updated Coordinated Framework for the Regulation of Biotechnology allows for a modernized regulatory system [178,183], and GE crops without recombinant DNA, pesticidal activity, or food safety attributes are exempt from the regulation [178]. Recently, genome-edited mustard greens (*Brassica juncea*) approved by the Department of Agriculture (USDA) had been released to market [184]. In Japan, the GE products with SDN (site-directed nuclease)-1 type of modifications, which are characterized as mutations without using a DNA template, are not regarded as "living modified organisms" [185]. GE tomato with a high content of γ-aminobutyric acid was approved by the Japan government in 2021 [178,186]. As well-known wine countries, France and Italy, for example, are bound by European Union (EU) GM legislation. In July 2018, the European Court of Justice ruled that plants generated by NGTs (new genomic techniques) should be strictly regulated according to the Directive 2001/18/EC, which is previously applied to GM organisms [187]. This judgment is totally contrary to what EU scientific and breeder community expect, and as a consequence, research into plant genome editing in Europe has been greatly hampered in recent years [182]. However, the good news is that a regulatory proposal was published by the European Commission in July 2023, and Category 1 NGT plants (NGT1) are considered to be equivalent to conventionally bred plants [187]. Even though there is still a long way to go before the proposal is officially approved to be a law, exciting changes have emerged in how GE plants, including non-crops, are to be regulated in the future.

6. Future Perspectives

6.1. *Taking Advantage of the Model System*

A stable and efficient system for studying grapevine genome editing is always important and indispensable. Notably, grapevine has a relatively long juvenile stage, and it is impractical to harvest grape berries in a short time after plant regeneration, which undoubtedly restricts the investigation of the genes of interest associated with fruit quality traits. In this case, the grape microvine, a *gai1* mutant allele that confers a dwarf stature, short generation cycles and continuous flowering [188], could be an ideal model system for grapevine genome editing. Using the microvine model, many grapevine fruit attributes can be investigated easily. The great potential of multiplex genome editing could be further exploited by targeting multiple genes involving different traits of not only stress resistance but also fruit quality.

6.2. *Exploiting the Precise Genome Editing*

Most of the state-of-the-art CRISPR tools have not yet been applied in grapevine, and some efficient genome editing methods like virus-induced plant genome editing are not applicable to grapevine due to the lack of valid virus vectors. All these issues should be addressed by researchers in the near future. The CRISPR toolbox in grapevine needs to be expanded, and precise genome editors like BEs and PEs should be used more for grapevine genome editing. Moreover, the knockout of a developmentally important gene could sometimes result in developmental defect or even death. The conditional or tissue/cell-type-specific knockout of genes of interest could serve as tailored solutions to address specific biological questions [189,190]. Genome editing in grapevine could be more flexible and versatile, and the target regions can be expanded to promoters or uORFs to regulate the gene expression level or protein abundance [146,147]. Novel alleles can be created by manipulating the specific elements in gene promoters [146]. Precise knock-in of large DNA fragment has been accomplished by using PE-based tools like twinPE [99] and GRAND [101]. The targeted insertion of DNA fragments at a designed site in the grapevine genome could be tested with these tools in the future, and a robust promoter or regulatory element may be integrated into the grape genome to regulate the expression of the genes of interest instead of cloning gene coding sequences laboriously. The application of precise genome editing would simplify the gene functional research in grapevine.

6.3. *Accelerating the Improvement of Grapevine Traits*

With the development of sequencing technology, increasing data of grapevine genomes, resequencing, and transcriptomes are available now and provide us with a lot of candidate genes based on bioinformatics-assisted analyses. For example, over 44 grape genomes have been released to date and whole-genome resequencing projects containing almost 5000 accessions have been published; more than 900 genes involved in grapevine resistance, quality, and development were identified from these datasets using the genome-wide association studies (GWASs) and other methods [191,192]. CRISPR/Cas editing, combined with a grape transformation system, could serve as a rapid and efficient approach to preliminarily confirm the functions of selected candidate genes without the cloning of their coding sequences (Figure 5). The verified results enable researchers to improve the grapevine traits more accurately and rapidly. In the future, new grapevine elite cultivars with improved fruit quality and increased tolerance to challenging environments such as cold, high temperatures, and drought should facilitate the sustainable development of viticulture and wine production.

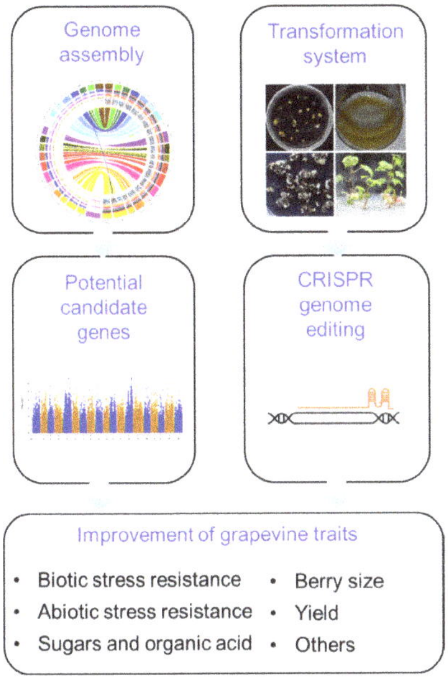

Figure 5. A strategy for accelerated improvement of grapevine traits. Bioinformatics analysis based on sequencing data and grapevine phenome provides candidate genes, and CRISPR editing, together with grape transformation system, enables a rapid verification of gene functions. The results pave the way for improvement of different grapevine traits.

6.4. Generating Transgene-Free, Edited Grapevine Plants

Considering that CRISPR editing, particularly conventional plasmid-mediated genome editing, may raise the concern of GM vines, which are morally unacceptable by consumers according to the online survey results conducted in the USA recently [193]. Furthermore, grapevine is in general vegetatively propagated, and it is unlikely to remove transgenic elements via genetic segregation. The use of the inducible Flp/FRT system or Cas9-based synthetic cleavage target sites (CTS) close to T-DNA borders can produce transgene-free genetically modified grapevine plants [194]. To guarantee the removal of transgenes, FRT sequences or Cas9 CTS are put next to T-DNA borders in designed binary vectors, and site-specific recombinase Flippase or Cas9 is driven by a heat-shock inducible promoter; the integrated T-DNA could be removed from the grapevine genome after the editing by heat-shock treatment [194]. It should be noted that a minimal trace of exogenous DNA sequence was still left in grapevine genome after the excision by using this strategy [194]. DNA-free genome editing using CRISPR/Cas9 RNPs followed by protoplast regeneration is a promising approach to obtain transgene-free, genome-edited grapevine plants [136,158,167,195]. The CRISPR/Cas9 components, sgRNA and Cas9, were expressed and preassembled *in vitro*, avoiding the use of DNA (plasmid) during the transformation [136,158,167]. This direct delivery of preassembled CRISPR/Cas9 RNPs into protoplasts may address public concerns about GM vines. Notably, it is almost impossible to get transgene-free grapevine plants in all grape varieties at present, because only limited varieties have been successfully used for the induction of EC, which serve as a source of protoplasts for transformation.

7. Conclusions

The emergence of CRISPR/Cas genome editing has initiated a new era in which genetic manipulation becomes precise and predictable. The application of CRISPR/Cas has advanced the pace of research in plants, including the grapevine. Targeted modifications of grapevine traits are highlighted by the improvement of plant resistance to the major diseases, namely, powdery mildew, gray mold, and downy mildew. Plant development, which includes shoot branching, stomata development, and female flower morphology, has also been modified in grapevine by using CRISPR/Cas technologies. The accumulation of important secondary metabolites such as anthocyanins and flavonoids was also improved through the targeted editing of specific genes. Even if in the past 8 years there have been some advances in grapevine genome editing, researchers are still expected to promote the application of CRISPR/Cas technologies in grapevine. We believe that CRISPR technology will play an increasingly important role in grapevine genome engineering due to the predicted innovation in CRISPR technology in the decade ahead.

Author Contributions: C.R. and Z.L. conceived the paper; C.R., M.S.M.M. and N.A. wrote the manuscript; C.R., N.A. and Y.K. collected the data; C.R. and Z.L. revised the manuscript. All authors have read and agreed to the published version of the manuscript.

Funding: This research is funded by grants from the Agricultural Breeding Project of Ningxia Hui Autonomous Region (NXNYYZ202101) and the Youth Innovation Promotion Association CAS (2022078).

Conflicts of Interest: The authors declare no conflicts of interest.

References

1. Ishino, Y.; Shinagawa, H.; Makino, K.; Amemura, M.; Nakata, A. Nucleotide sequence of the Iap gene, responsible for alkaline phosphatase isozyme conversion in *Escherichia coli*, and identification of the gene product. *J. Bacteriol.* **1987**, *169*, 5429–5433. [CrossRef] [PubMed]
2. Barrangou, R.; Fremaux, C.; Deveau, H.; Richards, M.; Boyaval, P.; Moineau, S.; Romero, D.A.; Horvath, P. CRISPR provides acquired resistance against viruses in prokaryotes. *Science* **2007**, *315*, 1709–1712. [CrossRef] [PubMed]
3. Marraffini, L.A.; Sontheimer, E.J. CRISPR interference limits horizontal gene transfer in staphylococci by targeting DNA. *Science* **2008**, *322*, 1843–1845. [CrossRef]
4. Brouns, S.J.; Jore, M.M.; Lundgren, M.; Westra, E.R.; Slijkhuis, R.J.; Snijders, A.P.; Dickman, M.J.; Makarova, K.S.; Koonin, E.V.; van der Oost, J. Small CRISPR RNAs guide antiviral defense in prokaryotes. *Science* **2008**, *321*, 960–964. [CrossRef] [PubMed]
5. Jinek, M.; Chylinski, K.; Fonfara, I.; Hauer, M.; Doudna, J.A.; Charpentier, E. A programmable dual-RNA-guided DNA endonuclease in adaptive bacterial immunity. *Science* **2012**, *337*, 816–821. [CrossRef]
6. Gasiunas, G.; Barrangou, R.; Horvath, P.; Siksnys, V. Cas9-crRNA ribonucleoprotein complex mediates specific DNA cleavage for adaptive immunity in bacteria. *Proc.Natl. Acad. Sci. USA* **2012**, *109*, E2579–E2586. [CrossRef] [PubMed]
7. Cong, L.; Ran, F.A.; Cox, D.; Lin, S.; Barretto, R.; Habib, N.; Hsu, P.D.; Wu, X.; Jiang, W.; Marraffini, L.A.; et al. Multiplex genome engineering using CRISPR/Cas systems. *Science* **2013**, *339*, 819–823. [CrossRef]
8. Urnov, F.D.; Rebar, E.J.; Holmes, M.C.; Zhang, H.S.; Gregory, P.D. Genome editing with engineered zinc finger nucleases. *Nat. Rev. Genet.* **2010**, *11*, 636–646. [CrossRef]
9. Boch, J.; Scholze, H.; Schornack, S.; Landgraf, A.; Hahn, S.; Kay, S.; Lahaye, T.; Nickstadt, A.; Bonas, U. Breaking the code of DNA binding specificity of TAL-type III effectors. *Science* **2009**, *326*, 1509–1512. [CrossRef]
10. Joung, J.K.; Sander, J.D. TALENs: A widely applicable technology for targeted genome editing. *Nat. Rev. Mol. Cell Biol.* **2013**, *14*, 49–55. [CrossRef]
11. Wang, H.; La Russa, M.; Qi, L.S. CRISPR/Cas9 in Genome Editing and Beyond. *Annu. Rev. Biochem.* **2016**, *85*, 227–264. [CrossRef] [PubMed]
12. Xie, K.; Yang, Y. RNA-guided genome editing in plants using a CRISPR-Cas system. *Mol. Plant* **2013**, *6*, 1975–1983. [CrossRef] [PubMed]
13. Jiang, W.; Zhou, H.; Bi, H.; Fromm, M.; Yang, B.; Weeks, D.P. Demonstration of CRISPR/Cas9/sgRNA-mediated targeted gene modification in Arabidopsis, tobacco, sorghum and rice. *Nucleic Acids Res.* **2013**, *41*, e188. [CrossRef] [PubMed]
14. Brooks, C.; Nekrasov, V.; Lippman, Z.B.; Van Eck, J. Efficient gene editing in tomato in the first generation using the clustered regularly interspaced short palindromic repeats/CRISPR-associated9 system. *Plant Physiol.* **2014**, *166*, 1292–1297. [CrossRef]
15. Jia, H.; Wang, N. Targeted genome editing of sweet orange using Cas9/sgRNA. *PLoS ONE* **2014**, *9*, e93806. [CrossRef]
16. Ren, C.; Liu, X.; Zhang, Z.; Wang, Y.; Duan, W.; Li, S.; Liang, Z. CRISPR/Cas9-mediated efficient targeted mutagenesis in Chardonnay (*Vitis vinifera* L.). *Sci. Rep.* **2016**, *6*, 32289. [CrossRef]
17. Malnoy, M.; Viola, R.; Jung, M.H.; Koo, O.J.; Kim, S.; Kim, J.S.; Velasco, R.; Nagamangala Kanchiswamy, C. DNA-Free Genetically Edited Grapevine and Apple Protoplast Using CRISPR/Cas9 Ribonucleoproteins. *Front. Plant Sci.* **2016**, *7*, 1904. [CrossRef]
18. Ren, C.; Li, H.; Liu, Y.; Li, S.; Liang, Z. Highly efficient activation of endogenous gene in grape using CRISPR/dCas9-based transcriptional activators. *Hortic. Res.* **2022**, *9*, uhab037. [CrossRef] [PubMed]

19. Yang, Y.; Wheatley, M.; Meakem, V.; Galarneau, E.; Gutierrez, B.; Zhong, G.Y. Editing *VvDXS1* for the creation of muscat flavour in *Vitis vinifera* cv. Scarlet Royal. *Plant Biotechnol. J.* **2024**, *22*, 1610–1621. [CrossRef]
20. Makarova, K.S.; Wolf, Y.I.; Alkhnbashi, O.S.; Costa, F.; Shah, S.A.; Saunders, S.J.; Barrangou, R.; Brouns, S.J.; Charpentier, E.; Haft, D.H.; et al. An updated evolutionary classification of CRISPR-Cas systems. *Nat. Rev. Microbiol.* **2015**, *13*, 722–736. [CrossRef]
21. Anzalone, A.V.; Koblan, L.W.; Liu, D.R. Genome editing with CRISPR-Cas nucleases, base editors, transposases and prime editors. *Nat. Biotechnol.* **2020**, *38*, 824–844. [CrossRef] [PubMed]
22. Shamshirgaran, Y.; Liu, J.; Sumer, H.; Verma, P.J.; Taheri-Ghahfarokhi, A. Tools for Efficient Genome Editing; ZFN, TALEN, and CRISPR. *Methods Mol. Biol.* **2022**, *2495*, 29–46. [PubMed]
23. Taheri-Ghahfarokhi, A.; Taylor, B.J.M.; Nitsch, R.; Lundin, A.; Cavallo, A.L.; Madeyski-Bengtson, K.; Karlsson, F.; Clausen, M.; Hicks, R.; Mayr, L.M.; et al. Decoding non-random mutational signatures at Cas9 targeted sites. *Nucleic Acids Res.* **2018**, *46*, 8417–8434. [CrossRef]
24. Ceccaldi, R.; Rondinelli, B.; D'Andrea, A.D. Repair pathway choices and consequences at the double-strand break. *Trends Cell Biol.* **2016**, *26*, 52–64. [CrossRef] [PubMed]
25. Ran, F.A.; Hsu, P.D.; Lin, C.Y.; Gootenberg, J.S.; Konermann, S.; Trevino, A.; Scott, D.A.; Inoue, A.; Matoba, S.; Zhang, Y.; et al. Double nicking by RNA-guided CRISPR Cas9 for enhanced genome editing specificity. *Cell* **2013**, *154*, 1380–1389. [CrossRef]
26. Schiml, S.; Fauser, F.; Puchta, H. The CRISPR/Cas9 system can be used as nuclease for *in planta* gene targeting and as paired nickases for directed mutagenesis in Arabidopsis resulting in heritable progeny. *Plant J.* **2014**, *80*, 1139–1150. [CrossRef]
27. Qi, L.S.; Larson, M.H.; Gilbert, L.A.; Doudna, J.A.; Weissman, J.S.; Arkin, A.P.; Lim, W.A. Repurposing CRISPR as an RNA-guided platform for sequence-specific control of gene expression. *Cell* **2013**, *152*, 1173–1183. [CrossRef]
28. Gilbert, L.A.; Larson, M.H.; Morsut, L.; Liu, Z.; Brar, G.A.; Torres, S.E.; Stern-Ginossar, N.; Brandman, O.; Whitehead, E.H.; Doudna, J.A.; et al. CRISPR-mediated modular RNA-guided regulation of transcription in eukaryotes. *Cell* **2013**, *154*, 442–451. [CrossRef] [PubMed]
29. Rusk, N. CRISPRs and epigenome editing. *Nat. Methods* **2014**, *11*, 28. [CrossRef]
30. Hsu, P.D.; Scott, D.A.; Weinstein, J.A.; Ran, F.A.; Konermann, S.; Agarwala, V.; Li, Y.; Fine, E.J.; Wu, X.; Shalem, O.; et al. DNA targeting specificity of RNA-guided Cas9 nucleases. *Nat. Biotechnol.* **2013**, *31*, 827–832. [CrossRef]
31. Kleinstiver, B.P.; Prew, M.S.; Tsai, S.Q.; Topkar, V.V.; Nguyen, N.T.; Zheng, Z.; Gonzales, A.P.; Li, Z.; Peterson, R.T.; Yeh, J.R.; et al. Engineered CRISPR/Cas9 nucleases with altered PAM specificities. *Nature* **2015**, *523*, 481–485. [CrossRef] [PubMed]
32. Hu, J.H.; Miller, S.M.; Geurts, M.H.; Tang, W.; Chen, L.; Sun, N.; Zeina, C.M.; Gao, X.; Rees, H.A.; Lin, Z.; et al. Evolved Cas9 variants with broad PAM compatibility and high DNA specificity. *Nature* **2018**, *556*, 57–63. [CrossRef] [PubMed]
33. Kim, H.K.; Lee, S.; Kim, Y.; Park, J.; Min, S.; Choi, J.W.; Huang, T.P.; Yoon, S.; Liu, D.R.; Kim, H.H. High-throughput analysis of the activities of xCas9, SpCas9-NG and SpCas9 at matched and mismatched target sequences in human cells. *Nat. Biomed. Eng.* **2020**, *4*, 111–124. [CrossRef]
34. Nishimasu, H.; Shi, X.; Ishiguro, S.; Gao, L.; Hirano, S.; Okazaki, S.; Noda, T.; Abudayyeh, O.O.; Gootenberg, J.S.; Mori, H.; et al. Engineered CRISPR/Cas9 nuclease with expanded targeting space. *Science* **2018**, *361*, 1259–1262. [CrossRef] [PubMed]
35. Walton, R.T.; Christie, K.A.; Whittaker, M.N.; Kleinstiver, B.P. Unconstrained genome targeting with near-PAMless engineered CRISPR/Cas9 variants. *Science* **2020**, *368*, 290–296. [CrossRef]
36. Ran, F.A.; Cong, L.; Yan, W.X.; Scott, D.A.; Gootenberg, J.S.; Kriz, A.J.; Zetsche, B.; Shalem, O.; Wu, X.; Makarova, K.S.; et al. *In vivo* genome editing using *Staphylococcus aureus* Cas9. *Nature* **2015**, *520*, 186–191. [CrossRef]
37. Müller, M.; Lee, C.M.; Gasiunas, G.; Davis, T.H.; Cradick, T.J.; Siksnys, V.; Bao, G.; Cathomen, T.; Mussolino, C. *Streptococcus thermophilus* CRISPR/Cas9 systems enable specific editing of the human genome. *Mol. Ter.* **2016**, *24*, 636–644. [CrossRef]
38. Hirano, H.; Gootenberg, J.S.; Horii, T.; Abudayyeh, O.O.; Kimura, M.; Hsu, P.D.; Nakane, T.; Ishitani, R.; Hatada, I.; Zhang, F.; et al. Structure and engineering of *Francisella novicida* Cas9. *Cell* **2016**, *164*, 950–961. [CrossRef]
39. Kim, E.; Koo, T.; Park, S.W.; Kim, D.; Kim, K.; Cho, H.Y.; Song, D.W.; Lee, K.J.; Jung, M.H.; Kim, S.; et al. *In vivo* genome editing with a small Cas9 orthologue derived from *Campylobacter jejuni*. *Nat. Commun.* **2017**, *8*, 14500. [CrossRef]
40. Chatterjee, P.; Jakimo, N.; Jacobson, J.M. Minimal PAM specificity of a highly similar SpCas9 ortholog. *Sci. Adv.* **2018**, *4*, eaau0766. [CrossRef]
41. Edraki, A.; Mir, A.; Ibraheim, R.; Gainetdinov, I.; Yoon, Y.; Song, C.Q.; Cao, Y.; Gallant, J.; Xue, W.; Rivera-Pérez, J.A.; et al. A compact, high-accuracy Cas9 with a dinucleotide PAM for *in vivo* genome editing. *Mol. Cell* **2019**, *73*, 714–726.e4. [CrossRef]
42. Slaymaker, I.M.; Gao, L.; Zetsche, B.; Scott, D.A.; Yan, W.X.; Zhang, F. Rationally engineered Cas9 nucleases with improved specificity. *Science* **2016**, *351*, 84–88. [CrossRef] [PubMed]
43. Kleinstiver, B.P.; Pattanayak, V.; Prew, M.S.; Tsai, S.Q.; Nguyen, N.T.; Zheng, Z.; Joung, J.K. High-fidelity CRISPR/Cas9 nucleases with no detectable genome-wide of-target effects. *Nature* **2016**, *529*, 490–495. [CrossRef] [PubMed]
44. Kulcsár, P.I.; Tálas, A.; Huszár, K.; Ligeti, Z.; Tóth, E.; Weinhardt, N.; Fodor, E.; Welker, E. Crossing enhanced and high fidelity SpCas9 nucleases to optimize specificity and cleavage. *Genome Biol.* **2017**, *18*, 190. [CrossRef] [PubMed]
45. Chen, J.S.; Dagdas, Y.S.; Kleinstiver, B.P.; Welch, M.M.; Sousa, A.A.; Harrington, L.B.; Sternberg, S.H.; Joung, J.K.; Yildiz, A.; Doudna, J.A. Enhanced proofreading governs CRISPR-Cas9 targeting accuracy. *Nature* **2017**, *550*, 407–410. [CrossRef] [PubMed]
46. Casini, A.; Olivieri, M.; Petris, G.; Montagna, C.; Reginato, G.; Maule, G.; Lorenzin, F.; Prandi, D.; Romanel, A.; Demichelis, F.; et al. A highly specifc SpCas9 variant is identified by *in vivo* screening in yeast. *Nat. Biotechnol.* **2018**, *36*, 265–271. [CrossRef]

47. Lee, J.K.; Jeong, E.; Lee, J.; Jung, M.; Shin, E.; Kim, Y.H.; Lee, K.; Jung, I.; Kim, D.; Kim, S.; et al. Directed evolution of CRISPR/Cas9 to increase its specificity. *Nat. Commun.* **2018**, *9*, 3048. [CrossRef]
48. Zetsche, B.; Gootenberg, J.S.; Abudayyeh, O.O.; Slaymaker, I.M.; Makarova, K.S.; Essletzbichler, P.; Volz, S.E.; Joung, J.; van der Oost, J.; Regev, A.; et al. Cpf1 is a single RNA-guided endonuclease of a class 2 CRISPR-Cas system. *Cell* **2015**, *163*, 759–771. [CrossRef]
49. Zetsche, B.; Heidenreich, M.; Mohanraju, P.; Fedorova, I.; Kneppers, J.; DeGennaro, E.M.; Winblad, N.; Choudhury, S.R.; Abudayyeh, O.O.; Gootenberg, J.S.; et al. Multiplex gene editing by CRISPR-Cpf1 using a single crRNA array. *Nat. Biotechnol.* **2017**, *35*, 31–34. [CrossRef]
50. Endo, A.; Masafumi, M.; Kaya, H.; Toki, S. Efficient targeted mutagenesis of rice and tobacco genomes using Cpf1 from *Francisella novicida*. *Sci. Rep.* **2016**, *6*, 38169. [CrossRef]
51. Tang, X.; Lowder, L.G.; Zhang, T.; Malzahn, A.A.; Zheng, X.; Voytas, D.F.; Zhong, Z.; Chen, Y.; Ren, Q.; Li, Q.; et al. A CRISPR-Cpf1 system for efficient genome editing and transcriptional repression in plants. *Nat. Plants* **2017**, *3*, 17103. [CrossRef] [PubMed]
52. Hu, X.; Wang, C.; Liu, Q.; Fu, Y.; Wang, K. Targeted mutagenesis in rice using CRISPR-Cpf1 system. *J. Genet. Genom.* **2017**, *44*, 71–73. [CrossRef] [PubMed]
53. Zhong, Z.; Zhang, Y.; You, Q.; Tang, X.; Ren, Q.; Liu, S.; Yang, L.; Wang, Y.; Liu, X.; Liu, B.; et al. Plant genome editing using FnCpf1 and LbCpf1 nucleases at redefined and altered PAM sites. *Mol. Plant* **2018**, *11*, 999–1002. [CrossRef] [PubMed]
54. Pietralla, J.; Capdeville, N.; Schindele, P.; Puchta, H. Optimizing ErCas12a for efficient gene editing in *Arabidopsis thaliana*. *Plant Biotechnol. J.* **2024**, *22*, 401–412. [CrossRef]
55. Zhang, Y.; Ren, Q.; Tang, X.; Liu, S.; Malzahn, A.A.; Zhou, J.; Wang, J.; Yin, D.; Pan, C.; Yuan, M.; et al. Expanding the scope of plant genome engineering with Cas12a orthologs and highly multiplexable editing systems. *Nat. Commun.* **2021**, *12*, 1944. [CrossRef]
56. Zhang, L.; Li, G.; Zhang, Y.; Cheng, Y.; Roberts, N.; Glenn, S.E.; DeZwaan-McCabe, D.; Rube, H.T.; Manthey, J.; Coleman, G. Boosting genome editing efficiency in human cells and plants with novel LbCas12a variants. *Genome Biol.* **2023**, *24*, 102. [CrossRef]
57. Huang, H.; Huang, G.; Tan, Z.; Hu, Y.; Shan, L.; Zhou, J.; Zhang, X.; Ma, S.; Lv, W.; Huang, T.; et al. Engineered Cas12a-Plus nuclease enables gene editing with enhanced activity and specificity. *BMC Biol.* **2022**, *20*, 91. [CrossRef] [PubMed]
58. Huang, T.K.; Armstrong, B.; Schindele, P.; Puchta, H. Efficient gene targeting in *Nicotiana tabacum* using CRISPR/SaCas9 and temperature tolerant LbCas12a. *Plant Biotechnol. J.* **2021**, *19*, 1314–1324. [CrossRef] [PubMed]
59. Kleinstiver, B.P.; Sousa, A.A.; Walton, R.T.; Tak, Y.E.; Hsu, J.Y.; Clement, K.; Welch, M.M.; Horng, J.E.; Malagon-Lopez, J.; Scarfò, I.; et al. Engineered CRISPR-Cas12a variants with increased activities and improved targeting ranges for gene, epigenetic and base editing. *Nat. Biotechnol.* **2019**, *37*, 276–282. [CrossRef]
60. Tak, Y.E.; Kleinstiver, B.P.; Nuñez, J.K.; Hsu, J.Y.; Horng, J.E.; Gong, J.; Weissman, J.S.; Joung, J.K. Inducible and multiplex gene regulation using CRISPR-Cpf1-based transcription factors. *Nat. Methods* **2017**, *14*, 1163–1166. [CrossRef]
61. Li, X.; Wang, Y.; Liu, Y.; Yang, B.; Wang, X.; Wei, J.; Lu, Z.; Zhang, Y.; Wu, J.; Huang, X.; et al. Base editing with a Cpf1-cytidine deaminase fusion. *Nat. Biotechnol.* **2018**, *36*, 324–327. [CrossRef] [PubMed]
62. Saito, M.; Xu, P.; Faure, G.; Maguire, S.; Kannan, S.; Altae-Tran, H.; Vo, S.; Desimone, A.; Macrae, R.K.; Zhang, F. Fanzor is a eukaryotic programmable RNA-guided endonuclease. *Nature* **2023**, *620*, 660–668. [CrossRef] [PubMed]
63. Karvelis, T.; Druteika, G.; Bigelyte, G.; Budre, K.; Zedaveinyte, R.; Silanskas, A.; Kazlauskas, D.; Venclovas, Č.; Siksnys, V. Transposon-associated TnpB is a programmable RNA-guided DNA endonuclease. *Nature* **2021**, *599*, 692–696. [CrossRef] [PubMed]
64. Komor, A.C.; Kim, Y.B.; Packer, M.S.; Zuris, J.A.; Liu, D.R. Programmable editing of a target base in genomic DNA without double-stranded DNA cleavage. *Nature* **2016**, *533*, 420–424. [CrossRef]
65. Gaudelli, N.M.; Komor, A.C.; Rees, H.A.; Packer, M.S.; Badran, A.H.; Bryson, D.I.; Liu, D.R. Programmable base editing of A*T to G*C in genomic DNA without DNA cleavage. *Nature* **2017**, *551*, 464–471. [CrossRef]
66. Kurt, I.C.; Zhou, R.; Iyer, S.; Garcia, S.P.; Miller, B.R.; Langner, L.M.; Grunewald, J.; Joung, J.K. CRISPR C-to-G base editors for inducing targeted DNA transversions in human cells. *Nat. Biotechnol.* **2021**, *39*, 41–46. [CrossRef]
67. Komor, A.C.; Zhao, K.T.; Packer, M.S.; Gaudelli, N.M.; Waterbury, A.L.; Koblan, L.W.; Kim, Y.B.; Badran, A.H.; Liu, D.R. Improved base excision repair inhibition and bacteriophage Mu Gam protein yields C:G-to-T:A base editors with higher efficiency and product purity. *Sci. Adv.* **2017**, *3*, eaao4774. [CrossRef]
68. Ren, B.; Yan, F.; Kuang, Y.; Li, N.; Zhang, D.; Zhou, X.; Lin, H.; Zhou, H. Improved base editor for efficiently inducing genetic variations in rice with CRISPR/Cas9-guided hyperactive hAID mutant. *Mol. Plant* **2018**, *11*, 623–626. [CrossRef] [PubMed]
69. Shimatani, Z.; Kashojiya, S.; Takayama, M.; Terada, R.; Arazoe, T.; Ishii, H.; Teramura, H.; Yamamoto, T.; Komatsu, H.; Miura, K.; et al. Targeted base editing in rice and tomato using a CRISPR-Cas9 cytidine deaminase fusion. *Nat. Biotechnol.* **2017**, *35*, 441–443. [CrossRef]
70. Zong, Y.; Song, Q.; Li, C.; Jin, S.; Zhang, D.; Wang, Y.; Qiu, J.L.; Gao, C. Efficient C-to-T base editing in plants using a fusion of nCas9 and human APOBEC3A. *Nat. Biotechnol.* **2018**, *36*, 950–954. [CrossRef]
71. Neugebauer, M.E.; Hsu, A.; Arbab, M.; Krasnow, N.A.; McElroy, A.N.; Pandey, S.; Doman, J.L.; Huang, T.P.; Raguram, A.; Banskota, S.; et al. Evolution of an adenine base editor into a small, efficient cytosine base editor with low off-target activity. *Nat. Biotechnol.* **2023**, *41*, 673–685. [CrossRef] [PubMed]

72. Chen, L.; Zhu, B.; Ru, G.; Meng, H.; Yan, Y.; Hong, M.; Zhang, D.; Luan, C.; Zhang, S.; Wu, H.; et al. Re-engineering the adenine deaminase TadA-8e for efficient and specific CRISPR-based cytosine base editing. *Nat. Biotechnol.* **2023**, *41*, 663–672. [CrossRef]
73. Hua, K.; Tao, X.; Han, P.; Wang, R.; Zhu, J.K. Genome engineering in rice using Cas9 variants that recognize NG PAM sequences. *Mol. Plant* **2019**, *12*, 1003–1014. [CrossRef] [PubMed]
74. Zhong, Z.; Sretenovic, S.; Ren, Q.; Yang, L.; Bao, Y.; Qi, C.; Yuan, M.; He, Y.; Liu, S.; Liu, X.; et al. Improving plant genome editing with high-fidelity xCas9 and non-canonical PAM-targeting Cas9-NG. *Mol. Plant* **2019**, *12*, 1027–1036. [CrossRef]
75. Wang, M.; Xu, Z.; Gosavi, G.; Ren, B.; Cao, Y.; Kuang, Y.; Zhou, C.; Spetz, C.; Yan, F.; Zhou, X.; et al. Targeted base editing in rice with CRISPR/ScCas9 system. *Plant Biotechnol. J.* **2020**, *18*, 1645–1647. [CrossRef]
76. Koblan, L.W.; Doman, J.L.; Wilson, C.; Levy, J.M.; Tay, T.; Newby, G.A.; Maianti, J.P.; Raguram, A.; Liu, D.R. Improving cytidine and adenine base editors by expression optimization and ancestral reconstruction. *Nat. Biotechnol.* **2018**, *36*, 843–846. [CrossRef]
77. Hua, K.; Tao, X.; Yuan, F.; Wang, D.; Zhu, J.K. Precise A·T to G·C base editing in the rice genome. *Mol. Plant* **2018**, *11*, 627–630. [CrossRef]
78. Li, C.; Zong, Y.; Wang, Y.; Jin, S.; Zhang, D.; Song, Q.; Zhang, R.; Gao, C. Expanded base editing in rice and wheat using a Cas9-adenosine deaminase fusion. *Genome Biol.* **2018**, *19*, 59. [CrossRef] [PubMed]
79. Gaudelli, N.M.; Lam, D.K.; Rees, H.A.; Sola-Esteves, N.M.; Barrera, L.A.; Born, D.A.; Edwards, A.; Gehrke, J.M.; Lee, S.J.; Liquori, A.J.; et al. Directed evolution of adenine base editors with increased activity and therapeutic application. *Nat. Biotechnol.* **2020**, *38*, 892–900. [CrossRef]
80. Richter, M.F.; Zhao, K.T.; Eton, E.; Lapinaite, A.; Newby, G.A.; Thuronyi, B.W.; Wilson, C.; Koblan, L.W.; Zeng, J.; Bauer, D.E.; et al. Phage-assisted evolution of an adenine base editor with improved Cas domain compatibility and activity. *Nat. Biotechnol.* **2020**, *38*, 883–891. [CrossRef]
81. Wei, C.; Wang, C.; Jia, M.; Guo, H.X.; Luo, P.Y.; Wang, M.G.; Zhu, J.K.; Zhang, H. Efficient generation of homozygous substitutions in rice in one generation utilizing an rABE8e base editor. *J. Integr. Plant Biol.* **2021**, *63*, 1595–1599. [CrossRef] [PubMed]
82. Tan, J.; Zeng, D.; Zhao, Y.; Wang, Y.; Liu, T.; Li, S.; Xue, Y.; Luo, Y.; Xie, X.; Chen, L.; et al. PhieABEs: A PAM-less/free high-efficiency adenine base editor toolbox with wide target scope in plants. *Plant Biotechnol. J.* **2022**, *20*, 934–943. [CrossRef]
83. Yan, D.; Ren, B.; Liu, L.; Yan, F.; Li, S.; Wang, G.; Sun, W.; Zhou, X.; Zhou, H. High-efficiency and multiplex adenine base editing in plants using new TadA variants. *Mol. Plant* **2021**, *14*, 722–731. [CrossRef]
84. Zhao, D.; Li, J.; Li, S.; Xin, X.; Hu, M.; Price, M.A.; Rosser, S.J.; Bi, C.; Zhang, X. Glycosylase base editors enable C-to-A and C-to-G base changes. *Nat. Biotechnol.* **2021**, *39*, 35–40. [CrossRef] [PubMed]
85. Tian, Y.; Shen, R.; Li, Z.; Yao, Q.; Zhang, X.; Zhong, D.; Tan, X.; Song, M.; Han, H.; Zhu, J.K.; et al. Efficient C-to-G editing in rice using an optimized base editor. *Plant Biotechnol. J.* **2022**, *20*, 1238–1240. [CrossRef]
86. Tong, H.; Wang, X.; Liu, Y.; Liu, N.; Li, Y.; Luo, J.; Ma, Q.; Wu, D.; Li, J.; Xu, C.; et al. Programmable A-to-Y base editing by fusing an adenine base editor with an N-methylpurine DNA glycosylase. *Nat Biotechnol.* **2023**, *41*, 1080–1084. [CrossRef]
87. Tong, H.; Liu, N.; Wei, Y.; Zhou, Y.; Li, Y.; Wu, D.; Jin, M.; Cui, S.; Li, H.; Li, G.; et al. Programmable deaminase-free base editors for G-to-Y conversion by engineered glycosylase. *Natl. Sci. Rev.* **2023**, *10*, nwad143. [CrossRef] [PubMed]
88. Zhang, X.; Zhu, B.; Chen, L.; Xie, L.; Yu, W.; Wang, Y.; Li, L.; Yin, S.; Yang, L.; Hu, H.; et al. Dual base editor catalyzes both cytosine and adenine base conversions in human cells. *Nat. Biotechnol.* **2020**, *38*, 856–860. [CrossRef] [PubMed]
89. Li, C.; Zong, Y.; Jin, S.; Zhu, H.; Lin, D.; Li, S.; Qiu, J.L.; Wang, Y.; Gao, C. SWISS: Multiplexed orthogonal genome editing in plants with a Cas9 nickase and engineered CRISPR RNA scaffolds. *Genome Biol.* **2020**, *21*, 141. [CrossRef]
90. Liang, Y.; Xie, J.; Zhang, Q.; Wang, X.; Gou, S.; Lin, L.; Chen, T.; Ge, W.; Zhuang, Z.; Lian, M.; et al. AGBE: A dual deaminase-mediated base editor by fusing CGBE with ABE for creating a saturated mutant population with multiple editing patterns. *Nucleic Acids Res.* **2022**, *50*, 5384–5399. [CrossRef]
91. Anzalone, A.V.; Randolph, P.B.; Davis, J.R.; Sousa, A.A.; Koblan, L.W.; Levy, J.M.; Chen, P.J.; Wilson, C.; Newby, G.A.; Raguram, A.; et al. Search-and-replace genome editing without double-strand breaks or donor DNA. *Nature* **2019**, *576*, 149–157. [CrossRef] [PubMed]
92. Chen, P.J.; Hussmann, J.A.; Yan, J.; Knipping, F.; Ravisankar, P.; Chen, P.F.; Chen, C.; Nelson, J.W.; Newby, G.A.; Sahin, M.; et al. Enhanced prime editing systems by manipulating cellular determinants of editing outcomes. *Cell* **2021**, *184*, 5635–5652. [CrossRef] [PubMed]
93. Li, J.; Zhang, C.; He, Y.; Li, S.; Yan, L.; Li, Y.; Zhu, Z.; Xia, L. Plant base editing and prime editing: The current status and future perspectives. *J. Integr. Plant Biol.* **2023**, *65*, 444–467. [CrossRef]
94. Lin, Q.; Jin, S.; Zong, Y.; Yu, H.; Zhu, Z.; Liu, G.; Kou, L.; Wang, Y.; Qiu, J.L.; Li, J.; et al. High-efficiency prime editing with optimized, paired pegRNAs in plants. *Nat. Biotechnol.* **2021**, *39*, 923–927. [CrossRef]
95. Liu, Y.; Yang, G.; Huang, S.; Li, X.; Wang, X.; Li, G.; Chi, T.; Chen, Y.; Huang, X.; Wang, X. Enhancing prime editing by Csy4-mediated processing of pegRNA. *Cell Res.* **2021**, *31*, 1134–1136. [CrossRef]
96. Li, X.; Zhou, L.; Gao, B.Q.; Li, G.; Wang, X.; Wang, Y.; Wei, J.; Han, W.; Wang, Z.; Li, J.; et al. Highly efficient prime editing by introducing same-sense mutations in pegRNA or stabilizing its structure. *Nat. Commun.* **2022**, *13*, 1669. [CrossRef]
97. Nelson, J.W.; Randolph, P.B.; Shen, S.P.; Everette, K.A.; Chen, P.J.; Anzalone, A.V.; An, M.; Newby, G.A.; Chen, J.C.; Hsu, A.; et al. Engineered pegRNAs improve prime editing efficiency. *Nat. Biotechnol.* **2022**, *40*, 402–410. [CrossRef] [PubMed]
98. Choi, J.; Chen, W.; Suiter, C.C.; Lee, C.; Chardon, F.M.; Yang, W.; Leith, A.; Daza, R.M.; Martin, B.; Shendure, J. Precise genomic deletions using paired prime editing. *Nat. Biotechnol.* **2022**, *40*, 218–226. [CrossRef]

99. Anzalone, A.V.; Gao, X.D.; Podracky, C.J.; Nelson, A.T.; Koblan, L.W.; Raguram, A.; Levy, J.M.; Mercer, J.A.M.; Liu, D.R. Programmable deletion, replacement, integration and inversion of large DNA sequences with twin prime editing. *Nat. Biotechnol.* **2022**, *40*, 731–740. [CrossRef]
100. Jiang, T.; Zhang, X.O.; Weng, Z.; Xue, W. Deletion and replacement of long genomic sequences using prime editing. *Nat. Biotechnol.* **2022**, *40*, 227–234. [CrossRef]
101. Wang, J.; He, Z.; Wang, G.; Zhang, R.; Duan, J.; Gao, P.; Lei, X.; Qiu, H.; Zhang, C.; Zhang, Y.; et al. Efficient targeted insertion of large DNA fragments without DNA donors. *Nat. Methods* **2022**, *19*, 331–340. [CrossRef] [PubMed]
102. O'Connell, M. Molecular Mechanisms of RNA Targeting by Cas13-containing Type VI CRISPR-Cas Systems. *J. Mol. Biol.* **2019**, *431*, 66–87. [CrossRef] [PubMed]
103. Abudayyeh, O.O.; Gootenberg, J.S.; Konermann, S.; Joung, J.; Slaymaker, I.M.; Cox, D.B.; Shmakov, S.; Makarova, K.S.; Semenova, E.; Minakhin, L.; et al. C2c2 is a single-component programmable RNA-guided RNA-targeting CRISPR effector. *Science* **2016**, *353*, aaf5573. [CrossRef]
104. Abudayyeh, O.O.; Gootenberg, J.S.; Essletzbichler, P.; Han, S.; Joung, J.; Belanto, J.J.; Verdine, V.; Cox, D.B.T.; Kellner, M.J.; Regev, A.; et al. RNA targeting with CRISPR-Cas13. *Nature* **2017**, *550*, 280–284. [CrossRef] [PubMed]
105. East-Seletsky, A.; O'Connell, M.R.; Knight, S.C.; Burstein, D.; Cate, J.H.; Tjian, R.; Doudna, J.A. Two distinct RNase activities of CRISPR-C2c2 enable guide-RNA processing and RNA detection. *Nature* **2016**, *538*, 270–273. [CrossRef]
106. Cox, D.B.T.; Gootenberg, J.S.; Abudayyeh, O.O.; Franklin, B.; Kellner, M.J.; Joung, J.; Zhang, F. RNA editing with CRISPR-Cas13. *Science* **2017**, *358*, 1019–1027. [CrossRef] [PubMed]
107. Gootenberg, J.S.; Abudayyeh, O.O.; Lee, J.W.; Essletzbichler, P.; Dy, A.J.; Joung, J.; Verdine, V.; Donghia, N.; Daringer, N.M.; Freije, C.A.; et al. Nucleic acid detection with CRISPR-Cas13a/C2c2. *Science* **2017**, *356*, 438–442. [CrossRef]
108. Terns, M.P. CRISPR-based Technologies: Impact of RNA-targeting Systems. *Mol. Cell* **2018**, *72*, 404–412. [CrossRef]
109. Kavuri, N.R.; Ramasamy, M.; Qi, Y.; Mandadi, K. Applications of CRISPR/Cas13-Based RNA Editing in Plants. *Cells* **2022**, *11*, 2665. [CrossRef]
110. O'Connell, M.R.; Oakes, B.L.; Sternberg, S.H.; East-Seletsky, A.; Kaplan, M.; Doudna, J.A. Programmable RNA recognition and cleavage by CRISPR/Cas9. *Nature* **2014**, *516*, 263–266.
111. Liu, Y.; Chen, Z.; He, A.; Zhan, Y.; Li, J.; Liu, L.; Wu, H.; Zhuang, C.; Lin, J.; Zhang, Q.; et al. Targeting cellular mRNAs translation by CRISPR/Cas9. *Sci. Rep.* **2016**, *6*, 29652. [CrossRef] [PubMed]
112. Price, A.A.; Sampson, T.R.; Ratner, H.K.; Grakoui, A.; Weiss, D.S. Cas9-mediated targeting of viral RNA in eukaryotic cells. *Proc. Natl. Acad. Sci. USA* **2015**, *112*, 6164–6169. [CrossRef]
113. Strutt, S.C.; Torrez, R.M.; Kaya, E.; Negrete, O.A.; Doudna, J.A. RNA-dependent RNA targeting by CRISPR/Cas9. *eLife* **2018**, *7*, e32724. [CrossRef] [PubMed]
114. Dugar, G.; Leenay, R.T.; Eisenbart, S.K.; Bischler, T.; Aul, B.U.; Beisel, C.L.; Sharma, C.M. CRISPR RNA-Dependent Binding and Cleavage of Endogenous RNAs by the *Campylobacter jejuni* Cas9. *Mol. Cell* **2018**, *69*, 893–905.e7. [CrossRef] [PubMed]
115. Wang, Y.; Liu, X.; Ren, C.; Zhong, G.Y.; Yang, L.; Li, S.; Liang, Z. Identification of genomic sites for CRISPR/Cas9-based genome editing in the *Vitis vinifera* genome. *BMC Plant Biol.* **2016**, *16*, 96. [CrossRef]
116. Osakabe, Y.; Liang, Z.; Ren, C.; Nishitani, C.; Osakabe, K.; Wada, M.; Komori, S.; Malnoy, M.; Velasco, R.; Poli, M.; et al. CRISPR-Cas9-mediated genome editing in apple and grapevine. *Nat. Protoc.* **2018**, *13*, 2844–2863. [CrossRef]
117. Nakajima, I.; Ban, Y.; Azuma, A.; Onoue, N.; Moriguchi, T.; Yamamoto, T.; Toki, S.; Endo, M. CRISPR/Cas9-mediated targeted mutagenesis in grape. *PLoS ONE* **2017**, *12*, e0177966. [CrossRef]
118. Wang, X.; Tu, M.; Wang, D.; Liu, J.; Li, Y.; Li, Z.; Wang, Y.; Wang, X. CRISPR/Cas9-mediated efficient targeted mutagenesis in grape in the first generation. *Plant Biotechnol. J.* **2018**, *16*, 844–855. [CrossRef]
119. Wan, D.; Guo, Y.; Cheng, Y.; Hu, Y.; Xiao, S.; Wang, Y.; Wen, Y.Q. CRISPR/Cas9-mediated mutagenesis of *VvMLO3* results in enhanced resistance to powdery mildew in grapevine (*Vitis vinifera*). *Hortic. Res.* **2020**, *7*, 116. [CrossRef]
120. Olivares, F.; Loyola, R.; Olmedo, B.; Miccono, M.L.Á.; Aguirre, C.; Vergara, R.; Riquelme, D.; Madrid, G.; Plantat, P.; Mora, R.; et al. CRISPR/Cas9 Targeted Editing of Genes Associated with Fungal Susceptibility in *Vitis vinifera* L. cv. Thompson Seedless Using Geminivirus-Derived Replicons. *Front. Plant Sci.* **2021**, *12*, 791030. [CrossRef]
121. Li, M.Y.; Jiao, Y.T.; Wang, Y.T.; Zhang, N.; Wang, B.B.; Liu, R.Q.; Yin, X.; Xu, Y.; Liu, G.T. CRISPR/Cas9-mediated *VvPR4b* editing decreases downy mildew resistance in grapevine (*Vitis vinifera* L.). *Hortic. Res.* **2020**, *7*, 149. [CrossRef] [PubMed]
122. Djennane, S.; Gersch, S.; Le-Bohec, F.; Piron, M.C.; Baltenweck, R.; Lemaire, O.; Merdinoglu, D.; Hugueney, P.; Nogué, F.; Mestre, P. CRISPR/Cas9 editing of *Downy mildew resistant 6* (*DMR6-1*) in grapevine leads to reduced susceptibility to *Plasmopara viticola*. *J. Exp. Bot.* **2024**, *75*, 2100–2112. [CrossRef]
123. Giacomelli, L.; Zeilmaker, T.; Giovannini, O.; Salvagnin, U.; Masuero, D.; Franceschi, P.; Vrhovsek, U.; Scintilla, S.; Rouppe van der Voort, J.; Moser, C. Simultaneous editing of two *DMR6* genes in grapevine results in reduced susceptibility to downy mildew. *Front. Plant Sci.* **2023**, *14*, 1242240. [CrossRef]
124. Ren, F.; Ren, C.; Zhang, Z.; Duan, W.; Lecourieux, D.; Li, S.; Liang, Z. Efficiency Optimization of CRISPR/Cas9-Mediated Targeted Mutagenesis in Grape. *Front. Plant Sci.* **2019**, *10*, 612. [CrossRef]
125. Medvedieva, M.O.; Blume, Y.B. Legal regulation of plant genome editing with the CRISPR/Cas9 technology as an example. *Cytol. Genet.* **2018**, *52*, 204–212. [CrossRef]

126. Su, W.; Xu, M.; Radani, Y.; Yang, L. Technological development and application of plant genetic transformation. *Int. J. Mol. Sci.* **2023**, *24*, 10646. [CrossRef]
127. Ren, C.; Guo, Y.; Kong, J.; Lecourieux, F.; Dai, Z.; Li, S.; Liang, Z. Knockout of *VvCCD8* gene in grapevine affects shoot branching. *BMC Plant Biol.* **2020**, *20*, 47. [CrossRef] [PubMed]
128. Sunitha, S.; Rock, C.D. CRISPR/Cas9-mediated targeted mutagenesis of *TAS4* and *MYBA7* loci in grapevine rootstock 101-14. *Transgenic Res.* **2020**, *29*, 355–367. [CrossRef]
129. Ren, C.; Liu, Y.; Guo, Y.; Duan, W.; Fan, P.; Li, S.; Liang, Z. Optimizing the CRISPR/Cas9 system for genome editing in grape by using grape promoters. *Hortic. Res.* **2021**, *8*, 52. [CrossRef]
130. Iocco-Corena, P.; Chaïb, J.; Torregrosa, L.; Mackenzie, D.; Thomas, M.R.; Smith, H.M. VviPLATZ1 is a major factor that controls female flower morphology determination in grapevine. *Nat. Commun.* **2021**, *12*, 6995. [CrossRef]
131. Tu, M.; Fang, J.; Zhao, R.; Liu, X.; Yin, W.; Wang, Y.; Wang, X.; Wang, X.; Fang, Y. CRISPR/Cas9-mediated mutagenesis of *VvbZIP36* promotes anthocyanin accumulation in grapevine (*Vitis vinifera*). *Hortic. Res.* **2022**, *9*, uhac022. [CrossRef] [PubMed]
132. Huang, P.; Lu, M.; Li, X.; Sun, H.; Cheng, Z.; Miao, Y.; Fu, Y.; Zhang, X. An Efficient Agrobacterium rhizogenes-Mediated Hairy Root Transformation Method in a Soybean Root Biology Study. *Int. J. Mol. Sci.* **2022**, *23*, 12261. [CrossRef]
133. Yang, Y.; Ke, J.; Han, X.; Wuddineh, W.A.; Song, G.Q.; Zhong, G.Y. Removal of a 10-kb *Gret1* transposon from *VvMybA1* of *Vitis vinifera* cv. Chardonnay. *Hortic. Res.* **2022**, *9*, uhac201. [CrossRef]
134. Nakajima, I.; Kawahigashi, H.; Nishitani, C.; Azuma, A.; Haji, T.; Toki, S.; Endo, M. Targeted deletion of grape retrotransposon associated with fruit skin color via CRISPR/Cas9 in *Vitis labrascana* 'Shine Muscat'. *PLoS ONE* **2023**, *18*, e0286698. [CrossRef]
135. Clemens, M.; Faralli, M.; Lagreze, J.; Bontempo, L.; Piazza, S.; Varotto, C.; Malnoy, M.; Oechel, W.; Rizzoli, A.; Dalla Costa, L. *VvEPFL9-1* Knock-Out via CRISPR/Cas9 Reduces Stomatal Density in Grapevine. *Front. Plant Sci.* **2022**, *13*, 878001. [CrossRef]
136. Scintilla, S.; Salvagnin, U.; Giacomelli, L.; Zeilmaker, T.; Malnoy, M.A.; Rouppe van der Voort, J.; Moser, C. Regeneration of non-chimeric plants from DNA-free edited grapevine protoplasts. *Front. Plant Sci.* **2022**, *13*, 1078931. [CrossRef] [PubMed]
137. Villette, J.; Lecourieux, F.; Bastiancig, E.; Héloir, M.C.; Poinssot, B. New improvements in grapevine genome editing: High efficiency biallelic homozygous knock-out from regenerated plantlets by using an optimized zCas9i. *Plant Methods* **2024**, *20*, 45. [CrossRef] [PubMed]
138. Li, Y.; Mansour, H.; Wang, T.; Poojari, S.; Li, F. Naked-Eye Detection of Grapevine Red-Blotch Viral Infection Using a Plasmonic CRISPR Cas12a Assay. *Anal. Chem.* **2019**, *91*, 11510–11513. [CrossRef] [PubMed]
139. Ng, H.; Dean, N. Dramatic improvement of CRISPR/Cas9 editing in Candida albicans by increased single guide RNA expression. *mSphere* **2017**, *2*, e00385-16. [CrossRef]
140. Long, L.; Guo, D.D.; Gao, W.; Yang, W.W.; Hou, L.P.; Ma, X.N.; Miao, Y.C.; Botella, J.R.; Song, C.P. Optimization of CRISPR/Cas9 genome editing in cotton by improved sgRNA expression. *Plant Methods* **2018**, *14*, 85. [CrossRef]
141. Li, B.; Fu, C.; Zhou, J.; Hui, F.; Wang, Q.; Wang, F.; Wang, G.; Xu, Z.; Che, L.; Yuan, D.; et al. Highly Efficient Genome Editing Using Geminivirus-Based CRISPR/Cas9 System in Cotton Plant. *Cells* **2022**, *11*, 2902. [CrossRef] [PubMed]
142. Zhang, Z.; Mao, Y.; Ha, S.; Liu, W.; Botella, J.R.; Zhu, J.K. A multiplex CRISPR/Cas9 platform for fast and efficient editing of multiple genes in Arabidopsis. *Plant Cell Rep.* **2016**, *35*, 1519–1533. [CrossRef] [PubMed]
143. He, Y.; Zhang, T.; Yang, N.; Xu, M.; Yan, L.; Wang, L.; Wang, R.; Zhao, Y. Self-cleaving ribozymes enable the production of guide RNAs from unlimited choices of promoters for CRISPR/Cas9 mediated genome editing. *J. Genet. Genom.* **2017**, *44*, 469–472.
144. Tang, X.; Ren, Q.; Yang, L.; Bao, Y.; Zhong, Z.; He, Y.; Liu, S.; Qi, C.; Liu, B.; Wang, Y.; et al. Single transcript unit CRISPR 2.0 systems for robust Cas9 and Cas12a mediated plant genome editing. *Plant Biotechnol. J.* **2019**, *17*, 1431–1445. [CrossRef]
145. Moradpour, M.; Abdulah, S.N.A. CRISPR/dCas9 platforms in plants: Strategies and applications beyond genome editing. *Plant Biotechnol. J.* **2020**, *18*, 32–44. [CrossRef]
146. Shi, L.; Su, J.; Cho, M.J.; Song, H.; Dong, X.; Liang, Y.; Zhang, Z. Promoter editing for the genetic improvement of crops. *J. Exp. Bot.* **2023**, *74*, 4349–4366. [CrossRef]
147. Xing, S.; Chen, K.; Zhu, H.; Zhang, R.; Zhang, H.; Li, B.; Gao, C. Fine-tuning sugar content in strawberry. *Genome Biol.* **2020**, *21*, 230. [CrossRef]
148. Zinelabidine, L.H.; Torres-Pérez, R.; Grimplet, J.; Baroja, E.; Ibáñez, S.; Carbonell-Bejerano, P.; Martínez-Zapater, J.M.; Ibáñez, J.; Tello, J. Genetic variation and association analyses identify genes linked to fruit set-related traits in grapevine. *Plant Sci.* **2021**, *306*, 110875.
149. Tello, J.; Torres-Pérez, R.; Flutre, T.; Grimplet, J.; Ibáñez, J. VviUCC1 Nucleotide Diversity, Linkage Disequilibrium and Association with Rachis Architecture Traits in Grapevine. *Genes* **2020**, *11*, 598. [CrossRef]
150. Ren, C.; Lin, Y.; Li, H.; Li, S.; Liang, Z. Targeted genome editing in grape using multiple CRISPR-guided editing systems. *bioRxiv* **2022**. [CrossRef]
151. Ren, C.; Gathunga, E.K.; Li, X.; Li, H.; Kong, J.; Dai, Z.; Liang, Z. Efficient genome editing in grapevine using CRISPR/LbCas12a system. *Mol. Hortic.* **2023**, *3*, 21. [CrossRef] [PubMed]
152. Jiao, B.; Hao, X.; Liu, Z.; Liu, M.; Wang, J.; Liu, L.; Liu, N.; Song, R.; Zhang, J.; Fang, Y.; et al. Engineering CRISPR immune systems conferring GLRaV-3 resistance in grapevine. *Hortic. Res.* **2022**, *9*, uhab023. [CrossRef]
153. Wang, Y.; Lecourieux, F.; Zhang, R.; Dai, Z.; Lecourieux, D.; Li, S.; Liang, Z. Data Comparison and Software Design for Easy Selection and Application of CRISPR-based Genome Editing Systems in Plants. *Genom. Proteom. Bioinf.* **2021**, *19*, 937–948. [CrossRef]

154. Liu, H.; Ding, Y.; Zhou, Y.; Jin, W.; Xie, K.; Chen, L.L. CRISPR-P 2.0: An Improved CRISPR-Cas9 Tool for Genome Editing in Plants. *Mol. Plant* **2017**, *10*, 530–532. [CrossRef] [PubMed]
155. Xie, X.; Ma, X.; Zhu, Q.; Zeng, D.; Li, G.; Liu, Y.G. CRISPR-GE: A Convenient Software Toolkit for CRISPR-Based Genome Editing. *Mol. Plant.* **2017**, *10*, 1246–1249. [CrossRef]
156. Lowder, L.G.; Zhou, J.; Zhang, Y.; Malzahn, A.; Zhong, Z.; Hsieh, T.F.; Voytas, D.F.; Zhang, Y.; Qi, Y. Robust Transcriptional Activation in Plants Using Multiplexed CRISPR-Act2.0 and mTALE-Act Systems. *Mol. Plant* **2018**, *11*, 245–256. [CrossRef] [PubMed]
157. Trinidad, J.L.; Longkumer, T.; Kohli, A. Rice Protoplast Isolation and Transfection for Transient Gene Expression Analysis. *Methods Mol. Biol.* **2021**, *2238*, 313–324.
158. Najafi, S.; Bertini, E.; D'Incà, E.; Fasoli, M.; Zenoni, S. DNA-free genome editing in grapevine using CRISPR/Cas9 ribonucleoprotein complexes followed by protoplast regeneration. *Hortic. Res.* **2022**, *10*, uhac240. [CrossRef]
159. Kong, Q.; Li, J.; Wang, S.; Feng, X.; Shou, H. Combination of Hairy Root and Whole-Plant Transformation Protocols to Achieve Efficient CRISPR/Cas9 Genome Editing in Soybean. *Plants* **2023**, *12*, 1017. [CrossRef]
160. Hosseini, S.M.; Bahramnejad, B.; Douleti Baneh, H.; Emamifar, A.; Goodwin, P.H. Hairy root culture optimization and resveratrol production from *Vitis vinifera* subsp. sylvesteris. *World J. Microbiol. Biotechnol.* **2017**, *33*, 67. [CrossRef]
161. Tisserant, L.P.; Aziz, A.; Jullian, N.; Jeandet, P.; Clément, C.; Courot, E.; Boitel-Conti, M. Enhanced Stilbene Production and Excretion in *Vitis vinifera* cv Pinot Noir Hairy Root Cultures. *Molecules* **2016**, *21*, 1703. [CrossRef] [PubMed]
162. Franks, T.K.; Powell, K.S.; Choimes, S.; Marsh, E.; Iocco, P.; Sinclair, B.J.; Ford, C.M.; van Heeswijck, R. Consequences of transferring three sorghum genes for secondary metabolite (cyanogenic glucoside) biosynthesis to grapevine hairy roots. *Transgenic Res.* **2006**, *15*, 181–195. [CrossRef] [PubMed]
163. Cutanda-Perez, M.C.; Ageorges, A.; Gomez, C.; Vialet, S.; Terrier, N.; Romieu, C.; Torregrosa, L. Ectopic expression of *VlmybA1* in grapevine activates a narrow set of genes involved in anthocyanin synthesis and transport. *Plant Mol. Biol.* **2009**, *69*, 633–648. [CrossRef] [PubMed]
164. Gambino, G.; Ruffa, P.; Vallania, R.; Gribaudo, I. Somatic embryogenesis from whole flowers, anthers and ovaries of grapevine (*Vitis* spp.). *Plant Cell Tiss. Org. Cult.* **2007**, *90*, 79–83. [CrossRef]
165. Dai, L.; Zhou, Q.; Li, R.; Du, Y.; He, J.; Wang, D.; Cheng, S.; Zhang, J.; Wang, Y. Establishment of a picloram-induced somatic embryogenesis system in *Vitis vinifera* cv. chardonnay and genetic transformation of a stilbene synthase gene from wild-growing *Vitis* species. *Plant Cell Tiss. Org. Cult.* **2015**, *121*, 397–412. [CrossRef]
166. Iocco, P.; Franks, T.; Thomas, M. Genetic transformation of major wine grape cultivars of *Vitis vinifera* L. *Transgenic Res.* **2001**, *10*, 105–112. [CrossRef]
167. Gambino, G.; Nuzzo, F.; Moine, A.; Chitarra, W.; Pagliarani, C.; Petrelli, A.; Boccacci, P.; Delliri, A.; Velasco, R.; Nerva, L.; et al. Genome editing of a recalcitrant wine grape genotype by lipofectamine-mediated delivery of CRISPR/Cas9 ribonucleoproteins to protoplasts. *Plant J.* **2024**, *119*, 404–412. [CrossRef] [PubMed]
168. Yan, T.; Hou, Q.; Wei, X.; Qi, Y.; Pu, A.; Wu, S.; An, X.; Wan, X. Promoting genotype-independent plant transformation by manipulating developmental regulatory genes and/or using nanoparticles. *Plant Cell Rep.* **2023**, *42*, 1395–1417. [CrossRef]
169. Vidal, J.R.; Kikkert, J.R.; Wallace, P.G.; Reisch, B.I. High-efficiency biolistic co-transformation and regeneration of 'Chardonnay' (*Vitis vinifera* L.) containing npt-II and antimicrobial peptide genes. *Plant Cell Rep.* **2003**, *22*, 252–260. [CrossRef]
170. Vidal, J.R.; Kikkert, J.R.; Donzelli, B.D.; Wallace, P.G.; Reisch, B.I. Biolistic transformation of grapevine using minimal gene cassette technology. *Plant Cell Rep.* **2006**, *25*, 807–814. [CrossRef]
171. Lowe, K.; Wu, E.; Wang, N.; Hoerster, G.; Hastings, C.; Cho, M.J.; Scelonge, C.; Lenderts, B.; Chamberlin, M.; Cushatt, J.; et al. Morphogenic Regulators Baby boom and Wuschel Improve Monocot Transformation. *Plant Cell* **2016**, *28*, 1998–2015. [CrossRef] [PubMed]
172. Lowe, K.; La Rota, M.; Hoerster, G.; Hastings, C.; Wang, N.; Chamberlin, M.; Wu, E.; Jones, T.; Gordon-Kamm, W. Rapid genotype "independent" *Zea mays* L. (maize) transformation via direct somatic embryogenesis. *In Vitro Cell. Dev. Biol. Plant* **2018**, *54*, 240–252. [CrossRef]
173. Chen, J.; Tomes, S.; Gleave, A.P.; Hall, W.; Luo, Z.; Xu, J.; Yao, J.L. Significant improvement of apple (*Malus domestica* Borkh.) transgenic plant production by pre-transformation with a Baby boom transcription factor. *Hortic. Res.* **2022**, *9*, uhab014. [CrossRef] [PubMed]
174. Debernardi, J.M.; Tricoli, D.M.; Ercoli, M.F.; Hayta, S.; Ronald, P.; Palatnik, J.F.; Dubcovsky, J. A GRF-GIF chimeric protein improves the regeneration efficiency of transgenic plants. *Nat. Biotechnol.* **2020**, *38*, 1274–1279. [CrossRef] [PubMed]
175. Maher, M.F.; Nasti, R.A.; Vollbrecht, M.; Starker, C.G.; Clark, M.D.; Voytas, D.F. Plant gene editing through de novo induction of meristems. *Nat. Biotechnol.* **2020**, *38*, 84–89. [CrossRef]
176. Kwak, S.Y.; Lew, T.T.S.; Sweeney, C.J.; Koman, V.B.; Wong, M.H.; Bohmert-Tatarev, K.; Snell, K.D.; Seo, J.S.; Chua, N.H.; Strano, M.S. Chloroplast-selective gene delivery and expression in planta using chitosan-complexed single-walled carbon nanotube carriers. *Nat. Nanotechnol.* **2019**, *14*, 447–455. [CrossRef]
177. Wang, B.; Huang, J.; Zhang, M.; Wang, Y.; Wang, H.; Ma, Y.; Zhao, X.; Wang, X.; Liu, C.; Huang, H.; et al. Carbon dots enable efficient delivery of functional DNA in plants. *ACS Appl. Bio Mater.* **2020**, *3*, 8857–8864. [CrossRef]
178. Caradus, J.R. Processes for regulating genetically modified and gene edited plants. *GM Crops Food* **2023**, *14*, 1–41. [CrossRef]

179. Eckerstorfer, M.F.; Engelhard, M.; Heissenberger, A.; Simon, S.; Teichmann, H. Plants developed by new genetic modification techniques-comparison of existing regulatory frameworks in the EU and NonEU countries. *Front. Bioeng. Biotechnol.* **2019**, *7*, 26. [CrossRef]
180. McHughen, A. A critical assessment of regulatory triggers for products of biotechnology: Product vs. process. *GM Crops Food* **2016**, *7*, 125–158. [CrossRef]
181. Broll, H.; Braeuning, A.; Lampen, A. European Court of Justice decision for genome editing: Consequences on food/feed risk assessment and detection. *Food Control* **2019**, *104*, 288–291. [CrossRef]
182. Puchta, H. Regulation of gene-edited plants in Europe: From the valley of tears into the shining sun? *Abiotech* **2023**, *5*, 231–238. [CrossRef] [PubMed]
183. Turnbull, C.; Lillemo, M.; Hvoslef-Eide, T.A.K. Global regulation of genetically modified crops aimed the gene edited crop boom-A review. *Front. Plant Sci.* **2021**, *12*, 630396. [CrossRef] [PubMed]
184. Molla, K.A.; Chakravorty, N.; Bansal, K.C. Genome editing for food, nutrition, and health. *Nucleus* **2024**, *67*, 1–4. [CrossRef]
185. Tsuda, M.; Watanabe, K.N.; Ohsawa, R. Regulatory status of genome-edited organisms under the Japanese cartagena act. *Front. Bioeng. Biotechnol.* **2019**, *7*, 387. [CrossRef]
186. Lemke, S.L. Gene editing in plants. a nutrition professional's guide to the science, regulatory, and social considerations. *Nutr. Today* **2022**, *57*, 57–63. [CrossRef]
187. Bohle, F.; Schneider, R.; Mundorf, J.; Zühl, L.; Simon, S.; Engelhard, M. Where does the EU-path on new genomic techniques lead us? *Front. Genome Ed.* **2024**, *6*, 1377117. [CrossRef]
188. Chaïb, J.; Torregrosa, L.; Mackenzie, D.; Corena, P.; Bouquet, A.; Thomas, M.R. The grape microvine -a model system for rapid forward and reverse genetics of grapevines. *Plant J.* **2010**, *62*, 1083–1092. [CrossRef]
189. Dai, X.; Chen, X.; Fang, Q.; Li, J.; Bai, Z. Inducible CRISPR genome-editing tool: Classifications and future trends. *Crit. Rev. Biotechnol.* **2018**, *38*, 573–586. [CrossRef]
190. Wang, X.; Ye, L.; Lyu, M.; Ursache, R.; Löytynoja, A.; Mähönen, A.P. An inducible genome editing system for plants. *Nat. Plants* **2020**, *6*, 766–772. [CrossRef]
191. Dong, Y.; Duan, S.; Xia, Q.; Liang, Z.; Dong, X.; Margaryan, K.; Musayev, M.; Goryslavets, S.; Zdunić, G.; Bert, P.F.; et al. Dual domestications and origin of traits in grapevine evolution. *Science* **2023**, *379*, 892–901. [CrossRef] [PubMed]
192. Wang, Y.; Ding, K.; Li, H.; Kuang, Y.; Liang, Z. Biography of *Vitis* genomics: Recent advances and prospective. *Hortic. Res.* **2024**, *11*, uhae128. [CrossRef] [PubMed]
193. Uddin, A.; Gallardo, R.K.; Rickard, B.; Alston, J.; Sambucci, O. Consumer acceptance of new plant-breeding technologies: An application to the use of gene editing in fresh table grapes. *PLoS ONE* **2022**, *17*, e0270792. [CrossRef] [PubMed]
194. Dalla Costa, L.; Piazza, S.; Pompili, V.; Salvagnin, U.; Cestaro, A.; Moffa, L.; Vittani, L.; Moser, C.; Malnoy, M. Strategies to produce T-DNA free CRISPRed fruit trees via *Agrobacterium tumefaciens* stable gene transfer. *Sci. Rep.* **2020**, *10*, 20155. [CrossRef]
195. Tricoli, D.M.; Debernardi, J.M. An efficient protoplast-based genome editing protocol for *Vitis* species. *Hortic. Res.* **2023**, *11*, uhad266. [CrossRef]

horticulturae

MDPI

Article

Effects of Different Biostimulants on Growth and Development of Grapevine Seedlings under High-Temperature Stress

Jiuyun Wu [1,2,†], Haixia Zhong [1,2,†], Yaning Ma [1], Shijian Bai [1,3], Vivek Yadav [2], Chuan Zhang [1,2], Fuchun Zhang [1,2], Wei Shi [4], Riziwangguli Abudureheman [1,*] and Xiping Wang [1,5,*]

1 Turpan Research Institute of Agricultural Sciences, Xinjiang Academy of Agricultural Sciences, Xinjiang Grape Engineering Technology Research Center, Turpan 838000, China; kobewjy@163.com (J.W.); zhonghaixia1@sina.cn (H.Z.); 18599491754@163.com (Y.M.); bsj19861103@163.com (S.B.); 2016204005@njau.edu.cn (C.Z.); zhangfc@xaas.ac.cn (F.Z.)

2 The State Key Laboratory of Genetic Improvement and Germplasm Innovation of Crop Resistance in Arid Desert Regions (Preparation), Key Laboratory of Genome Research and Genetic Improvement of Xinjiang Characteristic Fruits and Vegetables, Institute of Horticultural Crops, Xinjiang Academy of Agricultural Sciences, Urumqi 830091, China; vivekyadav@nwafu.edu.cn

3 Xinjiang Uighur Autonomous Region of Grapes and Melons Research Institution, Shanshan 838200, China

4 Turpan Eremophytes Botanic Graden, Xinjiang Institute of Ecology and Geography, Chinese Academy of Sciences, Urumqi 830091, China; shiwei@ms.xjb.ac.cn

5 Colleges of Horticulture, Northwest A&F University, Yangling 712100, China

* Correspondence: 13289951672@163.com (R.A.); wangxiping@nwsuaf.edu.cn (X.W.); Tel.: +86-29-87082129 (X.W.)

† These authors contributed equally to this work.

Abstract: High temperatures significantly affect the growth and development of grapevines, cause irreversible damage to plants, and severely impact grape production and quality. Biostimulants can promote the growth of plants and enhance their resistance to adverse stress. However, the effects of biostimulants on grapevines under high temperatures have not been studied in detail. To analyze the effects of various biostimulants on the growth and development of grape seedlings under high temperatures, we measured chlorophyll fluorescence parameters with observed seedling phenotypes under high temperatures in open field conditions in Turpan. We conducted a comprehensive analysis of the effects of different biostimulants on the growth, development, and photosynthesis of grapevine seedlings. Our study aimed to provide scientific evidence to improve cultivation methods for grapevines under high-temperature stress. The results revealed that biostimulants have a positive effect on promoting the growth of grapevine seedlings under high-temperature stress conditions. They also positively affect the accumulation of chlorophyll components in grapevine leaves, inhibiting chlorophyll degradation and maintaining photosynthesis. However, the effects of different biostimulants were inconsistent. A comprehensive analysis revealed the following effectiveness order: T2 > T1 > T3 > Control. These findings suggest that T2 is the most effective in alleviating high-temperature stress and promoting grapevine growth. We recommend the use of T2 to improve the cultivation of grapevine seedlings during high-temperature periods. This has implications for grape production in hot and arid climatic areas.

Keywords: grapevine; heat stress; biostimulant; JIP-test

1. Introduction

Grapevines (*Vitis vinifera* L.) are sessile organisms that cannot change their location in the field and inevitably face various biological and abiotic stresses during their growth and development. The grape is an economically important crop species worldwide, but its quality and production are often limited by high-temperature stress. With global warming, high temperatures have become one of the primary abiotic stress factors that restrict the yield and quality of grapes [1–3]. Turpan, one of the most important grape production areas in Xinjiang, China, covers an area of 36,253 hectares with a yield of 1,447,800 tons. 'Thompson Seedless' is the main cultivated grape variety in Turpan, with a planting area of 32,640 hectares (https://www.tlf.gov.cn/tlfs/sjkf/tlfdata.shtml, accessed on 28 November 2023). However, due to the distinctive geography of the Turpan region, its temperatures always exceed 40 °C for more than 35 days each summer. This intense heat can cause grapevines to lose water, wilt, and suffer cell damage. It can also disrupt their photosynthesis process. This seriously affects the growth and development of grapevine seedlings and causes irreversible damage to plants, resulting in stunted growth and even death of the plants [4–6].

Biostimulants are agronomic products that have become highly important in agriculture because they are formulated with substances capable of stimulating physiological and biochemical processes in plants, which help them adapt to different detrimental environmental conditions [7]. In general, a biostimulant is an agrochemical product formulated with mixtures of natural substances or microorganisms, which is used for enhancing nutrition efficiency and crop quality traits [8]. Moreover, biostimulants have varying effects on different plant species or cultivars, supporting the activity of microorganisms and serving as substrates for the formation of biologically active substances by these microorganisms. They can enhance the tolerance to stress in plants, and it has been determined that using biostimulants can improve the efficiency of mineral nutrition, resistance to abiotic stress (drought, high temperatures, salinity, heavy metals, etc.), and crop yield, or enhance quality characteristics, regardless of its content in essential mineral nutrients for plants, to maintain a good agronomic yield and quality of harvest under these conditions [9]. Furthermore, recent studies have evidenced the important role of biostimulants in minimizing cellular oxidative stress in plants. For instance, plant-based modified biostimulants (copper chlorophyllin) were used for plants under salinity stress, and findings showed that cellular oxidative stress was decreased in Arabidopsis thaliana plants [10]. A recent study also showed that seaweed extract biostimulants affected the sugarcane morphology and physiology through significant changes in oxidative stress [11].

Photosynthesis is one of the most sensitive physiological processes highly affected by temperature conditions. Leaves, being crucial organs for plant photosynthesis and gas exchange, exhibit the most obvious signs in response to heat stress [10]. High temperatures not only damage the tissue structure of grapevine leaves but also inhibit photosynthesis and nutrient metabolism. Additionally, it affects the growth and development of grapevine plants, leading to an imbalance in energy metabolism and material transformation processes. Eventually, this can also result in heat damage, malaise, or even the death of the plant [12–14]. A number of studies have reported that biostimulants have a significant impact on the growth and development of crops such as Capsicum, Malus, Nicotiana, and others. Additionally, the use of biostimulants can alleviate damage caused by heat stress. Biostimulants are widely used in crop cultivation due to their high efficiency, environmental friendliness, and absence of residues [15–19]. However, the effects of biostimulants on grapevines under high-temperature stress remain poorly understood. Therefore, we conducted research on the effects of different biostimulants on the growth and development of grapevine seedlings under natural high temperatures. Our aim is to provide a basis for reinforcing the cultivation of grape seedlings, offering a reference for alleviating heat stress, and exploring pathways for stress-resistant cultivation.

2. Materials and Methods

2.1. Plant Material and Treatments

'Thompson Seedless' (*V. vinifera*) is the main cultivated variety in Turpan grape production area, which covers a planting area of 32,640 hectares. We selected 'Thompson Seedless' (*V. vinifera*) grape plants as an experimental material and cultivated them in plastic pots at the grapevine garden of the Turpan Research Institute of Agricultural Sciences, Xinjiang Academy of Agricultural Sciences (XAAS). The garden is located at 89°11′ E, 42°56′ N, at an altitude of 0 m [10]. Uniform grapevine seedlings aged one year (1a) were selected and transplanted into pots. For the experiment, 1-year-old grapevine seedlings with almost the same height and stem diameter were cultivated in plastic pots containing soil mixed uniformly with substrate, nutrient soil, and field soil in a 1:1:1 ($v/v/v$) ratio. The plastic pots, with dimensions of 37.5 cm in diameter and 40.5 cm in height, were used for the cultivation. To ensure the optimal growth of the seedlings during observation and sampling, all containers were mixed with the same soil uniformly, and the soil moisture was carefully maintained at a moderate level.

Three treatments were planned, and the grapevine seedlings were treated with biostimulants through a combination of foliar spraying and root irrigation. The plants were sprayed and irrigated approximately once every 7 days. Water was used as a control (C), following the same cultivation conditions as the test treatments. Each treatment is sprayed with a watering can (200 mL) filled with biostimulants when there is no wind and the air temperature is below 30 °C (9:30–10:30). The upper and lower leaves are wet as is the standard to ensure all leaves are evenly sprayed. The plants are fully irrigated by mixing every 100 mL of biological agent with 1000 mL of water. Each experiment utilized every single plant in the container, with each treatment having three grapevine seedlings and three biological replicates. The biostimulant products with patents were researched and provided by Turpan Eremophytes Botanic Garden, Xinjiang Institute of Ecology and Geography, Chinese Academy of Sciences. The biostimulants were named T1 (Main compounds: β-Myrcene; Patent No.: ZL202010410895.1), T2 (Main compounds: BaZFP924 protein; Patent No.: ZL202110900671.3), and T3 (Main compounds: Aspergone; Patent No.: ZL202210880877.9), respectively.

2.2. Temperature Measurement Tested

Throughout the period of high-temperature conditions in the grape research farm, we employed the MicroLite USB Temperature Data Logger (manufactured by Fourier Systems, Fourtec-Fourier Technologies, Ltd., San Francisco, CA, USA) to monitor temperature fluctuations at a specific location. Temperature readings were recorded and logged every hour, covering the entirety of the high-temperature span from 1 May to 31 August 2022.

2.3. Plant, Leaf and Root Observation and Tested

The height of the plants, the length of the internodes (3rd, 4th, 5th, 7th, and 9th internodes in the branch of the vine), the stem diameter, and the root length (cm) were measured using a ruler. Additionally, the dry weight and fresh weight of the roots (g) were determined using an electronic balance. Three healthy, fully opened leaves from the 5th to 9th positions from the top were randomly selected and labeled on each plant. Chlorophyll (Chl) levels were measured using a TYS-B tester (Zhejiang Tuopu Yunnong Technology, Ltd., Hangzhou, Zhejiang, China) from 9:00 to 11:00, expressed by SPAD. The leaves were wiped clean before determination, and measurements were taken every 10 days.

For each grapevine seedling, three leaves were selected from 5th to 9th nodes from the top. The leaf area (mm^2) was measured using a leaf area meter. Subsequently, the fresh leaf mass (g) was determined using an electronic scale. The leaves were then placed in a room for drying, and the dry leaf mass (g) was measured after complete drying.

2.4. Measurement of Polyphasic Chl Fluorescence Transient OJIP

We selected fully opened grapevine leaves from the 5th to 9th positions from the top at 12:00 to 16:00 on a sunny day with no wind or clouds for measuring Chl, a fluorescence parameter, using the OJIP test. The parameters were measured using the Fluor Pen FP110 (Ecotech Ecological Technology Co., Ltd., Drásov, Czech Republic). Before measurements, leaves were acclimated to darkness for 20 min; the parameters included F_0, F_v, F_m, F_v/F_0, F_v/F_m, Ψ_0, φE_0, PI_{ABS}, ABS/RC, TR_0/RC, ET_0/RC, DI_0/RC, etc. to provide information on the photochemical activity of PSII and the status of the PQ-pool [20].

2.5. Data Analysis

The test data were used for variance analysis, employing the LSD method for multiple comparisons and assessing the significance of differences. A significance level of $p < 0.05$ was considered for differences, while $p < 0.01$ denoted an exceptionally significant distinction. ANOVA, PCA, and figure generation were conducted using GraphPad Prism ver.9.0 (October 2020, Dotmatics Corporation, Boston, MA, USA).

3. Results

3.1. Temperature Dynamics in the Field

In May, the average temperature in the field was 30.52 °C, with the maximum temperature reaching 42.65 °C. By late June, the region had transitioned into a high-temperature phase, experiencing a peak of 44.29 °C, featuring 26 days surpassing 35 °C and 12 days exceeding 40 °C. July witnessed a continuous temperature rise, registering an average temperature of 32.92 °C and a maximum of 44.29 °C. During this period, 30 days surpassed 35 °C, with 14 days exceeding 40 °C. August marked the onset of an extremely high-temperature period, reaching a maximum of 45.67 °C. This scorching phase included 26 days beyond 35 °C and a notable 18 days surpassing 40 °C (refer to Table 1). Throughout the recorded high-temperature period spanning June to August, the maximum temperature soared to 45.67 °C, while the lowest dipped to 18.90 °C, resulting in an average temperature of 32.67 °C. Turpan experienced the initiation of the high-temperature period in early June 2022 (Figure 1), culminating in an extreme high-temperature phase in August 2022. The average temperature during this period was 32.13 °C, with a maximum temperature soaring to a remarkable 45.67 °C.

Table 1. Temperature variation during high-temperature period (2022).

Air Temperature	May	June	July	August
Maximum Temperature/°C	42.65	44.29	44.28	45.67
Minimum Temperature/°C	16.82	22.27	22.10	18.90
Average Temperature/°C	30.52	32.92	33.62	31.49
≥35.00 °C/Day	27.00	26.00	30.00	26.00
≥40.00 °C/Day	9.00	12.00	14.00	18.00

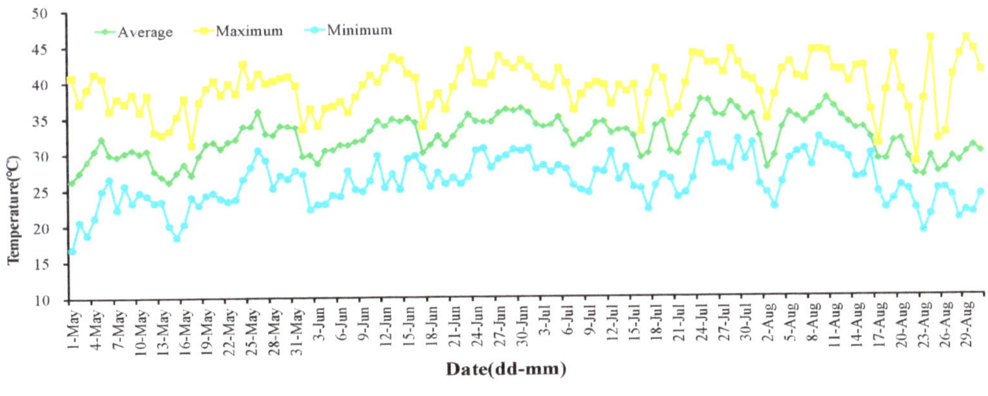

Figure 1. The temperature of the viticultural region of XAAS in summer in Turpan.

3.2. Plants and Roots Distribution

During the plant growth and development stage (30-May to 12-September), the control group always exhibited the smallest plant type and the shortest plant height, which was significantly lower than that of the treated groups ($p < 0.05$, 12-September). The root distribution in the control group was relatively smaller, with the shortest root length, indicating weaker overall plant growth compared to the treated groups (see Figures 2 and 3a).

Figure 2. Effects of different biostimulants on plants and roots of grapevine.

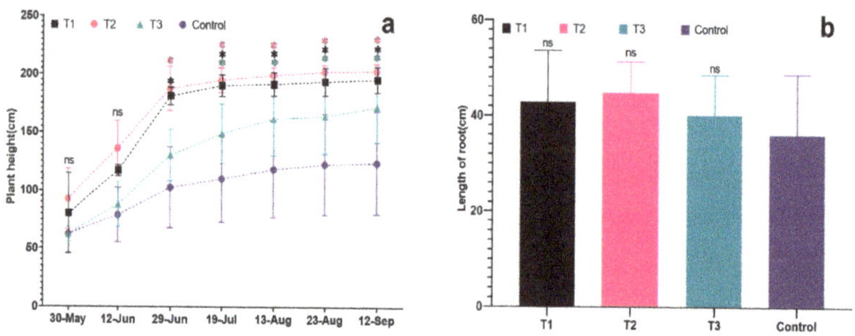

Figure 3. Effects of different biostimulants on plant height and root length of grapevine: (**a**) Plant height of grapevine; (**b**) Root length of grapevine. Graphs are plotted with the mean ± SE of three replications. * represent significant differences in expression levels at $p < 0.05$ and $p < 0.01$. "ns" is used to show non-significant differences.

Among the treated groups, T2 showed significantly better plant growth than the other treatments, and the seedlings' growth followed the order: T2 > T1 > T3 > Control. The results revealed a trend of fast growth in the early stage and slow growth in the later stage for grapevine seedlings. The average plant height for each treatment was 195.90 cm, 203.00 cm, and 172.20 cm, respectively. The control group had the shortest average plant height (only 123.20 cm), significantly lower than the treated groups ($p < 0.05$). Additionally, the control group exhibited a relatively smaller root distribution, the shortest root length, and weaker overall plant growth compared to the treated groups. In contrast,

T2 had an average plant height of 203.00 cm, significantly higher than the other treated groups ($p < 0.05$). T2 plants demonstrated the greatest strength, with more extensive root distribution, better growth of fresh branches, and overall plant development (see Figures 2 and 3a).

The root length of each treatment was significantly greater than that of the control (Figure 3b), with the root lengths ranked as follows: T2 > T1 > T3 > Control. The respective root lengths were 42.87 cm, 44.73 cm, 40.17 cm, and 36.10 cm. Notably, T2 exhibited the longest root length at 44.73 cm. The fresh root weight of T1 was the highest (66.0 g), followed by T2 (51.0 g) and T3 (28.3 g). In comparison, the control group had the lowest fresh root weight (24.7 g), significantly lower than each treatment ($p < 0.05$) and notably lower than T1 and T2 ($p < 0.01$). This indicates that all three biostimulants had a positive impact on the root growth of grapevine seedlings. Examining dry root weight, T2 had the highest value (17.7 g), followed by T3 (16.0 g) and T1 (15.0 g). The control group exhibited the lowest dry root weight (13.7 g), significantly lower than T2 ($p < 0.01$). The order of dry root weight was T2 > T1 > T3 > Control, with the dry root weights of each treatment being significantly higher than that of the control ($p < 0.05$).

3.3. Growth of Grapevine Seedling

All grapevine seedlings exhibited certain differences in height during the initial measurement on 30 May (see Figure 3a), These height growth for each treatment was 37.17 cm, 36.83 cm, 30.40 cm, and 38.83 cm, respectively, although none of these differences were deemed significant ($p < 0.05$). However, vines began to grow rapidly in June, the plant height has grown in 22.93 cm to 63.83 cm, height of T1 increased is the greatest, control is the lowest, was significantly lower than other treatments (see Table 2). By the last measurement on 12 September, the plant heights for each treatment were recorded as 195.90 cm, 203.00 cm, 172.20 cm, and 123.20 cm for the control group. These values represented a notable increase compared to the initial measurements on 30 May. The recorded growth for each treatment was 115.90 cm, 103.80 cm, and 114.40 cm, respectively. In contrast, the control group only exhibited a growth of 83.00 cm, which was significantly lower than that of T1 ($p < 0.01$). Upon analyzing the internode length of different nodes, all grapevine seedlings showed certain differences in internode length during the initial measurement on 30 May (see Figure 4b). Although there were no significant differences, it was observed that the internode length of each node in the control group was lower than that of the treated groups. Furthermore, the internode length of T1 was the maximum, consistently surpassing the lengths observed in other treatments, which is significantly greater than the control in the seventh and ninth internodes. This indicates that T1 had a more pronounced effect on the height growth of grapevine seedlings (Figure 4b).

Table 2. Effects of different biostimulants on grapevine growth.

Group	T1	T2	T3	Control
30-May	37.17 ± 37.59 ^{aA}	36.83 ± 5.11 ^{aA}	30.40 ± 33.19 ^{aA}	38.83 ± 32.37 ^{aA}
12-June	63.83 ± 10.49 ^{aA}	51.40 ± 12.61 ^{abA}	42.17 ± 16.00 ^{abA}	22.93 ± 21.03 ^{bA}
29-June	9.17 ± 4.37 ^{aA}	7.43 ± 8.79 ^{aA}	18.50 ± 22.15 ^{aA}	7.23 ± 12.36 ^{aA}
13-August	1.33 ± 1.04 ^{aA}	4.33 ± 4.01 ^{aA}	12.83 ± 16.64 ^{aA}	8.00 ± 7.76 ^{aA}
23-August	2.43 ± 2.53 ^{aA}	3.00 ± 1.00 ^{aA}	2.50 ± 2.00 ^{aA}	4.50 ± 3.60 ^{aA}
12-September	1.93 ± 1.44 ^{aA}	0.83 ± 0.58 ^{bAB}	8.00 ± 4.82 ^{bAB}	1.50 ± 0.87 ^{bB}

The average values are means ± S.E. Different lowercase letters in each column indicate a significant difference at $p < 0.05$; Uppercase letters in each column indicate a significant difference at $p < 0.01$.

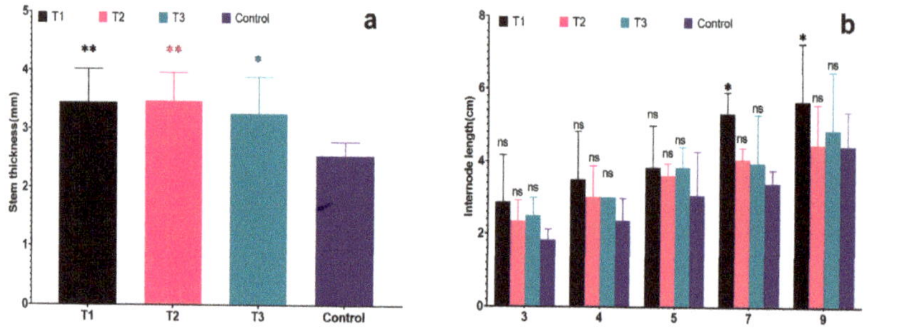

Figure 4. Effects of different biostimulants on stem diameter and internode length of grapevine: (**a**) Stem thickness; (**b**) Internode length. Graphs are plotted with the mean ± SE of three replications. * and ** represent significant differences in expression levels at $p < 0.05$ and $p < 0.01$. "ns" is used to show non-significant differences among the treatments.

In addition, the stem diameters of the seedlings in each treatment were 3.45 cm, 3.47 cm, 3.26 cm, and 2.55 cm, respectively. The stem diameters of the seedlings in each treatment were significantly greater than the control ($p < 0.05$). Moreover, the stem diameters of T1 were significantly greater than those of the control ($p < 0.01$). This suggests that all of the biostimulants had a certain effect on promoting the shoot growth of grapevine seedlings. Notably, T2 had a more pronounced effect on the stem diameter of grapevine seedlings (Figure 4a).

3.4. Leaves and Chl Content

In Figure 5a,b, it is evident that the dry and fresh leaf weights of grapevine seedlings in each treatment exceeded those of the control. Notably, T2 exhibited the highest fresh leaf weight at 1.10 g, followed by T1 (1.03 g) and T3 (0.96 g). In comparison, the control group had the smallest average fresh leaf weight at 0.95 g, which was obviously lower than T2 ($p < 0.05$). In leaf dry weight, T2 had the highest value (0.31 g), followed by T1 and T3 at 0.29 g and 0.28 g, respectively. The control had the lowest dry leaf weight, measuring 0.27 g, which was significantly lower than T2 ($p < 0.05$). In terms of leaf area, the treated plants showed a larger leaf area compared to the control group (see Figure 5c). The ranking of leaf area for the treatments was as follows: T1 > T2 > T3 > Control. Specifically, the leaf areas were 95.76 cm^2 for T1, 90.61 cm^2 for T2, 84.83 cm^2 for T3, and 79.38 cm^2 for the control. There was a significant difference in leaf area between T1 and the control, with T1 having the largest leaf area and the control having the smallest ($p < 0.05$). The chlorophyll (Chl) levels directly reflected the nutritional and photosynthetic capacity of plants. In each treatment, there was an initial increase in Chl, followed by a decrease and a slight rise again. Initially (on 30 May), the Chl levels were almost the same for all treatments. However, as the plants grew and developed, and with temperature changes, the Chl levels in the leaves began to vary. Notably, T2 consistently showed the highest Chl levels for most of the observation period. In the early stages, the control had slightly higher Chl levels than T2 and T3, but as time progressed, it became lower than each treatment, particularly during the middle of the high-temperature period (after 19 July). In particular, the Chl content of T3 was always lower than other treatments in the early stage, but it was significantly higher than others treated in the later stage, indicating that all three biostimulants had a positive effect on the Chl accumulation of grapevine leaves or alleviating the degradation of Chl, and the Chl of grapevine leaves was always maintained at a high level in treatment, which was facilitating the growth of grapevine seedlings under the high temperature.

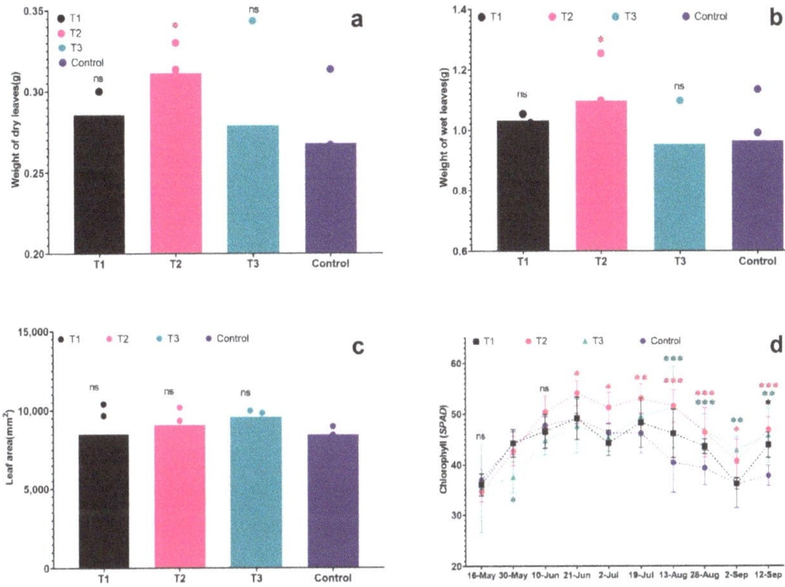

Figure 5. Effects of different biostimulants on grape leaves: (**a**) Weight of dry leaves; (**b**) Weight of fresh leaves; (**c**) Leaf area; (**d**) Chl content. Colour represent treatment. Graphs are plotted with the mean ± SE of three replications. *, ** and *** represent significant differences in expression levels at $p < 0.05$, $p < 0.01$ and $p < 0.001$. "ns" is used for non-significant differences.

3.5. Polyphasic Chl Fluorescence Transient OJIP

There was no significant difference between the treatments in the OJIP normalization curve from June to August (Figure 6a,c,e). However, it is evident that the OJIP curve of the control was slightly higher than that of the treated group in June (Figure 6a). Results of the high-temperature period (July and August) showed that the OJIP curve of the control became significantly lower than that of each treated group (Figure 6c,e). Therefore, Through the double normalization of the OJIP curve data, it becomes evident that the O-J phase, J-I phase, and I-P phase of each treatment were significantly lower than those of the control, except for the time in the I-P phase of T1 (Figure 6b). This suggests that the photosynthetic efficiency of the control was significantly higher than that of the other treatments in the high-temperature early stage (June).

However, during the high-temperature period in July, except for T1, which was slightly lower than the control in the O-J phase for a brief period, the photosynthetic efficiency of the three treatments in the O-J and J-I phases was consistently higher than that of the control. This indicates that the photosynthetic efficiency of the seedlings in the three treatments at this stage was significantly higher than that of the control (Figure 6d). After the high-temperature period in August, both T2 and T1 exhibited significantly higher values during the O-J phase and J-I phase. However, in the I-P phase, they were lower than the control values. Specifically, T1 was significantly lower than the control in the O-J phase, J-I phase, and I-P phase, as illustrated in Figure 6f.

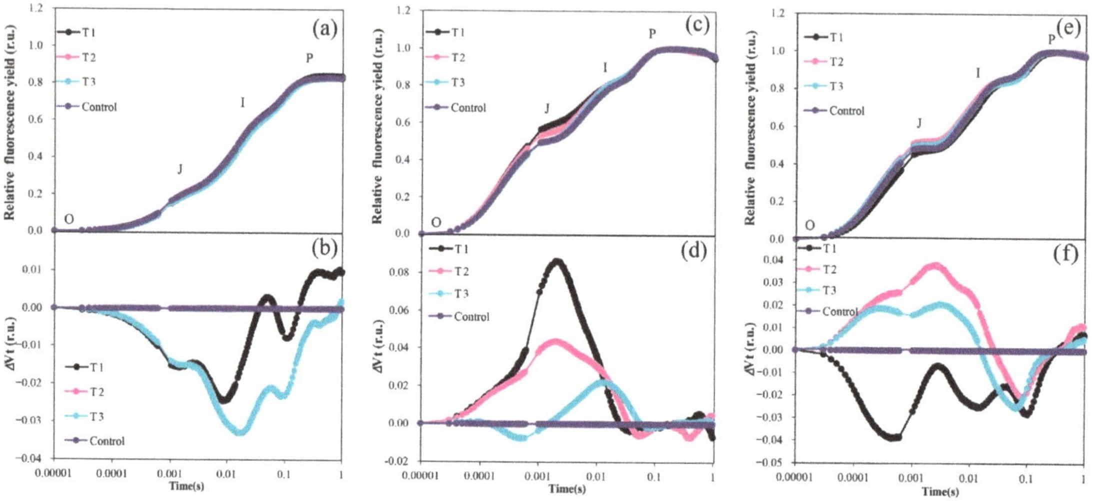

Figure 6. OJIP curves of grapevine for different treatments under high−temperature conditions: (**a**) Normalized OJIP curve for June; (**b**) Double normalized OJIP curve for June month; (**c**) Normalized OJIP curve for July (**d**) Double normalized OJIP curve for July month; (**e**) Normalized OJIP curve for August; (**f**) Double normalized OJIP curve for August.

3.6. *Chl a Fluorescence Parameters*

F_0 reflected the degree of damage to the thylakoid membrane. A higher F_0 value indicates more severe damage to the thylakoid membrane. The maximum fluorescence F_m reflects the electron transport from Photosystem II (PSII), and a lower Fm value indicates a higher degree of heat damage. The F_v/F_0 represents the potential activity of PSII, reflecting the activity of the PSII center. The maximum photochemical quantum yield, F_v/F_m, characterizes the photoenergy conversion efficiency of the PSII center. Ψ_0 represented the ratio of electron transport optical quantum flux in the captured light quantum flux. φE_0 represents the photoquantum yield for electron transport, reflecting the proportion of absorbed light quanta that transport electrons to other downstream electron acceptors. PI_{ABS} is the photochemical performance index based on absorbed light energy and ABS/RC is the absorbed light quantum flux at the reaction center of PSII. TR_0/RC is the initial (or maximum) captured light quantum flux of the reaction center, ET_0/RC is the initial electron transport light quantum flux of the reaction center, and DI_0/RC reflects the ratio of energy dissipated by the PSII reaction center as thermal energy.

The results revealed that the control group exhibited lower values compared to the treated groups in terms of F_v/F_m, F_v/F_0, F_v, φE_0, PI_{ABS}, and other fluorescence parameters. Additionally, DI_0/RC was significantly higher in the control group than in each treated group. This indicates that during the high-temperature period (August), the light energy conversion efficiency, activity of the PSII center, optical quantum yield, and photochemical properties of the control group were lower than those of each treated group. Furthermore, the heat dissipation ratios in the control group were significantly higher than in all the treated groups. These findings suggest that all three biostimulants had a certain effect on the photosynthetic efficiency of grapevine seedlings under high-temperature conditions. T2 was significantly higher than that of others treated in aspects of Ψ_0, F_v/F_m, F_v/F_0, φE_0, and PI_{ABS} (Figure 7a), while parameters of F_0, TR_0/RC, and ABS/RC were significantly lower than those of other treatments (Figure 7b). This indicated that the degree of heat damage of T2 was relatively weak under high-temperature conditions, and it explained that T2 could inhibit or alleviate the heat damage, maintain the relatively high photosynthesis ability of seedlings, and facilitate the growth and development of grapevine seedlings.

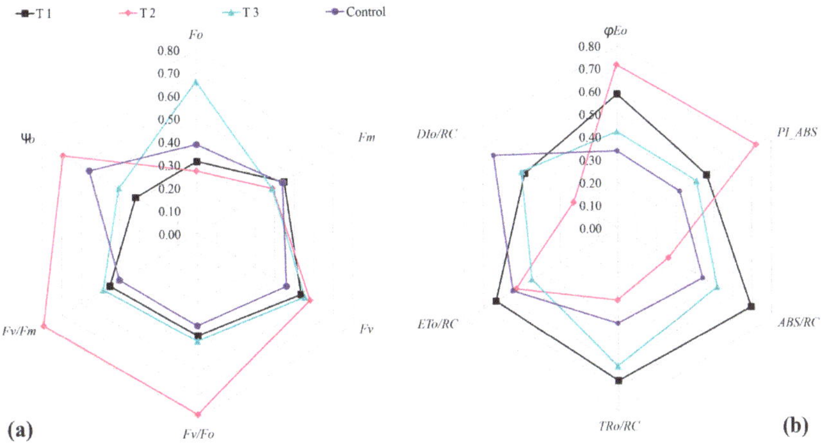

Figure 7. Comparison of Chl a fluorescence parameter in different treatments. (**a**) F_0, F_v, F_m, F_v/F_0, F_v/F_m, and Ψ_0; (**b**) φE_0, PI_{ABS}, ABS/RC, TR_0/RC, ET_0/RC, and DI_0/RC.

4. Discussion

With the global climate changing, high-temperature conditions are becoming more frequent and lasting longer, significantly affecting plant growth, development, and crop yield. Grapevines, constantly exposed in the field, must withstand various unavoidable abiotic stresses during their growth and development. With the continuous increase in global temperatures, heat stress has gradually become a major limiting factor in the development of the grape industry [2,21]. Heat stress (>35 °C) can significantly impede the photosynthesis and nutrient metabolism of grapes, leading many high-quality varieties to lose their inherent characteristics. This directly affects the commodity and market value of grapes [3,22]. More importantly, dealing with heat stress will pose an unavoidable challenge for the global grape industry [15,23–25]. Turpan, Xinjiang, has become one of the most renowned grape-producing areas in China, which is attributed to its superior light and heat conditions that provide a unique environment for viticulture growth. However, the distinctive geographical condition also results in an atmosphere where temperatures equal or exceed 35 °C for over 100 days a year. The prolonged high-temperature conditions can affect the photosynthetic process, limiting the synthesis and transfer of photosynthetic products and significantly impacting its growth and development. When heat stress surpasses the plant's regulatory capacity, it manifests as injury symptoms in the phenotype, ultimately leading to wilting and, in severe cases, the death of the plants [5,26,27].

The 'Thompson Seedless' cultivar plays a pivotal role in Turpan, covering a planting area of 32,640 hectares, which accounts for 90.03% of the viticulture in the region [28]. High temperatures have frequently impacted the photosynthesis of plants, limited grape production, and impeded the growth of grapevine seedlings. This significantly affects the overall growth, development, and yield of grapevines. Biostimulants have been widely employed in crops due to their safety, non-polluting nature, low residue levels, and their ability to facilitate the growth and development of crops [29–31]. However, it remains uncertain whether they effectively work on grapevine seedlings under naturally high temperatures. Simulating high temperatures indoors may not fully and accurately reflect the growth of seedlings, particularly when considering new production methods [32–34]. Therefore, we planned to analyze the effects of various biostimulants on the growth and development of grape seedlings, as well as the fluorescence characteristics of chlorophyll, under natural high temperatures in Turpan. This study may not only offer insights into enhancing the cultivation of grapevine seedlings but also explore methods for cultivation in high-temperature conditions. Additionally, it holds reference significance for grape production in thermal and arid areas.

High temperature has a deleterious impact on plant osmotic adjustment by increasing evapotranspiration, which has an irreversible impact on solute generation, which is required for stress tolerance. Grapevines are regarded as a model perennial fruit crop for heat tolerance research, revealing that high temperatures pose various challenges to the development of high-quality grape berries [35]. We found that all biostimulants have diverse effects on the growth and development parameters of grapevine seedlings. The grapevine seedlings treated with biostimulants exhibited better performance than those in the control group in terms of plant height, stem diameter, and root development. This indicates that all three biostimulants have a positive effect on the growth of grapevine seedlings, which is consistent with findings from previous studies [36,37]. Moreover, the dry leaf weight, fresh leaf weight, leaf area, and chlorophyll content of each treatment were higher than those of the control. This indicates that the biostimulants also had a positive effect on leaf growth, consistent with findings from previous studies on *Pyrus* [38], *Gossypium* [39], *Maize* [40], and many other crops.

We also observed that as the duration of high temperature prolonged, the control group exhibited lower values compared to each treatment in features such as F_v/F_m, F_v/F_0, F_v, φE_0, and PI_{ABS}. In contrast, DI_0/RC was significantly higher in the control group than in any of the treated groups. Among the biostimulants, T2 was significantly higher than that of others treated in aspects of Ψ_0, F_v/F_m, F_v/F_0, φE_0, and PI_{ABS} (Figure 7a), while parameters of F_0, TR_0/RC, and ABS/RC were significantly lower than those of other treatments. Some other parameters, like the dry and fresh weight of leaves, were comparatively higher in T2 when compared with any other treatments. The overall findings of various parameters showed that the degree of heat damage of T2 was relatively weak under high-temperature conditions. Overall findings indicate that the light energy conversion efficiency, activity of the PSII center, optical quantum yield, and photochemical properties of the control group were lower than those of each treated group, and the heat dissipation ratios in the control group were significantly higher than in all the treated groups during the high-temperature period. Furthermore, these suggest that three biostimulants had a positive effect on maintaining high levels of photosynthetic efficiency in grapevine seedlings under high temperatures [26]. Moreover, the results indicate that biostimulants could inhibit or alleviate damage to the photosynthesis of seedlings under high temperatures, thus enabling the maintenance of a relatively high level of photosynthesis. This conclusion is consistent with previous studies [41–44]. Furthermore, three biostimulants promoted the accumulation of chlorophyll or inhibited its decomposition in grapevine leaves. They assisted in maintaining a consistently high level of chlorophyll, facilitating the growth of grapevine seedlings during periods of high temperature. This indicates that biostimulants are not only beneficial to vine growth but also contribute to enhancing the heat tolerance of grapevine seedlings. And also, we should enhance our ability to comprehensively understand the various variables influencing heat stress and develop strategies for heat-tolerance grapevine breeding and cultivation [35,45].

5. Conclusions

Biostimulants play a crucial role in enhancing the growth and development of the roots, leaves, and overall health of vine seedlings. They contribute to promoting the accumulation of chlorophyll (Chl) and inhibiting its degradation in leaves. This ensures that photosynthetic efficiency remains at a high level, which is beneficial for the growth of grapevine seedlings, particularly under high-temperature conditions. Among these biostimulants, '6-B-2′ (treated 2) stands out as particularly advantageous for the growth and development of vine seedlings in thermal and arid areas. It demonstrates the most effective results in alleviating heat stress and facilitating seedling growth.

Author Contributions: Conceptualization, J.W. and X.W.; methodology, R.A., J.W. and W.S.; software, J.W., C.Z., V.Y. and F.Z.; validation, J.W., H.Z. and X.W.; investigation, R.A., S.B. and Y.M.; writing—review and editing: J.W., H.Z. and X.W.; visualization, J.W. and V.Y.; supervision, H.Z. and X.W.; funding acquisition, J.W. and X.W. All authors have read and agreed to the published version of the manuscript.

Funding: This research was funded by the Natural Science Foundation of Xinjiang Uygur Autonomous Region (2023D01A96), Xinjiang Uygur Autonomous Region Tianshan Talents Training Program (2022TSYCJC0036), Xinjiang Uygur Autonomous Region Innovation Environment Construction Special Program (PT2314), and Xinjiang Uygur Autonomous Region Tianchi Talent—Special Expert Project (Xiping Wang, 2022), and the Hunan Innovative Province Construction Program (2023WK4008-4).

Data Availability Statement: The data that support the findings of this study are available upon request from the corresponding author.

Acknowledgments: We would like to thank the supportive and scientific staff in the research lab and grape research farm of Xinjiang Academy of Agriculture Sciences. Thanks to Wei Shi from the Turpan Eremophytes Botanical Garden, the Xinjiang Institute of Ecology and Geography, and the Chinese Academy of Sciences, who provided the biostimulant products and funding for testing.

Conflicts of Interest: The authors declare that they have no conflicts of interest.

References

1. Hayes, S.; Schachtschabel, J.; Mishkind, M.; Munnik, T.; Arisz, S.A. Hot topic: Thermosensing in plants. *Plant Cell Environ.* **2021**, *44*, 2018–2033. [CrossRef]
2. Kim, J.S.; Jeon, B.W.; Kim, J. Signaling peptides regulating abiotic stress responses in plants. *Front. Plant Sci.* **2021**, *12*, 704490. [CrossRef]
3. Upadhyay, A.; Upadhyay, A.K. Global transcriptome analysis of heat stress response of grape variety 'Fantasy Seedless' under different irrigation regimens. *Vitis* **2021**, *60*, 143–151. [CrossRef]
4. Wu, J.Y.; Lian, W.J.; Zeng, X.Y.; Liu, Z.G.; Mao, L.; Liu, Y.X.; Jiang, J.F. Physiological response to high temperature and heat tolerance evaluation of different grape cultivars. *Acta Bot. Boreali-Occident. Sin.* **2019**, *39*, 1075–1084. [CrossRef]
5. Zha, Q.; Yin, X.; Xi, X.; Jiang, A. Colored shade nets can relieve abnormal fruit softening and premature leaf senescence of 'Jumeigui' grapes during ripening under greenhouse conditions. *Plants* **2022**, *11*, 1227. [CrossRef]
6. Liu, Y.; Jiang, J.; Fan, X.; Zhang, Y.; Wu, J.; Wang, L.; Liu, C. Identification of heat tolerance in chinese wildgrape germplasm resources. *Horticulturae* **2020**, *6*, 68. [CrossRef]
7. García-Sánchez, F.; Simón-Grao, S.; Navarro-Pérez, V.; Alfosea-Simón, M. Scientific Advances in Biostimulation Reported in the 5th Biostimulant World Congress. *Horticulturae* **2022**, *8*, 665. [CrossRef]
8. Du Jardin, P. Plant biostimulants: Definition, concept, main categories and regulation. *Sci. Hortic.* **2015**, *196*, 3–14. [CrossRef]
9. Tariq, M.; Khan, A.; Asif, M.; Khan, F.; Ansari, T.; Shariq, M.; Siddiqui, M.A. Biological control: A sustainable and practical approach for plant disease management. *Acta Agric. Scand. Sect. B—Soil Plant Sci.* **2020**, *70*, 507–524. [CrossRef]
10. Islam, M.T.; Ckurshumova, W.; Fefer, M.; Liu, J.; Uddin, W.; Rosa, C. A plant based modified biostimulant (Copper Chlorophyllin), mediates defense response in *Arabidopsis thaliana* under salinity stress. *Plants* **2021**, *10*, 625. [CrossRef] [PubMed]
11. Jacomassi, L.M.; Viveiros, J.O.; Oliveira, M.P.; Momesso, L.; de Siqueira, G.F.; Crusciol, C.A.C. A seaweed extract-based biostimulant mitigates drought stress in sugarcane. *Front. Plant Sci.* **2022**, *13*, 865291. [CrossRef]
12. Wu, J.; Abudureheman, R.; Zhong, H.; Yadav, V.; Zhang, C.; Ma, Y.; Liu, X.; Zhang, F.; Zha, Q.; Wang, X. The impact of high temperatures in the field on leaf tissue structure in different grape cultivars. *Horticulturae* **2023**, *9*, 731. [CrossRef]
13. Zha, Q.; Xi, X.J.; He, Y.N.; Jiang, A.L. Comprehensive evaluation of heat resistance in 68 *Vitis* germplasm resources. *Vitis* **2018**, *57*, 75–81. [CrossRef]
14. Asproudi, A.; Petrozziello, M.; Cavalletto, S.; Guidoni, S. Grape aroma precursors in cv. Nebbiolo as affected by vine microclimate. *Food Chem.* **2016**, *211*, 947–956. [CrossRef]
15. Khan, W.; Rayirath, U.P.; Subramanian, S.; Jithesh, M.N.; Rayorath, P.; Hodges, D.M.; Critchley, A.T.; Craigie, J.S.; Norrie, J.; Prithiviraj, B. Seaweed Extracts as Biostimulants of Plant Growth and Development. *J. Plant Growth Regul.* **2009**, *28*, 386–399. [CrossRef]
16. Yang, Y.; Li, Q.; Tang, X.N.; Han, X.M.; Zhang, J.K. Combined effect of organic fertilizer and biological agent on grape production. *China Fruit Veg.* **2020**, *40*, 101–104. [CrossRef]
17. Zheng, L.; Huang, S.; Huang, J.; Deng, Y.; Wu, Z.; Jiang, Z.; Yu, G. Biological control agents colonize litchi fruit during storage and stimulate physiological responses to delay pericarp browning. *Front. Microbiol.* **2023**, *13*, 1093699. [CrossRef] [PubMed]
18. Leneveu-Jenvrin, C.; Charles, F.; Barba, F.J.; Remize, F. Role of biological control agents and physical treatments in maintaining the quality of fresh and minimally-processed fruit and vegetables. *Crit. Rev. Food Sci. Nutr.* **2019**, *60*, 2837–2855. [CrossRef]

19. Candido, V.; Campanelli, G.; Addabbo, T.D.; Castronuovo, D.; Renco, M.; Camele, I. Growth and yield promoting effect of articificial mycorrhizationcombined with different fertiliser rates on field-grown tomato. *Ital. J. Agron.* **2013**, *8*, 168–174.
20. He, Y.; Yadav, V.; Bai, S.; Wu, J.; Zhou, X.; Zhang, W.; Han, S.; Wang, M.; Zeng, B.; Wu, X.; et al. Performance evaluation of new table grape varieties under high light intensity conditions based on the photosynthetic and chlorophyll fluorescence characteristics. *Horticulturae* **2023**, *9*, 1035. [CrossRef]
21. Venios, X.; Korkas, E.; Nisiotou, A.; Banilas, G. Grapevine responses to heat stress and global warming. *Plants* **2020**, *9*, 1754. [CrossRef]
22. IPCC. *Climate Change 2021: The Physical Science Basis*; Contribution of Working Group I to the Sixth Assessment Report of the Intergovernmental Panel on Climate Change; Cambridge University Press: Cambridge, UK, 2021.
23. Zha, Q.; Xi, X.J.; He, Y.N.; Jiang, A.L. High temperature in field: Effect on the leaf tissue structure of grape varieties. *Chin. Agric. Sci. Bull.* **2019**, *35*, 74–77. [CrossRef]
24. Chen, H.Y.; Liu, X.N.; Li, S.C.; Yuan, L.; Mu, H.Y.; Wang, Y.; Li, Y.; Duan, W.; Fan, P.G.; Liang, Z.C.; et al. The class B Heat Shock Factor HSFB1 regulates heat tolerance in grapevine. *Hortic Res.* **2023**, *10*, uhac001. [CrossRef]
25. Zhang, H.; Zhu, J.; Gong, Z.; Zhu, J.-K. Abiotic stress responses in plants. *Nat. Rev. Genet.* **2021**, *23*, 104–119. [CrossRef]
26. Zha, Q.; Xi, X.; He, Y.; Yin, X.; Jiang, A. Effect of short-time high-temperature treatment on the photosynthetic performance of different heat-tolerant grapevine cultivars. *Photochem. Photobiol.* **2021**, *97*, 763–769. [CrossRef]
27. Wu, J.; Zhang, F.; Liu, G.; Abudureheman, R.; Bai, S.; Wu, X.; Zhang, C.; Ma, Y.; Wang, X.; Zha, Q.; et al. Transcriptomeand coexpression network analysis reveals properties properties and candidate genes associated with grape (*Vitis vinifera* L.) heat tolerance. *Front. Plant Sci.* **2023**, *14*, 1270933. [CrossRef]
28. Hu, J.G.; Bai, S.J. Study on the Fruit Quality and Raisins Character of 'Thompson Seedless' and Its Lines. *Xinjiang Agric. Sci.* **2023**, 2751–2763. Available online: http://kns.cnki.net/kcms/detail/65.1097.S.20231114.1625.036.html (accessed on 28 November 2023).
29. Zhang, L.; Song, Y.; Li, J.; Liu, J.; Zhang, Z.; Xu, Y.; Fan, D.; Liu, M.; Ren, Y.; Xi, X.; et al. Development, identification and validation of a novel SSR molecular marker for heat resistance of grapes based on miRNA. *Horticulturae* **2023**, *9*, 931. [CrossRef]
30. Wang, Q.Z.; Xia, X.R.; Deng, T.; Li, X.C.; Zhou, Y.S.; Wu, S.J.; Yang, F.; Yu, S.S. Effects of three biological agents on disease control and quality improvement in the southern Xuanwei area. *Hunan Agric. Sci.* **2023**, *2*, 57–62. [CrossRef]
31. Yang, Y.Q. Effects of Exogenous Spermidine on Physiological and Biochemical Characteristics of Grape Seedlings under High Temperature Stress. Master's Thesis, Shandong Agricultural University, Taian, China, 21 July 2020.
32. Divi, U.K.; Rahman, T.; Krishna, P. Gene expression and functional analyses in brassinosteroid-mediated stress tolerance. *Plant Biotechnol. J.* **2015**, *14*, 419–432. [CrossRef] [PubMed]
33. Qiu, Y.M.; Wang, W.; Hu, X.; Sun, M.M.; Ni, J.Z. Realationship between leaf anatomical structure and heat resistance of *Rhododendron* simsii. *For. Environ. Sci.* **2021**, *37*, 69–81.
34. Bueno, A.; Alfarhan, A.; Arand, K.; Burghardt, M.; Deininger, A.-C.; Hedrich, R.; Leide, J.; Seufert, P.; Staiger, S.; Riederer, M. Effects of temperature on the cuticular transpiration barrier of two desert plants with water-spender and water-saver strategies. *J. Exp. Bot.* **2019**, *70*, 1613–1625. [CrossRef]
35. Yadav, V.; Zhong, H.; Patel, M.K.; Zhang, S.; Zhou, X.; Zhang, C.; Zhang, J.; Su, J.; Zhang, F.; Wu, X. Integrated omics-based exploration for temperature stress resilience: An approach to smart grape breeding strategies. *Plant Stress* **2024**, *11*, 100356. [CrossRef]
36. Sun, Y.; Gao, Y.; Wang, H.; Yang, X.; Zhai, H.; Du, Y. Stimulation of cyclic electron flow around PSI as a response to the combined stress of high light and high temperature in grape leaves. *Funct. Plant Biol.* **2018**, *45*, 1038–1045. [CrossRef] [PubMed]
37. Chun, J.; Sang, Y.S.; Chen, T.; Feng, Y.; Li, Q.; Zhang, Q.P. Effects of Seed-tuber dressing by *VDAL* and Harpin on potato growth and development. *Chin. Agric. Sci. Bull.* **2022**, *38*, 6–12.
38. Zhou, J.; Xiao, W.; Chen, X.D.; Gao, D.S.; Li, L. Effect of spraying amino acid-ca on photosynthetic characteristics, calcium content and quality of 'Whangkeumbae' Pear. *J. Shandong Agric. Univ. (Nat. Sci. Ed.)* **2018**, *49*, 551–555. [CrossRef]
39. Jin, R.; Li, J.; Liang, J.; Zhou, X.Y.; Zhang, X.Y.; Zeng, X.F.; Li, S.S. Effects of four plant growth regulators on growth and development, photosynthetic characteristics and yield of cotton. *Xinjiang Agric. Sci.* **2021**, *58*, 2035–2042. [CrossRef]
40. Wu, Q.; Ding, K.X.; Yu, M.L.; Huang, W.T.; Zuo, G.Q.; Feng, N.J.; Zheng, D.F. Effects of a new plant growth regulator B2 on photosynthetic fluorescence characteristics and yield of maize. *Crop Sci.* **2020**, *36*, 174–181.
41. Chen, K.; Li, G.J.; Bressan, R.A.; Song, C.P.; Zhu, J.K.; Zhao, Y. Abscisic acid dynamics, signaling, and functions in plants. *J. Integr. Plant Biol.* **2020**, *62*, 25–54. [CrossRef]
42. Raza, A.; Charagh, S.; Zahid, Z.; Mubarik, M.S.; Javed, R.; Siddiqui, M.H.; Hasanuzzaman, M. Jasmonic acid: A key frontier in conferring abiotic stress tolerance in plants. *Plant Cell Rep.* **2021**, *40*, 1513–1541. [CrossRef] [PubMed]
43. Ahammed, G.J.; Li, X.; Liu, A.; Chen, S. Brassinosteroids in plant tolerance to abiotic stress. *J. Plant Growth Regul.* **2020**, *39*, 1451–1464. [CrossRef]

44. Kim, S.; Hwang, G.; Kim, S.; Thi, T.N.; Kim, H.; Jeong, J.; Kim, J.; Kim, J.; Choi, G.; Oh, E. The epidermis coordinates thermoresponsive growth through the phyB-PIF4-auxin pathway. *Nat. Commun.* **2020**, *11*, 1053. [CrossRef] [PubMed]
45. Zha, Q.; Yin, X.; Xi, X.; Jiang, A. Heterologous *VvDREB2c* expression improves heat tolerance in *Arabidopsis* by inducing photoprotective responses. *Int. J. Mol. Sci.* **2023**, *24*, 5989. [CrossRef] [PubMed]

horticulturae

MDPI

Article

Distribution of *Plasmopara viticola* Causing Downy Mildew in Russian Far East Grapevines

Nikolay N. Nityagovsky [1,2], Alexey A. Ananev [1], Andrey R. Suprun [1], Zlata V. Ogneva [1], Alina A. Dneprovskaya [1,2], Alexey P. Tyunin [1], Alexandra S. Dubrovina [1], Konstantin V. Kiselev [1], Nina M. Sanina [2] and Olga A. Aleynova [1,*]

1 Laboratory of Biotechnology, Federal Scientific Center of the East Asia Terrestrial Biodiversity, Far Eastern Branch of the Russian Academy of Sciences, 690022 Vladivostok, Russia; niknit1996@gmail.com (N.N.N.); tyunin@biosoil.ru (A.P.T.); dubrovina@biosoil.ru (A.S.D.); kiselev@biosoil.ru (K.V.K.)

2 Institute of the World Ocean, Far Eastern Federal University, 690090 Vladivostok, Russia

* Correspondence: aleynova@biosoil.ru; Tel.: +7-4232-310718; Fax: +7-4232-310193

Abstract: Downy mildew is a severe disease that leads to significant losses in grape yields worldwide. It is caused by the oomycete *Plasmopara viticola*. The study of the distribution of this agent and the search for endophytic organisms that inhibit the growth of *P. viticola* are essential objectives to facilitate the transition to sustainable and high-yield agriculture, while respecting the environment. In this study, high-throughput sequencing of the *ITS* (ITS1f/ITS2 region) and *16S* (V4 region) amplicons was employed to analyze 80 samples of leaves and stems from different grapevine species and cultivars grown in the Russian Far East (*Vitis amurensis* Rupr., *Vitis coignetiae* Pulliat, and several grapevine cultivars). The analysis revealed the presence of *P. viticola* in 53.75% of the grape samples. The pathogen *P. viticola* was not detected in *V. amurensis* samples collected near Vladivostok and Russky Island. Among the *P. viticola*-affected samples, only two (out of the eighty analyzed grape samples) from the Makarevich vineyard in Primorsky Krai exhibited disease symptoms, while the majority appeared visually healthy. We also found six distinct *P. viticola* ASVs in our metagenomic data. Based on phylogenetic analysis, we hypothesize that the *P. viticola* population in the Russian Far East may have originated from the invasive *P. viticola* clade *aestivalis*, which has spread around the world from North America. To identify putative microbial antagonists of *P. viticola*, a differential analysis of high-throughput sequencing data was conducted using the DESeq2 method to compare healthy and *P. viticola*-affected samples. The in silico analysis revealed an increased representation of certain taxa in healthy samples compared to *P. viticola*-affected ones: fungi—*Kabatina* sp., *Aureobasidium* sp., and *Vishniacozyma* sp.; bacteria—*Hymenobacter* spp., *Sphingomonas* spp., *Massilia* spp., *Methylobacterium-Methylorubrum* spp., and *Chryseobacterium* spp. This in-silico-obtained information on the potential microbial antagonists of *P. viticola* serves as a theoretical basis for the development of biocontrol agents for grapevine downy mildew.

Keywords: endophytes; disease management; *Vitis amurensis*; *Vitis coignetiae*; associate microbiome

1. Introduction

Grapevine downy mildew, caused by oomycetes *Plasmopara viticola* (Berk. and M.A. Curtis) Berl. and de Toni, is an extremely destructive affliction that poses a significant threat to vineyards [1]. This pathogen can infect all green parts of the vine during the warmer and wetter periods of the growing season, causing significant losses in a short period of time [2]. In controlling oomycete fungi, including *P. viticola*, it is crucial to consistently administer fungicides. This proactive approach serves to safeguard against potential harm and mitigate substantial financial repercussions (up to 75% in humid grapevine-producing areas worldwide) [2,3].

Grapevine downy mildew, which is a common disease in North America, was first identified in 1889. Certain grapevines, such as *Muscadinia rotundifolia*, have shown resistance to this pathogen [4]. On the other hand, all major *Vitis vinifera* cultivars are highly susceptible to downy mildew [5]. To address this problem, the use of cultivars with natural disease resistance is a cost-effective and environmentally friendly alternative to the use of fungicides [6]. Several American and Asian *Vitis* species, such as *V. rupestris*, *V. rubra*, *V. candicans*, *V. amurensis*, *V. riparia*, *V. cinerea*, and *M. rotundifolia*, show different levels of resistance to *P. viticola*. This ranges from moderate resistance in some species to high resistance in others [7–9].

V. amurensis Rupr., native to East Asia, is mainly found in the southern Far East of Russia to northern Korea. *V. amurensis* shows numerous advantageous characteristics, such as its resilience to downy mildew [10], anthracnose, and white rot [11], and the ability to withstand cold temperatures [12]. Furthermore, *V. amurensis* contains valuable medicinal compounds, such as stilbenes, which are known for their antioxidant, anticancer, antibacterial, and antiaging properties [13]. The unique characteristics of this species have led grapevine breeders to incorporate it into their selective breeding programs. Through a comprehensive analysis of quantitative trait loci (QTL) on linkage group 14, a significant QTL controlling resistance to downy mildew resistance in *V. amurensis* was identified. This specific QTL, known as "Resistance to *Plasmopara viticola*" (*Rpv8*, *Rpv10*, and *Rpv12*), represents the first set of QTLs conferring resistance to *P. viticola* to be derived from an Asian *Vitis* species [14–16]. However, it is likely that *Rpv8*, *Rpv10*, and *Rpv12* are not present in every cultivar of this particular species. According to Wan et al. (2007), only one out of the nine wild *V. amurensis* accessions was found to be partially resistant in real cases of infection in China, while the rest were considered susceptible [17]. *V. coignetiae* Pulliat ex Planch., commonly known as crimson grapevine, is a deciduous climbing vine that is native to the temperate climates of East Asia. This includes regions such as Sakhalin Island in Russia, Japan, and Korea. This particular variety of grapevine is often used for its health juice and wine due to the abundance of polyphenols and anthocyanins found in its fruit [18,19]. However, according to a study conducted by Kim et al. in 2019, this grapevine species is susceptible to downy mildew [20].

As mentioned earlier, breeding resistant grape varieties is the most effective way to control downy mildew. However, the introduction of these varieties is a time-consuming and costly process. Given the favorable conditions for disease development, chemical control remains the most economically effective strategy for protecting crops from downy mildew. As an alternative to chemical fungicides, the use of biofungicides offers a biological approach to disease control [21]. Typically, endophytic microorganisms are used as biological control agents. Endophytes possess the ability to significantly impact host–pathogen dynamics, exerting their influence even prior to the emergence of disease. Notably, certain endophytes can instigate systemic resistance mechanisms within their host organisms, effectively stimulating the activation of defense genes targeting specific pathogens. [22]. For example, the *Bacillus velezensis* KOF112 showed biocontrol activities against downy mildew, inhibiting zoospore release from *P. viticola* zoosporangia [21]. Also, the endophytes of grape, such as the *Bacillus*, *Variovorax*, *Pantoea*, *Staphylococcus*, *Herbaspirillum*, and *Sphingomonas* bacterial genera, inhibited the mycelial growth of *Phytophthora infestans* used as a surrogate for *P. viticola* [23]. Moreover, dipeptides extracted from the grapevine endophyte *Alternaria alternate* showed efficacy in inhibiting *P. viticola* sporulation [24]. Culture filtrates obtained from the grape endophyte *Acremonium* spp. showed inhibitory activity against the *P. viticola* in vitro [22].

Therefore, the current study, using metagenome analysis, aimed to (I) detect the presence of *ITS P. viticola* sequences in wild *V. amurensis*, *V. coignetiae* grape, and cultivated grape of the Far East of Russia; and (II) perform a comparative analysis of the biodiversity of endophytic bacteria and fungi from healthy and mildew-infected grape samples in order to identify microorganisms that could theoretically be antagonists of *P. viticola*.

2. Materials and Methods

2.1. Plant Material and Surface Sterilization of Samples

To determine the presence of *ITS1* sequences of *P. viticola* in grapevines from the Far East of Russia, a total of 11 asymptomatic tissue samples from *V. amurensis*, 3 samples from *V. coignetiae*, and 4 samples from cultivated grapevines were collected. *V. amurensis* (Gh) has been carefully cultivated under special conditions in the greenhouse at the Federal Scientific Center of the East Asia Terrestrial Biodiversity in Vladivostok, Russia. Additionally, a visually healthy sample of *V. amurensis* (M) and a sample showing downy mildew symptoms (M-dm) were collected from the commercial vineyard "Makarevich". Finally, *V. amurensis* (S-Va) was sampled from the botanical garden on Sakhalin Island. The eight *V. amurensis* grapevines were collected from different non-protected natural populations. Two grapevines, P1 and P2, were found in close proximity to each other near Vladivostok, Russia, approximately 1 km apart. Another two grapevines, P3 and P4, were discovered on Russky and Rikord Islands in the southern Primorsky Territory of the Russian Far East. P5 and P6 were obtained from Ivanovka village and the Verkhne-Ussuriysky Research Station (SSA) of the Federal Scientific Center of the East Asia terrestrial. Lastly, two additional grapevines were collected from Litovko village (Kh-1) and Silinsky forest (Kh-2), situated in the southern Khabarovsk region of the Russian Far East. Additionally, one *V. coignetiae* grapevine was sampled from the botanical garden of Sakhalin Island (S-1). There were two additional *V. coignetiae* grapevines discovered within a natural population on Sakhalin Island, specifically near the cities Kholmsk (S-2) and Nevelsk (S-3). Furthermore, grapes from vineyards located in the Primorsky Territory of Russia were also gathered. Among the collected samples were (Ad) *V. vinifera* × *V. amurensis* cv. Adele (hybrid No. 82-41 F^3) and (Muk) *Vitis riparia* × *V. vinifera* cv. Mukuzani (with an unknown pedigree), which were obtained from the Makarevich vineyard. The samples of *Vitis labrusca* × *V. riparia* cv. Alfa (https://www.vivc.de/index.php?r=passport/view&id=346, accessed on 21 March 2024) (Alfa) and *Vitis* Elmer Swenson 2-7-13 cv. Prairie Star (https://www.vivc.de/index.php?r=passport/view&id=23087, accessed on 21 March 2024) (Pr-St) were selected in PRIM ORGANICA vineyard (Figure 1). It is important to note that all samples, except M-dm, were looking healthy, i.e., without symptoms of downy mildew.

Plant samples were collected on days with little cloud cover and no precipitation, specifically between 11 to 12 in the morning. The air temperature at the time of collection was between 18 and 20 °C. Each sample was promptly transported to the laboratory within a timeframe of 3 h to 1 day. To ensure a comprehensive analysis, a minimum of four biological replicates, consisting of two stems and two leaves, were obtained for each grapevine sample. These replicates were subjected to next-generation sequencing (NGS) for further analysis. Finally, a total of 52 biological replicates of *V. amurensis*, 12 biological replicates of *V. coignetiae*, and 16 biological replicates of cultivated grapevines were collected and thoroughly analyzed.

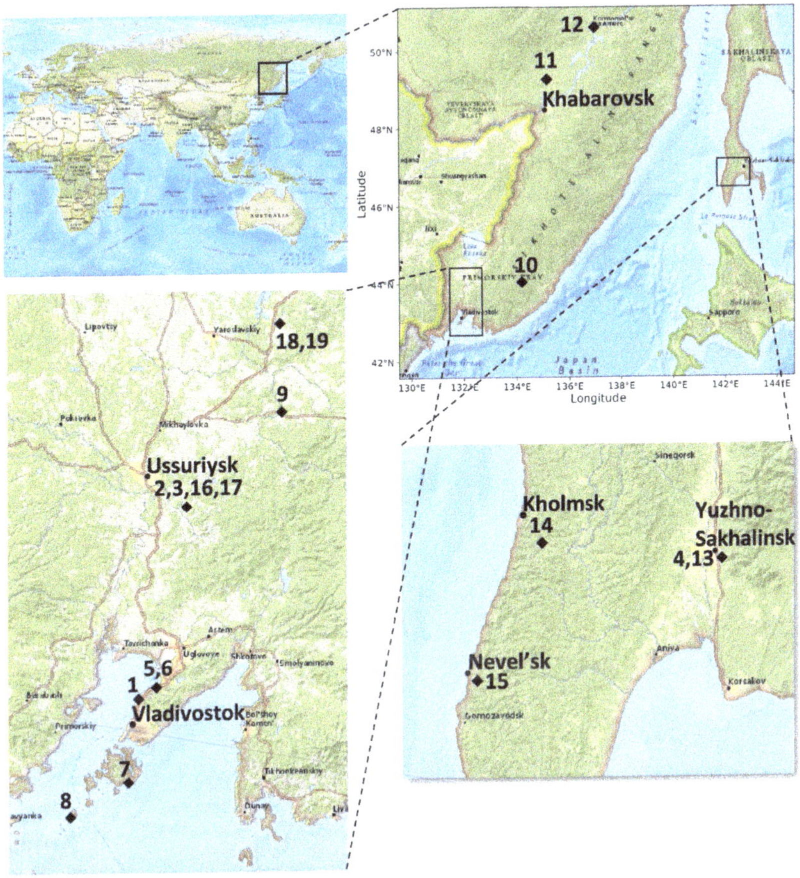

Figure 1. Plant material collection sites. The numbers indicate the collection sites of the plant material, which are listed in Supporting Information Table S1. The geographical map used: National Geographic World Map (esri) [25].

To prepare the grapevine tissues for further analysis, each grapevine sample was thoroughly washed with soap and subjected to a sequential sterilization process. First, they were immersed in 75% ethanol for 2 min, followed by a 1-min treatment with 10% hydrogen peroxide. Finally, they were rinsed five times with sterile water [26,27]. In order to assess the effectiveness of this surface sterilization method, a 100 µL sample of the final rinse water was cultured on R2A (PanReac, AppliChem, Darmstadt, Germany) and potato dextrose agar (PDA, Neogene, Watford, UK) plates to ascertain the absence of any bacterial or fungal colony growth originating from the external sources.

2.2. DNA Isolation, Library Preparation, and Illumina MiSeq Sequencing

To carry out NGS analysis, DNA was isolated from 200 mg of surface-sterilized grape leaves and stems using the CTAB-spin method, as previously reported [28]. The DNA samples were then sent to a reputable commercial organization Syntol (Moscow, Russia) for high-throughput sequencing using Illumina technology. To ensure the quality and quantity of the DNA, it underwent evaluation through the Nanodrop-1000 (Nanodrop, Wilmington, DE, USA) and Quantus Fluorometer (Promega, Madison, WI, USA). The libraries were meticulously prepared for sequencing, adhering precisely to the protocol outlined in the manual "16S Metagenomic Sequencing Library Preparation" (Part # 15,044,223 Rev. B;

Illumina). Bacterial *16S* rRNA regions were amplified from all samples using the primers 515F (5′GGTAATACGKAGGKKGCDAGC) and 806R (5′RTGGACTACCAGGGTATCTAA), specifically designed to target *Vitis* sp. plants. The primers ITS1f (5′CTTGGTCATTTAGAGG AAGTAA) and ITS2 (5′GCTGCGTTCTTCATCGATGC) were utilized to amplify the fungal *ITS1* rDNA regions in all of the samples. The Nextera® XT Index Kit reagents were used to index the amplicons. The library pool underwent sequencing on Illumina MiSeq platform, employing 2 × 250 paired-end reads, utilizing the MiSeq Reagent Kit v2, which allows for 500 cycles.

The accession numbers for *16S* and *ITS1* sequences have been successfully submitted and archived in the National Center for Biotechnology Information (NCBI) under the accession numbers PRJNA980748 and PRJNA998468 and in the database of laboratory Biotechnology, Federal Scientific Center of the East Asia Terrestrial Biodiversity, Far Eastern Branch of the Russian Academy of Sciences, Russia (https://biosoil.ru/downloads/biotech/Metagenoms/, accessed on 21 March 2024).

2.3. Data Processing

The Supporting Information Tables S2 and S3 provide an overview of the samples utilized in the bioinformatic analysis. Custom scripts based on the R and Bash languages were used to process the data obtained (https://github.com/niknit96/Nityagovsky_et.al.2024, accessed on 21 March 2024). The raw data were preprocessed using QIIME 2 [29] and DADA2 [30] programs. After eliminating the primers, PhiX reads and chimeric sequences, the paired-end reads were merged and arranged in a sorted order. Taxonomic identification of amplicon sequence variants (ASVs) was carried out utilizing the QIIME 2 Scikit-learn algorithm by employing the SILVA 138 pre-trained classifier for *16S* sequences (99% OTUs from the V4 region of the sequences) [31]. Additionally, for the *ITS* sequences, the UNITE pre-trained classifier was utilized (99% OTUs from the ITS1F/ITS2 region of sequences) [32].

The qiime2R [33], phyloseq [34], and tidyverse [35] R libraries were used in prefiltering and data preparation. Mitochondria, chloroplast, *Viridiplantae*, *Metazoa*, *Rhizaria*, *Protista*, *Alveolata*, and unidentified ASVs were deleted from the obtained data. Alpha diversity metrics in rarified samples to even sample depth were obtained using phyloseq R library [34]. The number of ASVs and Pielou's evenness index were used to characterize richness and evenness in microbial communities, respectively. The Wilcoxon rank sum test was performed to analyze the alpha diversity data between groups. For beta diversity analysis, data were transformed to even sampling depth. The calculation of Bray–Curtis beta diversity data was conducted by employing the Vegan package, a widely recognized tool in the field [36], and converted to nonmetric multidimensional scaling (NMDS). Statistical validation of beta diversity data was performed using the PERMANOVA test with 999 permutations. Data on the differential abundance ASVs between samples, in which *P. viticola* was found and in which the pathogen was absent, were obtained using the DESeq2 statistical tool with false discovery rate correction [37]. Visualization was conducted using the ggplot2 [35], tidyterra [38], sf [39], maptiles [40], and ggmagnify [41] R libraries.

Evolutionary analyses of the *P. viticola* ASVs in our dataset with the cryptic species described in [42,43] were performed in MEGA X [44]. The sequences were aligned using the Muscle algorithm [45]. Evolutionary history was inferred using the maximum likelihood (ML) method and the general time reversible (GTR) model [46]. We used the GTR model for the selected region of the *ITS* because it was the best model for this region in *P. viticola* data, as described in [42]. To estimate the percentage of trees where the associated taxa cluster together, we used the bootstrap test with 1000 replicates [47]. Initial trees for the heuristic search were automatically obtained by applying the maximum parsimony (MP) method. The *Phytophthora sojae ITS* sequence was used as the outgroup for phylogenetic analyses. The original sequences, aligned sequences, and MEGA tree session file are presented in the Supplementary Materials.

3. Results

3.1. Distribution of the Plasmopara Viticola ITS1 Sequences in Grape Samples

The Illumina NGS technology was employed to generate a substantial amount of data, resulting in a total of 16,315,902 *16S* and 5,192,469 *ITS1* paired-end reads. Extensive processing, including paired-end alignments, quality filtering, and the removal of unwanted sequences such as chimeric, mitochondria, chloroplast, *Viridiplantae*, *Metazoa*, *Rhizaria*, *Protista*, *Alveolata*, and unidentified sequences, led to the generation of 10,102,418 *16S* and 1,348,330 *ITS1* sequences from 80 grape samples (2–6 samples from each plant) (Supporting Information Table S2). Analyzing the *16S* data revealed that the average and median read numbers for the samples were 126,280 and 79,912, respectively. Similarly, for *ITS1* data, the average and median read numbers were 16,854 and 14,373, respectively (Supporting Information Table S3).

The geographical range of *P. viticola* in the collected grape samples was analyzed. It should be noted that *P. viticola* only causes downy mildew in the family Vitaceae. Therefore, in this study, it is acceptable to determine the *ITS1 P. viticola* sequences before the genus level. Furthermore, using the NCBI nucleotide BLAST (nucleotide–nucleotide BLAST) algorithm, the *P. viticola ITS1* sequences (Supporting Information Table S6) were determined to be *P. viticola* with high percentage identities (99–100%). It was shown that the greatest representation of *P. viticola ITS1* sequences was in samples collected in the Makarevich vineyard. The highest representation of the *P. viticola ITS1* sequence was 15.4–60.9% in grape sample M-dm, which had visible symptoms of downy mildew (Figure 2). In other samples without visible downy mildew symptoms, the percentage of *P. viticola ITS1* sequences was 0–48%. During the analysis of metagenomic data, a large percentage of downy mildew was found on Sakhalin Island and Rikord Island. The greatest representation of *P. viticola ITS1* sequences was in the botanical garden of Sakhalin Island, and the percentage ratio in *V. coignetiae* samples was higher (34–48%) compared to *V. amurensis* samples (0.3–1.4%). The representation of *P. viticola* sequences in samples collected on Rikord Island was 2–18%. Relatively small amounts of *ITS* of *P. viticola* sequences were detected in samples near the city of Nevelsk (0.2–14%) in the Silinsky forest of the Khabarovsky region (0.7–10%) (Figure 2). *P. viticola* sequences were present in trace amounts in the sample collected from a greenhouse at the Federal Scientific Center of the East Asia Terrestrial Biodiversity (0–2.2%), near the city of Kholmsk on Sakhalin Island (0.1–1.9%), the Verkhne-Ussuriysky Research Station (0–1.3%), Litovko village, the southern Khabarovsky region of the Russian Far East (0–0.4%), and PRIM ORGANICA vineyard (0–0.2%) (Figure 2). The grape samples collected near Vladivostok, on the Russky Island, and the village of Ivanovka did not have the *ITS1* sequence of *P. viticola* in their metagenome (Figure 2).

We found six *P. viticola* ASVs in our metagenomic data. We performed a phylogenetic analysis of these ASVs with the cryptic species described in our colleagues' works [42,43] (Figure 3). Based on the analysis, all of the ASVs clustered together in a well-supported branch, along with a representative *ITS* sequence of *P. viticola* clade *aestivalis*. Among the plants in which *P. viticola* ASVs were found are wild grapes *V. amurensis* (P-3, P-4, P-5, P-6, Kh-1, Kh-2, M, M-dm, and S-Va) and *V. coignetiae* (S-1, S-2, and S-3), as well as cultivated forms of grapes (Ad, Muk, Pr-St, and Alfa) (Table 1). According to our data, the most common *P. viticola* ASV is ASV 1, which was present in 37 out of 43 samples and had the highest mean relative abundance.

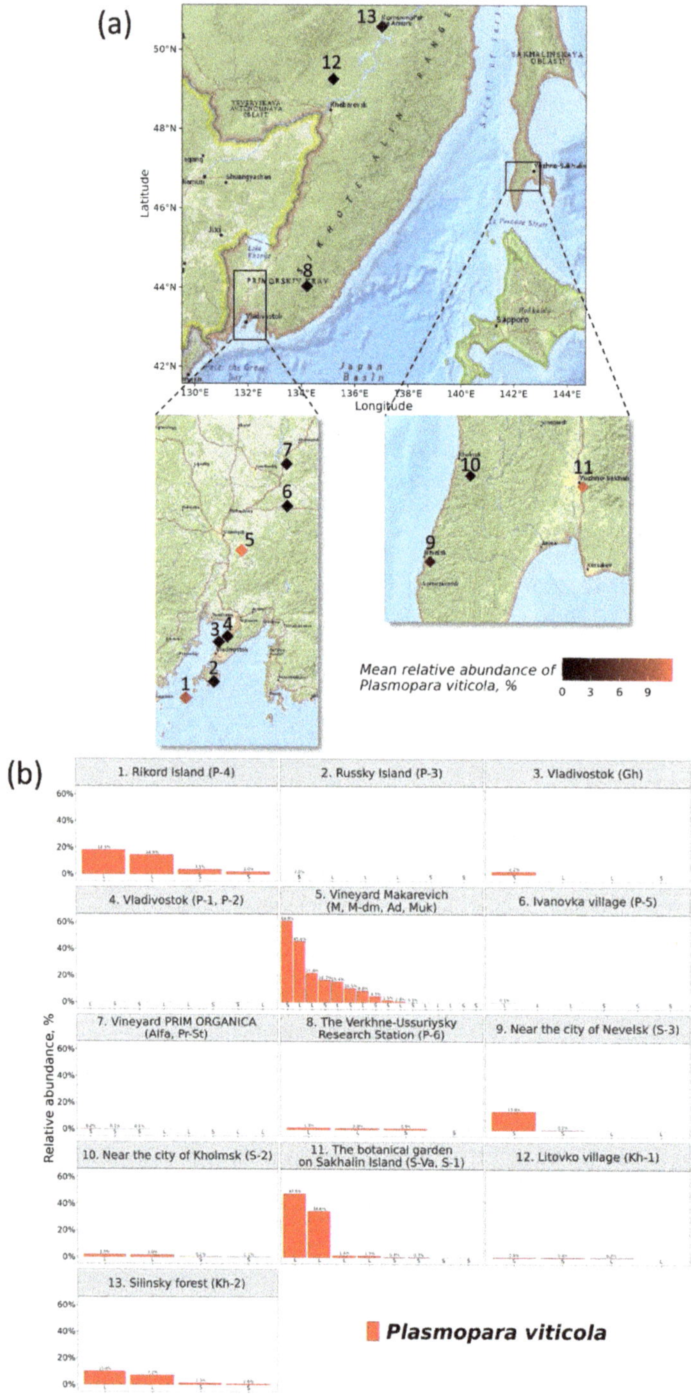

Figure 2. Relative representation of *ITS1* sequences of *Plasmopara viticola* in grape samples: (**a**) geographic map with mean relative abundance of *P. viticola* in sample locations; (**b**) relative

abundance of *P. viticola* in samples. The marks in the form of numbers on the map (**a**) correspond to the data in (**b**). The geographical map used: National Geographic World Map (esri) [25]. L—leaf; S—stem. Gh—*V. amurensis* in greenhouse at the Federal Scientific Center of the East Asia Terrestrial Biodiversity; M—*V. amurensis* in the commercial vineyard «Makarevich»; M-dm—*V. amurensis* with visible symptoms of *P. viticola* in «Makarevich» vineyard; S-Va—*V. amurensis* in the botanical garden on Sakhalin Island; P-1—*V. amurensis* in Vladivostok; P-2—*V. amurensis* in Vladivostok; P-3—*V. amurensis* in Russky Island; P-4—*V. amurensis* in Rikord Island; P-5—*V. amurensis* in Ivanovka village; P-6—*V. amurensis* in the Verkhne-Ussuriysky Research Station (SSA); Kh-1—*V. amurensis* in Litovko village, the southern Khabarovsky region of the Russian Far East; Kh-2—*V. amurensis* in the Silinsky forest; S-1—*V. coignetiae* in the botanical garden on Sakhalin Island; S-2—*V. coignetiae* near the city Kholmsk on Sakhalin Island; S-3—*V. coignetiae* near the city Nevelsk on Sakhalin Island; Pr-St—*Vitis* Elmer Swenson 2-7-13 cv. Prairie Star from commercial vineyard PRIM ORGANICA; Alfa- *Vitis labrusca* × *Vitis riparia* cv. Alfa from PRIM ORGANICA; Ad- *Vitis vinifera* × *V. amurensis* cv. Adele from commercial vineyard Makarevich; Muk—*V. riparia* × *V. vinifera* cv. Mukuzani.

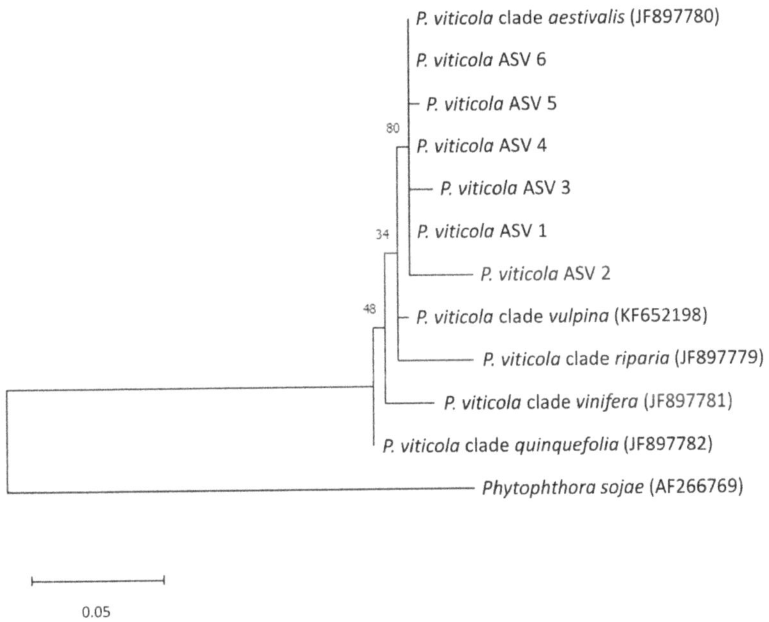

Figure 3. Evolutionary analysis of *Plasmopara viticola* ASVs in our NGS dataset with previously described cryptic species of *P. viticola* [42,43] using a maximum likelihood method. The ML method and the GTR model were utilized to deduce the evolutionary history. The tree with the highest log likelihood (−644,81) is shown. The branches display the percentage of trees in which the related taxa formed clusters, as determined by the bootstrap test (with 1000 replicates). Initial trees for the heuristic search were obtained automatically by applying the MP method. The tree is drawn to scale, with branch lengths measured in the number of substitutions per site. This analysis involved 12 nucleotide sequences. The final dataset consisted of a sum of 249 positions. The phylogenetic tree is rooted with the *Phytophthora sojae ITS* sequence. Evolutionary analyses were conducted in MEGA X. The original sequences, aligned sequences, and the MEGA tree session file are presented in the Supplementary Materials.

Table 1. Representation of *Plasmopara viticola* ASVs in NGS samples.

Name of ASV	Occurrence in the Plants	Mean Relative Abundance, %	Number of ASV-Affected Samples	Total *P. viticola*-Affected Samples
P. viticola ASV 1	P-4, P-5, P-6, Kh-1, Kh-2, M, M-dm, Ad, Muk, Pr-St, Alfa, S-Va, S-1, S-2, S-3	8.84	37	43
P. viticola ASV 2	P-3, P-4, P-6, M-dm, Ad, S-1, S-2	0.76	9	
P. viticola ASV 3	S-1, S-3, P-4, Kh-2	0.50	7	
P. viticola ASV 4	Gh, Kh-1, Muk, Ad, S-1	2.46	5	
P. viticola ASV 5	M-dm, S-1	1.75	2	
P. viticola ASV 6	S-1	0.28	1	

3.2. Comparative Analysis of the Biodiversity of Grape Endophytes in Grape Samples Affected by *P. viticola*

The outcomes of the alpha diversity analysis of the bacterial and fungal endophytic communities in the grape samples, categorized based on the occurrence of *P. viticola*, are visually presented in Figure 4a,b and Figure 4c,d, respectively. Based on the number of *16S* ASVs, samples affected by *P. viticola* are characterized by a reduced richness of the bacterial community compared to healthy samples ($p = 0.011$, Figure 4a). According to the Pielou's evenness index, the bacterial communities of *P. viticola*-affected and healthy samples have the same evenness ($p = 0.14$, Figure 4b). On the other hand, samples affected by *P. viticola* and healthy samples are not significantly different in terms of fungal community richness ($p = 0.26$, Figure 4c), but *P. viticola*-affected samples were characterized by a more even fungal community compared to healthy samples based on Pielou's evenness index ($p = 0.042$, Figure 4d).

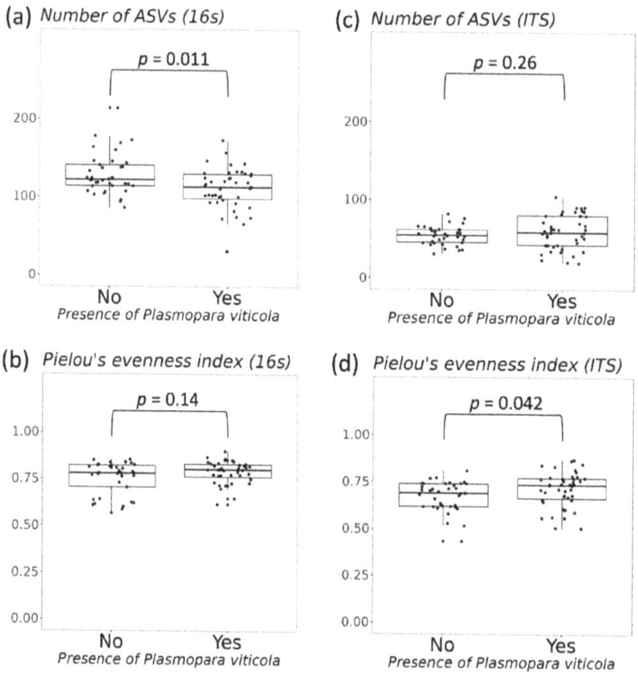

Figure 4. The alpha diversity metrics between samples, which are grouped based on the presence of *Plasmopara viticola*. (**a**,**b**) Number of ASVs and Pielou's evenness index for the endophytic bacterial community; (**c**,**d**) number of ASVs and Pielou's evenness index for the endophytic fungal community.

According to the NMDS ordination plots of beta diversity, *P. viticola*-affected and healthy endophytic bacterial or fungal communities overlap to a high degree, but fungal communities overlap more than bacterial communities (Figure 5a,b). The PERMANOVA test showed that the factor of the presence of *P. viticola* explained 4.4% of the differences between grape samples in the bacterial endophytic community (Figure 5a, Supporting Information Table S4), whereas this factor explained 3.2% of the differences between samples in the fungal endophytic community (Figure 5b, Supporting Information Table S4).

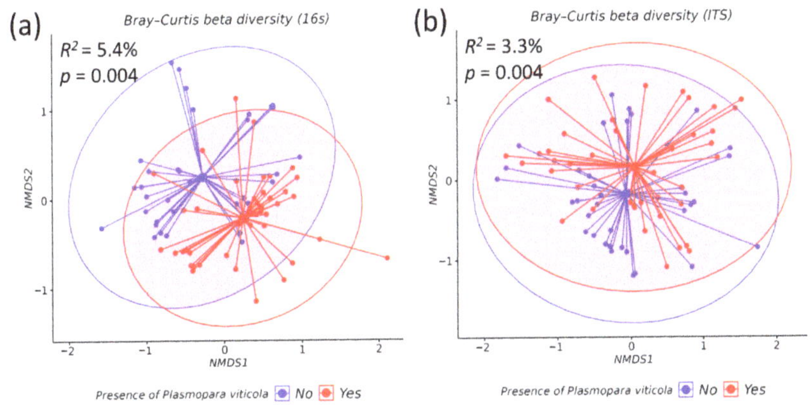

Figure 5. The comparison of endophytic bacterial and fungi communities of grapevines samples based on the presence of *Plasmopara viticola*: (**a**) Bray–Curtis beta diversity NMDS plot of grape endophytic bacteria; (**b**) Bray–Curtis beta diversity NMDS plot of grape endophytic fungi. The ellipses assume a multivariate normal distribution. The central points of ellipses are mean points.

3.3. The In Silico Analysis of Potential Microorganisms Plasmopara viticola Antagonists

According to the DESeq2 results, healthy samples were characterized by an increased abundance of 40 bacterial ASVs compared to *P. viticola*-affected samples (Figure 6, Supporting Information Table S5). These ASVs belonged to 4 taxa of the class level or 18 taxa of the genus level. The largest number of ASVs belonged to class Bacteroidia (18), followed by Alphaproteobacteria (11), Gammaproteobacteria (10), and Actinobacteria (1). At the genus level, the largest number of ASVs belonged to *Hymenobacter* (14), *Sphingomonas* (4), *Massilia* (4), *Methylobacterium-Methylorubrum* (2), and *Chryseobacterium* (2), followed by *Advenella*, *Microbacteriaceae* (ASV 7), *Brevundimonas*, *Devosia*, *Sphingomonadaceae* (ASV 7), *Spirosomaceae*, *Rhizobacter*, *Phyllobacterium*, *Xanthobacteraceae* (ASV 4), *Pedobacter*, *Nevskia*, *Pseudomonas*, *Escherichia-Shigella*, and *Polaromonas*. ASV, belonging to the genus *Cupriavidus*, was characterized by an increased abundance in *P. viticola*-affected samples compared to healthy samples (Figure 6, Supporting Information Table S5).

In the fungal community, healthy samples were characterized by an increased abundance of four ASVs compared to *P. viticola*-affected samples (Figure 7, Supporting Information Table S6). These ASVs belonged to two taxa at the class level or four taxa at the genus level, namely: *Dothideaceae* (ASV 2), *Kabatina*, and *Aureobasidium* of the class Dothideomycetes, and *Vishniacozyma* of the class Tremellomycetes. However, samples affected by *P. viticola* are characterized by an increased abundance of 2 ASVs belonging to *Ramularia* and 1 ASV belonging to *Taphrina* (Figure 7, Supporting Information Table S6).

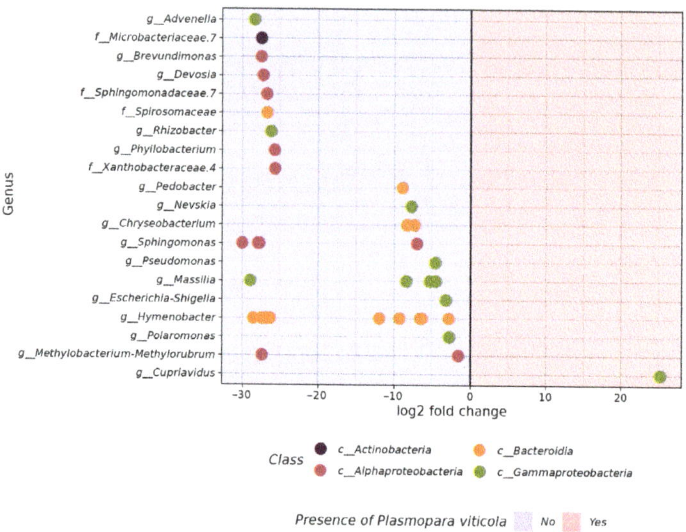

Figure 6. Significantly different abundant (adjusted $p < 0.01$) bacterial ASVs between grape samples, identified by the DESeq2 tool, which were grouped based on the presence of *Plasmopara viticola*. Dots mean ASVs, which were identified as genus-level taxa.

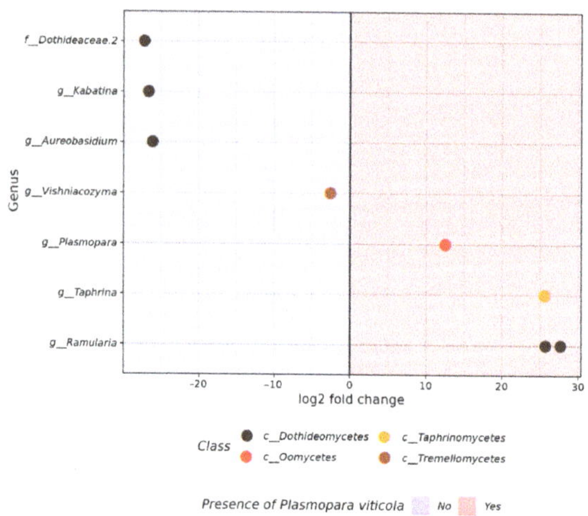

Figure 7. Significantly different abundant (adjusted $p < 0.01$) fungal ASVs between grape samples, identified by the DESeq2 tool, which were grouped based on the presence of *Plasmopara viticola*. Dots mean ASVs, which were identified as genus-level taxa.

4. Discussion

Grapevine downy mildew, caused by the pathogen known as *P. viticola*, is widely recognized as one of the most significant diseases affecting grapes worldwide. Throughout the growing season, this particular pathogen has the ability to infect any green component of the vine whenever the weather conditions are warm and wet. In regions with temperate climates, where the grapevine experiences dormancy, the pathogen adapts by forming oospores to ensure its survival in the absence of a host. The extensive application of fungicides has the potential to result in the development of *P. viticola* isolates that possess resistance to these chemical agents [48,49].

This paper examines the distribution of *P. viticola* in both unprotected areas and vineyards of the Russian Far East through the analysis of NGS grape samples. According to NGS data, the highest representation of *P. viticola ITS1* sequences was found in grape samples from the Makarevich vineyard. It is worth noting that only one of the collected grape samples showed visible symptoms of downy mildew. It is likely that *P. viticola* is circulating in the Magarach vineyard, and regular fungicide treatments only reduce the amount of the pathogen without destroying it. The fungicide treatment in the PRIM ORGANICA vineyard is the most effective in protecting against *P. viticola*, where the presence of this pathogen was minimal in the NGS samples. There was also a relatively high presence of *P. viticola* in *V. amurensis* samples collected on Rikord Island and in the Silinsky forest of the Khabarovsky region, as well as in *V. coignetiae* samples grown in the botanical garden on Sakhalin Island and near the town of Nevelsk. It is known that once a certain amount of time has passed, affected tissues of the host organism may display signs of downy mildew. The manifestation of these symptoms heavily relies on the specific environmental conditions of the region, as well as the susceptibility of the host to the disease [50]. *P. viticola* grows optimally under high relative humidity and mild temperatures [51]. It is likely that the climatic characteristics of northern regions and associated remote islands, namely, increased humidity, UV radiation, and low average temperatures, contribute to a more active distribution of *P. viticola*. The presence of *P. viticola ITS1* in samples of wild grapes without visible symptoms of downy mildew may indicate the resistance of these species to this pathogen. It is known that susceptible grapevine species infected by *P. viticola* produce mainly *trans*-resveratrol (3,5,4′-trihydroxystilbene) and *trans*- and *cis*-piceid (3-O-β-D-glucoside of resveratrol) [52], whereas resistant species produce *trans*-resveratrol, *trans*-pterostilbene (3,5-dimethoxy-4′-hydroxystilbene), and cyclic dehydrodimers of resveratrol *trans*-ε-viniferin and *trans*-δ-viniferin [53–55]. The wild grape *V. amurensis* is an important source of stilbenes [11], containing six main stilbenes [56]. Previous research has shown that UV-C-induced biosynthesis of stilbenes and flavonoids in grape leaves, especially resveratrol biosynthesis in grape leaves, is greatly increased in response to UV-C irradiation [57]. In addition, genetic studies have identified 33 loci of resistance to *P. viticola* (*Rpv*) in American and Asian *Vitis* spp. and in some *V. vinifera* cultivars [58]. Thus, the resistance loci of an individual grapevine, namely, the presence of the loci "resistance to *Plasmopara viticola*" (*Rpv8*, *Rpv10*, and *Rpv12*), or the higher content of stilbenes, or the stimulation of stilbene biosynthesis by UV radiation or other external factors, can influence the level of *P. viticola* representation in wild grape samples.

A recent study has proposed a possible scenario for the spread of *P. viticola* around the world [59]. Through analyzing sequences of nuclear and mitochondrial genes from invasive grapevine downy mildew populations, it was determined that all of these populations belonged to a single clade *aestivalis* of *P. viticola* [42,43,59]. This clade is known to infect the wild summer grape *V. aestivalis* in North America. The study suggests that the pathogen first spread from North America to Europe and then to other parts of the world. Our data partially support the close relationship between the *P. viticola* ASVs present in the Russian Far East and the invasive clade. However, the data are not sufficient to determine whether the invasion originated directly from Europe or if it is a secondary introduction from another country. Unfortunately, we only sequenced the *ITS* amplicon, which provides a low resolution for population studies of *P. viticola*. To draw more accurate conclusions about *P. viticola* populations in the Russian Far East, further research with a larger sample size and using additional marker genes used in phylogenetic analysis is necessary.

We also analyzed the microbiome of grape samples affected to different degrees by the oomycete *P. viticola*. It is known that many plant pathogens can form a specific microbiome that can also be an indicator of the pathogen. A comparative analysis of the microbiomes of grape samples without *P. viticola* representation and samples with high *P. viticola* representation allowed us to identify microorganisms that could hypothetically be antagonists of the downy mildew pathogen or associated with this oomycete. For example, a grape virus associated with *P. viticola* has recently been analyzed. It is likely that some viruses could act as new biocontrol agents for *P. viticola* [60].

In this work, we found both positive and negative correlations in the number of some endophytic microorganisms depending on the representation of *P. viticola ITS1* sequences in grape samples. For example, a high proportion of *P. viticola ITS1* sequences correlated positively with a high proportion of *16S* sequences from *Cupriavidus* endophytic bacteria in grape samples. There was early evidence that *Cupriavidus* species were associated with agricultural crops growing in alkaline soils [61]. It is possible that the association of these bacteria with *P. viticola* is related to the alkalinization of the internal tissues of grapes as a result of infection with *P. viticola*. The endophytic fungi genera *Ramularia* and *Taphrina* were also more abundant in grape samples with high levels of downy mildew. *Ramularia*, the white mold of plants, is a species-rich genus that harbors plant pathogens responsible for yield losses in many important crops, including barley, sugar beet, and strawberries [62]. It has been shown that barley plants with *mlo* resistance to downy mildew have an increased susceptibility to a new important disease—ramularia leaf spot [63]. In addition, *Taphrina* fungi are biotrophic plant pathogens that cause plant deformities [64]. Thus, infection with downy mildew leads to subsequent infection with other pathogenic fungi.

In addition to mildew-associated microorganisms, we found endophytes that are hypothetical antagonists of *P. viticola*. The presence of bacteria belonging to the genera *Hymenobacter*, *Sphingomonas*, *Massilia*, *Methylobacterium-Methylorubrum*, and *Chryseobacterium* was found to be significantly higher in grape samples with a low or absent content of *P. viticola ITS1* sequences compared to samples highly infected with downy mildew. Some species of genera *Hymenobacter* are known to be UV-resistant [65], and the pathogen *P. viticola* is very sensitive to UV radiation [66]. Therefore, the inversely proportional number of endophytic bacteria *Hymenobacter* spp. and *P. viticola* is likely to be related to the amount of UV exposure of individual grape samples. It is known that several *Methylobacterium* and *Sphingomonas* strains work against the proliferation of plant pahogene *Candidatus phytoplasma*, which is the primary cause of grapevine yellows. Additionally, it is worth highlighting that the presence of genera *Methylobacterium* and *Sphingomonas* in significant numbers is directly linked to the production of characteristic sensory compounds found in well-rounded wines [67]. Moreover, the presence of *Chryseobacterium* species contribute to enhanced plant growth through biocontrol activity against plant pathogens, including *Phytophthora capsici* [68]. Also, a low number of *ITS P. viticola* sequences correlated with high-percentage representation of endophytic fungi of the taxa *Dothideaceae.2*, *Kabatina*, *Aureobasidium*, and *Vishniacozyma* in grape samples. According to the literature, some species of fungi of the genus *Kabatina* can synthesize enfumafungin, a novel antifungal compound [69]. In addition, several species of *Aureobasidium* fungi possess the remarkable capacity to produce volatile organic compounds (VOCs) that display inhibitory effects on grape pathogens, most notably *Botrytis cinerea* [70,71]. *Vishniacozyma* is a versatile yeast genus that has been discovered in various ecological settings. Research has revealed that the population of *Vishniacozyma* sp. thrives during the berry ripening phase, showcasing its ability to flourish in conditions characterized by high sugar content and low moisture levels. It is noteworthy that the pathogenic fungus *P. viticola* typically flourishes in moist environments. This discrepancy in preferred habitats could potentially explain the inverse correlation between the prevalence of *P. viticola* and the presence of *Vishniacozyma* sp. Additionally, *Vishniacozyma* sp. has exhibited promising biocontrol properties against both blue molds and gray molds, which commonly infect pears [72]. Recent investigations by Zhu et al. in 2021 [73] have unveiled a potential antagonistic effect of *Vishniacozyma* sp. on *Erysisphe*. Nevertheless, the biocontrol potential of the endophytic bacteria *Methylobacterium* spp., *Sphingomonas* spp., and *Chryseobacterium* spp. and the fungi *Kabatina* sp., *Aureobasidium* sp., and *Vishniacozyma* sp. for grape downy mildew requires further analysis. Together, these endophytic antagonists represent a valuable resource that will undoubtedly be used in the foreseeable future to develop biocontrol or integrated programs to reduce chemical use against downy mildew.

5. Conclusions

In this study, NGS was utilized for the first time to analyze the distribution of downy mildew in wild *V. amurensis, V. coignetiae*, and cultivated grapes. Our data suggest that the population of *P. viticola* in the Russian Far East may be related to an invasive clade *aestivalis* of *P. viticola*, which has spread from North America to other parts of the world. Bioinformatic methods were also used to identify endophytic microorganisms associated or antagonistic to the downy mildew. The in silico analysis showed that certain genera of endophytic bacteria, namely, *Hymenobacter* spp., *Sphingomonas* spp., *Massilia* spp., *Methylobacterium-Methylorubrum* spp., and *Chryseobacterium* spp., and fungi, namely, *Kabatina* sp., *Aureobasidium* sp., and *Vishniacozyma* sp., could be hypothetical antagonists of *P. viticola*. The results obtained provide an important basis for the development of downy mildew biocontrol tools based on natural endophytic microorganisms.

Supplementary Materials: The following supporting information can be downloaded at https://www.mdpi.com/article/10.3390/horticulturae10040326/s1: Supporting Information Table S1: The code, location and GPS coordinates samples of grapevines of *Vitis amurensis, Vitis coignetiae* and cultivated grapes collected at July 2022; Supporting Information Table S2: *16s* data samples used in analysis; Supporting Information Table S3: *ITS* data samples used in analysis; Supporting Information Table S4: PERMANOVA results; Supporting Information Table S5: Identified by the DESeq2 tool significantly different abundant (adjusted $p < 0.01$) bacterial ASVs between grape samples which grouped based on presence of *Plasmopara viticola*; Supporting Information Table S6: Identified by the DESeq2 tool significantly different abundant (adjusted $p < 0.01$) fungal ASVs between grape samples which grouped based on presence of *Plasmopara viticola*; Original sequences.fasta: *ITS* sequences of *P. viticola* ASVs in our NGS dataset, *ITS* sequences of cryptic species of *P. viticola* and *Phytophthora sojae ITS* sequence; Aligned sequences.fasta: Aligned sequences used in phylogenetic analysis; MEGA tree session.mtsx: phylogenetic tree in MEGA tree session format.

Author Contributions: Conceptualization K.V.K. and O.A.A.; methodology and software N.N.N. and A.A.A.; validation N.M.S.; formal analysis Z.V.O., A.R.S. and A.A.D.; investigation N.N.N. and A.P.T.; resources, O.A.A.; data curation K.V.K.; writing—original draft preparation O.A.A.; writing—review and editing A.S.D.; visualization A.A.A.; supervision K.V.K.; project administration N.N.N.; funding acquisition O.A.A. All authors have read and agreed to the published version of the manuscript.

Funding: This work was supported by a grant from the Russian Science Foundation (grant number 22-74-10001, https://rscf.ru/project/22-74-10001) (accessed on 21 March 2024). The maintenance of the used microbial collection was carried out within the state assignment of the Ministry of Science and Higher Education of the Russian Federation (theme No. 124012200181-4).

Data Availability Statement: The data presented in this study are available within the article and Supplementary Materials.

Acknowledgments: We would like to thank the Sergei Makarevich, Alexander, and Anna Storozhenko for their primary contribution to the sample collection for the present study.

Conflicts of Interest: The authors declare no conflicts of interest.

References

1. Lafon, R.; Clerjeau, M. Downy Mildew. In *Compendium of Grape Diseases*; APS Press: St. Paul, MN, USA, 1988; pp. 11–13.
2. Koledenkova, K.; Esmaeel, Q.; Jacquard, C.; Nowak, J.; Clément, C.; Ait Barka, E. *Plasmopara viticola* the Causal Agent of Downy Mildew of Grapevine: From Its Taxonomy to Disease Management. *Front. Microbiol.* **2022**, *13*, 889472. [CrossRef]
3. Massi, F.; Torriani, S.F.F.; Borghi, L.; Toffolatti, S.L. Fungicide Resistance Evolution and Detection in Plant Pathogens: *Plasmopara viticola* as a Case Study. *Microorganisms* **2021**, *9*, 119. [CrossRef]
4. Feechan, A.; Anderson, C.; Torregrosa, L.; Jermakow, A.; Mestre, P.; Wiedemann-Merdinoglu, S.; Merdinoglu, D.; Walker, A.R.; Cadle-Davidson, L.; Reisch, B.; et al. Genetic Dissection of a TIR-NB-LRR Locus from the Wild North American Grapevine Species *Muscadinia rotundifolia* Identifies Paralogous Genes Conferring Resistance to Major Fungal and Oomycete Pathogens in Cultivated Grapevine. *Plant J.* **2013**, *76*, 661–674. [CrossRef]
5. Yin, L.; An, Y.; Qu, J.; Li, X.; Zhang, Y.; Dry, I.; Wu, H.; Lu, J. Genome Sequence of *Plasmopara viticola* and Insight into the Pathogenic Mechanism. *Sci. Rep.* **2017**, *7*, 46553. [CrossRef]

6. Delbac, L.; Delière, L.; Schneider, C.; Delmotte, F. Downy Mildew Is Able to Carry out Its Sexual Cycle on Resistant Grape Varieties: Sourced from the Research Article "Evidence for Sexual Reproduction and Fertile Oospore Production by *Plasmopara viticola* on the Leaves of Partially Resistant Grapevine Varieties" (ACTA Horticulturae, 1248. ISHS, 2019). Original Language of the Article: French. *IVES Tech. Rev. Vine Wine* **2020**. [CrossRef]
7. Olmo, H.P. *Vinifera rotundifolia* Hybrids as Wine Grapes. *Am. J. Enol. Vitic.* **1971**, *22*, 87–91. [CrossRef]
8. Staudt, G.; Kassemeyer, H.H. Evaluation of Downy Mildew Resistance in Various Accessions of Wild *Vitis* Species. *Vitis* **1995**, *34*, 225–228.
9. Díez-Navajas, A.M.; Wiedemann-Merdinoglu, S.; Greif, C.; Merdinoglu, D. Nonhost Versus Host Resistance to the Grapevine Downy Mildew, *Plasmopara viticola*, Studied at the Tissue Level. *Phytopathology* **2008**, *98*, 776–780. [CrossRef]
10. Wu, J.; Zhang, Y.; Zhang, H.; Huang, H.; Folta, K.M.; Lu, J. Whole Genome Wide Expression Profiles of *Vitis amurensis* Grape Responding to Downy Mildew by Using Solexa Sequencing Technology. *BMC Plant Biol.* **2010**, *10*, 234. [CrossRef]
11. Liu, L.; Li, H. Review: Research Progress in Amur Grape, *Vitis amurensis* Rupr. *Can. J. Plant Sci.* **2013**, *93*, 565–575. [CrossRef]
12. Ma, Y.Y.; Zhang, Y.L.; Shao, H.; Lu, J. Differential Physio-Biochemical Responses to Cold Stress of Cold-Tolerant and Non-Tolerant Grapes (*Vitis* L.) from China. *J. Agron. Crop Sci.* **2010**, *196*, 212–219. [CrossRef]
13. Chen, Q.; Diao, L.; Song, H.; Zhu, X. *Vitis amurensis* Rupr: A Review of Chemistry and Pharmacology. *Phytomedicine* **2018**, *49*, 111–122. [CrossRef]
14. Blasi, P.; Blanc, S.; Wiedemann-Merdinoglu, S.; Prado, E.; Rühl, E.H.; Mestre, P.; Merdinoglu, D. Construction of a Reference Linkage Map of *Vitis amurensis* and Genetic Mapping of *Rpv8*, a Locus Conferring Resistance to Grapevine Downy Mildew. *Theor. Appl. Genet.* **2011**, *123*, 43–53. [CrossRef]
15. Schwander, F.; Eibach, R.; Fechter, I.; Hausmann, L.; Zyprian, E.; Töpfer, R. *Rpv10*: A New Locus from the Asian *Vitis* Gene Pool for Pyramiding Downy Mildew Resistance Loci in Grapevine. *Theor. Appl. Genet.* **2012**, *124*, 163–176. [CrossRef]
16. Venuti, S.; Copetti, D.; Foria, S.; Falginella, L.; Hoffmann, S.; Bellin, D.; Cindrić, P.; Kozma, P.; Scalabrin, S.; Morgante, M.; et al. Historical Introgression of the Downy Mildew Resistance Gene *Rpv12* from the Asian Species *Vitis amurensis* into Grapevine Varieties. *PLoS ONE* **2013**, *8*, e61228. [CrossRef]
17. Wan, Y.; Schwaninger, H.; He, P.; Wang, Y. Comparison of Resistance to Powdery Mildew and Downy Mildew in Chinese Wild Grapes. *Vitis* **2007**, *46*, 132–136.
18. Takayama, F.; Nakamoto, K.; Kawasaki, H.; Mankura, M.; Egashira, T.; Ueki, K.; Hasegawa, A.; Okada, S.; Mori, A. Beneficial Effects of *Vitis coignetiae* Pulliat Leaves on Nonalcoholic Steatohepatitis in a Rat Model. *Acta Medica Okayama* **2009**, *63*, 105–111.
19. Kim, M.J.; Paramanantham, A.; Lee, W.S.; Yun, J.W.; Chang, S.H.; Kim, D.C.; Park, H.S.; Choi, Y.H.; Kim, G.S.; Ryu, C.H.; et al. Anthocyanins Derived from *Vitis coignetiae Pulliat* Contributes Anti-Cancer Effects by Suppressing NF-κB Pathways in Hep3B Human Hepatocellular Carcinoma Cells and in vivo. *Molecules* **2020**, *25*, 5445. [CrossRef]
20. Kim, B.R.; Kim, I.H.; Lee, J.S.; Choi, Y.J. First Report of Downy Mildew Caused by *Plasmopara viticola* on *Vitis coignetiae* in Korea. *Plant Dis.* **2019**, *103*, 1793. [CrossRef]
21. Hamaoka, K.; Aoki, Y.; Suzuki, S. Isolation and Characterization of Endophyte *Bacillus velezensis* KOF112 from Grapevine Shoot Xylem as Biological Control Agent for Fungal Diseases. *Plants* **2021**, *10*, 1815. [CrossRef]
22. Lo Piccolo, S.; Alfonzo, A.; Giambra, S.; Conigliaro, G.; Lopez-Llorca, L.V.; Burruano, S. Identification of *Acremonium* Isolates from Grapevines and Evaluation of Their Antagonism towards *Plasmopara viticola*. *Ann. Microbiol.* **2015**, *65*, 2393–2403. [CrossRef]
23. Bruisson, S.; Zufferey, M.; L'Haridon, F.; Trutmann, E.; Anand, A.; Dutartre, A.; De Vrieze, M.; Weisskopf, L. Endophytes and Epiphytes from the Grapevine Leaf Microbiome as Potential Biocontrol Agents Against Phytopathogens. *Front. Microbiol.* **2019**, *10*, 491406. [CrossRef] [PubMed]
24. Musetti, R.; Polizzotto, R.; Vecchione, A.; Borselli, S.; Zulini, L.; D'Ambrosio, M.; di Toppi, L.S.; Pertot, I. Antifungal Activity of Diketopiperazines Extracted from *Alternaria alternata* against *Plasmopara viticola*: An Ultrastructural Study. *Micron* **2007**, *38*, 643–650. [CrossRef]
25. National Geographic Basemap. Available online: https://www.esri.com/news/arcuser/0312/national-geographic-basemap.html (accessed on 21 March 2024).
26. Aleynova, O.A.; Nityagovsky, N.N.; Dubrovina, A.S.; Kiselev, K.V. The Biodiversity of Grapevine Bacterial Endophytes of *Vitis amurensis* Rupr. *Plants* **2022**, *11*, 1128. [CrossRef] [PubMed]
27. Aleynova, O.A.; Nityagovsky, N.N.; Suprun, A.R.; Ananev, A.A.; Dubrovina, A.S.; Kiselev, K.V. The Diversity of Fungal Endophytes from Wild Grape *Vitis amurensis* Rupr. *Plants* **2022**, *11*, 2897. [CrossRef] [PubMed]
28. Kiselev, K.V.; Nityagovsky, N.N.; Aleynova, O.A. A Method of DNA Extraction from Plants for Metagenomic Analysis Based on the Example of Grape *Vitis amurensis* Rupr. *Appl. Biochem. Microbiol.* **2023**, *59*, 361–367. [CrossRef]
29. Bolyen, E.; Rideout, J.R.; Dillon, M.R.; Bokulich, N.A.; Abnet, C.C.; Al-Ghalith, G.A.; Alexander, H.; Alm, E.J.; Arumugam, M.; Asnicar, F.; et al. Reproducible, Interactive, Scalable and Extensible Microbiome Data Science Using QIIME 2. *Nat. Biotechnol.* **2019**, *37*, 852–857. [CrossRef] [PubMed]
30. Callahan, B.J.; McMurdie, P.J.; Rosen, M.J.; Han, A.W.; Johnson, A.J.A.; Holmes, S.P. DADA2: High-Resolution Sample Inference from Illumina Amplicon Data. *Nat. Methods* **2016**, *13*, 581–583. [CrossRef] [PubMed]
31. Quast, C.; Pruesse, E.; Yilmaz, P.; Gerken, J.; Schweer, T.; Yarza, P.; Peplies, J.; Glöckner, F.O. The SILVA Ribosomal RNA Gene Database Project: Improved Data Processing and Web-Based Tools. *Nucleic Acids Res.* **2013**, *41*, D590–D596. [CrossRef] [PubMed]

32. Abarenkov, K.; Zirk, A.; Piirmann, T.; Pöhönen, R.; Ivanov, F.; Nilsson, R.H.; Kõljalg, U. *UNITE QIIME Release for Eukaryotes*; Version 29.11.2022; UNITE Community: London, UK, 2022. [CrossRef]
33. Bisanz, J. qiime2R: Importing QIIME2 Artifacts and Associated Data into R Sessions. 2018. Available online: https://github.com/jbisanz/qiime2R (accessed on 21 March 2024).
34. McMurdie, P.J.; Holmes, S. Phyloseq: An R Package for Reproducible Interactive Analysis and Graphics of Microbiome Census Data. *PLoS ONE* **2013**, *8*, e61217. [CrossRef]
35. Wickham, H.; Averick, M.; Bryan, J.; Chang, W.; McGowan, L.D.; François, R.; Grolemund, G.; Hayes, A.; Henry, L.; Hester, J.; et al. Welcome to the Tidyverse. *J. Open Source Softw.* **2019**, *4*, 1686. [CrossRef]
36. Oksanen, J.; Blanchet, F.G.; Friendly, M.; Kindt, R.; Legendre, P.; McGlinn, D.; Minchin, P.R.; O'Hara, R.B.; Simpson, G.L.; Solymos, P.; et al. Vegan: Community Ecology Package. 2020. Available online: https://CRAN.R-project.org/package=vegan (accessed on 21 March 2024).
37. Love, M.I.; Huber, W.; Anders, S. Moderated Estimation of Fold Change and Dispersion for RNA-Seq Data with DESeq2. *Genome Biol.* **2014**, *15*, 550. [CrossRef] [PubMed]
38. Hernangómez, D. Using the Tidyverse with Terra Objects: The Tidyterra Package. *J. Open Source Softw.* **2023**, *8*, 5751. [CrossRef]
39. Pebesma, E. Simple Features for R: Standardized Support for Spatial Vector Data. *R J.* **2018**, *10*, 439–446. [CrossRef]
40. Giraud, T. Maptiles: Download and Display Map Tiles. 2023. Available online: https://github.com/riatelab/maptiles (accessed on 21 March 2024).
41. Hugh-Jones, D. Ggmagnify: Create a Magnified Inset of Part of a Ggplot Object. 2023. Available online: https://github.com/hughjonesd/ggmagnify (accessed on 21 March 2024).
42. Rouxel, M.; Mestre, P.; Comont, G.; Lehman, B.L.; Schilder, A.; Delmotte, F. Phylogenetic and Experimental Evidence for Host-Specialized Cryptic Species in a Biotrophic Oomycete. *New Phytol.* **2013**, *197*, 251–263. [CrossRef]
43. Rouxel, M.; Mestre, P.; Baudoin, A.; Carisse, O.; Delière, L.; Ellis, M.A.; Gadoury, D.; Lu, J.; Nita, M.; Richard-Cervera, S.; et al. Geographic Distribution of Cryptic Species of *Plasmopara viticola* Causing Downy Mildew on Wild and Cultivated Grape in Eastern North America. *Phytopathology* **2014**, *104*, 692–701. [CrossRef]
44. Kumar, S.; Stecher, G.; Li, M.; Knyaz, C.; Tamura, K. MEGA X: Molecular Evolutionary Genetics Analysis across Computing Platforms. *Mol. Biol. Evol.* **2018**, *35*, 1547–1549. [CrossRef] [PubMed]
45. Edgar, R.C. MUSCLE: Multiple Sequence Alignment with High Accuracy and High Throughput. *Nucleic Acids Res.* **2004**, *32*, 1792–1797. [CrossRef] [PubMed]
46. Nei, M.; Kumar, S. *Molecular Evolution and Phylogenetics*; Oxford University Press: Oxford, UK, 2000; ISBN 978-0-19-513585-5.
47. Felsenstein, J. Confidence limits on phylogenies: An approach using the bootstrap. *Evolution* **1985**, *39*, 783–791. [CrossRef]
48. Toffolatti, S.L.; Serrati, L.; Sierotzki, H.; Gisi, U.; Vercesi, A. Assessment of QoI Resistance in *Plasmopara viticola* Oospores. *Pest Manag. Sci.* **2007**, *63*, 194–201. [CrossRef]
49. Huang, X.; Wang, X.; Kong, F.; van der Lee, T.; Wang, Z.; Zhang, H. Detection and Characterization of Carboxylic Acid Amide-Resistant *Plasmopara viticola* in China Using a TaqMan-MGB Real-Time PCR. *Plant Dis.* **2020**, *104*, 2338–2345. [CrossRef]
50. Yang, L.; Chu, B.; Deng, J.; Yuan, K.; Sun, Q.; Jiang, C.; Ma, Z. Use of a Real-Time PCR Method to Quantify the Primary Infection of *Plasmopara viticola* in Commercial Vineyards. *Phytopathol. Res.* **2023**, *5*, 19. [CrossRef]
51. Mouafo-Tchinda, R.A.; Beaulieu, C.; Fall, M.L.; Carisse, O. Effect of Temperature on Aggressiveness of *Plasmopara viticola* f. sp. *aestivalis* and *P. viticola* f. sp. *riparia* from Eastern Canada. *Can. J. Plant Pathol.* **2021**, *43*, 73–87. [CrossRef]
52. Pezet, R.; Pont, V.; Cuenat, P. Method to Determine Resveratrol and Pterostilbene in Grape Berries and Wines Using High-Performance Liquid Chromatography and Highly Sensitive Fluorimetric Detection. *J. Chromatogr. A* **1994**, *663*, 191–197. [CrossRef]
53. Langcake, P. Disease Resistance of *Vitis* spp. and the Production of the Stress Metabolites Resveratrol, ε-Viniferin, α-Viniferin and Pterostilbene. *Physiol. Plant Pathol.* **1981**, *18*, 213–226. [CrossRef]
54. Dercks, W.; Creasy, L.L. The Significance of Stilbene Phytoalexins in the *Plasmopara viticola*-Grapevine Interaction. *Physiol. Mol. Plant Pathol.* **1989**, *34*, 189–202. [CrossRef]
55. Pezet, R.; Perret, C.; Jean-Denis, J.B.; Tabacchi, R.; Gindro, K.; Viret, O. δ-Viniferin, a Resveratrol Dehydrodimer: One of the Major Stilbenes Synthesized by Stressed Grapevine Leaves. *J. Agric. Food Chem.* **2003**, *51*, 5488–5492. [CrossRef] [PubMed]
56. Kiselev, K.V.; Aleynova, O.A.; Grigorchuk, V.P.; Dubrovina, A.S. Stilbene Accumulation and Expression of Stilbene Biosynthesis Pathway Genes in Wild Grapevine *Vitis amurensis* Rupr. *Planta* **2017**, *245*, 151–159. [CrossRef] [PubMed]
57. Xi, H.-F.; Ma, L.; Wang, L.-N.; Li, S.-H.; Wang, L.-J. Differential Response of the Biosynthesis of Resveratrols and Flavonoids to UV-C Irradiation in Grape Leaves. *N. Z. J. Crop Hortic. Sci.* **2015**, *43*, 163–172. [CrossRef]
58. Possamai, T.; Wiedemann-Merdinoglu, S. Phenotyping for QTL Identification: A Case Study of Resistance to *Plasmopara viticola* and *Erysiphe necator* in Grapevine. *Front. Plant Sci.* **2022**, *13*, 930954. [CrossRef]
59. Fontaine, M.C.; Labbé, F.; Dussert, Y.; Delière, L.; Richart-Cervera, S.; Giraud, T.; Delmotte, F. Europe as a Bridgehead in the Worldwide Invasion History of Grapevine Downy Mildew, *Plasmopara viticola*. *Curr. Biol.* **2021**, *31*, 2155–2166.e4. [CrossRef]
60. Chiapello, M.; Rodríguez-Romero, J.; Nerva, L.; Forgia, M.; Chitarra, W.; Ayllón, M.A.; Turina, M. Putative New Plant Viruses Associated with *Plasmopara viticola*-Infected Grapevine Samples. *Ann. Appl. Biol.* **2020**, *176*, 180–191. [CrossRef]
61. Estrada-de los Santos, P.; Vacaseydel-Aceves, N.B.; Martínez-Aguilar, L.; Cruz-Hernández, M.A.; Mendoza-Herrera, A.; Caballero-Mellado, J. *Cupriavidus* and *Burkholderia* Species Associated with Agricultural Plants That Grow in Alkaline Soils. *J. Microbiol.* **2011**, *49*, 867–876. [CrossRef] [PubMed]

62. Videira, S.I.R.; Groenewald, J.Z.; Braun, U.; Shin, H.D.; Crous, P.W. All That Glitters Is Not *Ramularia*. *Stud. Mycol.* **2016**, *83*, 49–163. [CrossRef] [PubMed]
63. McGrann, G.; Stavrinides, A.; Russell, J.; Corbitt, M.; Booth, A.; Chartrain, L.; Thomas, W.; Brown, J. A Trade off between *mlo* Resistance to Powdery Mildew and Increased Susceptibility of Barley to a Newly Important Disease, Ramularia Leaf Spot. *J. Exp. Bot.* **2014**, *65*, 1025–1037. [CrossRef] [PubMed]
64. Tsai, I.J.; Tanaka, E.; Masuya, H.; Tanaka, R.; Hirooka, Y.; Endoh, R.; Sahashi, N.; Kikuchi, T. Comparative Genomics of *Taphrina* Fungi Causing Varying Degrees of Tumorous Deformity in Plants. *Genome Biol. Evol.* **2014**, *6*, 861–872. [CrossRef] [PubMed]
65. Maeng, S.; Kim, M.K.; Subramani, G. *Hymenobacter jejuensis* sp. nov., a UV Radiation-Tolerant Bacterium Isolated from Jeju Island. *Antonie Van Leeuwenhoek* **2020**, *113*, 553–561. [CrossRef]
66. Gadoury, D.M.; Sapkota, S.; Cadle-Davidson, L.; Underhill, A.; McCann, T.; Gold, K.M.; Gambhir, N.; Combs, D.B. Effects of Nighttime Applications of Germicidal Ultraviolet Light Upon Powdery Mildew (*Erysiphe necator*), Downy Mildew (*Plasmopara viticola*), and Sour Rot of Grapevine. *Plant Dis.* **2023**, *107*, 1452–1462. [CrossRef]
67. Wassermann, B.; Korsten, L.; Berg, G. Plant Health and Sound Vibration: Analyzing Implications of the Microbiome in Grape Wine Leaves. *Pathogens* **2021**, *10*, 63. [CrossRef] [PubMed]
68. Jung, H.; Lee, D.; Lee, S.; Kong, H.J.; Park, J.; Seo, Y.-S. Comparative Genomic Analysis of *Chryseobacterium* Species: Deep Insights into Plant-Growth-Promoting and Halotolerant Capacities. *Microb. Genom.* **2023**, *9*, 001108. [CrossRef]
69. Peláez, F.; Cabello, A.; Platas, G.; Díez, M.T.; del Val, A.G.; Basilio, A.; Martán, I.; Vicente, F.; Bills, G.F.; Giacobbe, R.A.; et al. The Discovery of Enfumafungin, a Novel Antifungal Compound Produced by an Endophytic *Hormonema* Species Biological Activity and Taxonomy of the Producing Organisms. *Syst. Appl. Microbiol.* **2000**, *23*, 333–343. [CrossRef]
70. Di Francesco, A.; Zajc, J.; Gunde-Cimerman, N.; Aprea, E.; Gasperi, F.; Placì, N.; Caruso, F.; Baraldi, E. Bioactivity of Volatile Organic Compounds by *Aureobasidium* Species against Gray Mold of Tomato and Table Grape. *World J. Microbiol. Biotechnol.* **2020**, *36*, 171. [CrossRef] [PubMed]
71. Yalage Don, S.M.; Schmidtke, L.M.; Gambetta, J.M.; Steel, C.C. *Aureobasidium pullulans* Volatilome Identified by a Novel, Quantitative Approach Employing SPME-GC-MS, Suppressed *Botrytis cinerea* and *Alternaria alternata* In Vitro. *Sci. Rep.* **2020**, *10*, 4498. [CrossRef] [PubMed]
72. Lutz, M.C.; Lopes, C.A.; Rodriguez, M.E.; Sosa, M.C.; Sangorrín, M.P. Efficacy and Putative Mode of Action of Native and Commercial Antagonistic Yeasts against Postharvest Pathogens of Pear. *Int. J. Food Microbiol.* **2013**, *164*, 166–172. [CrossRef] [PubMed]
73. Zhu, L.; Li, T.; Xu, X.; Shi, X.; Wang, B. Succession of Fungal Communities at Different Developmental Stages of Cabernet Sauvignon Grapes from an Organic Vineyard in Xinjiang. *Front. Microbiol.* **2021**, *12*, 718261. [CrossRef]

horticulturae

MDPI

Article

Negative Regulatory Role of Non-Coding RNA Vvi-miR3633a in Grapevine Leaves and Callus under Heat Stress

Lipeng Zhang [1], Yuanxu Teng [1], Junpeng Li [2], Yue Song [2], Dongying Fan [2], Lujia Wang [2], Zhen Zhang [2], Yuanyuan Xu [2], Shiren Song [2], Juan He [2], Yi Ren [3], Huaifeng Liu [1,*] and Chao Ma [1,2,*]

1 Key Laboratory of Special Fruits and Vegetables Cultivation Physiology and Germplasm Resources Utilization of Xinjiang Production and Construction Corps, Department of Horticulture, Agricultural College of Shihezi University, Shihezi 832003, China; zlp@stu.shzu.edu.cn (L.Z.)

2 Shanghai Collaborative Innovation Center of Agri-Seeds, School of Agriculture and Biology, Shanghai Jiao Tong University, Shanghai 200240, China

3 College of Landscape and Horticulture, Yunnan Agricultural University, Kunming 650201, China

* Correspondence: lhf_agr@shzu.edu.cn (H.L.); chaoma2015@sjtu.edu.cn (C.M.)

Abstract: The grapevine, a globally significant fruit and an essential fruit tree species in China, is vulnerable to the adverse effects of high temperatures. Understanding the roles of microRNA and transcription factors in plant development and stress resistance is crucial for mitigating the impact of high temperature on grape growth and yield. This study investigates the response of miRNA to high-temperature stress in grape leaves. The expression level of Vvi-miR3633a was found to be inhibited under heat treatment in both Thompson seedless and Shen yue varieties, while its potential target genes (*Vv-Atg36* and *Vv-GA3ox2*) were induced. Through transgenic overexpression experiments, it was demonstrated that Vvi-miR3633a plays a role in thermal response by affecting the expression of target genes. Furthermore, under heat stress conditions, overexpression of Vvi-miR3633a in grape callus decreased heat resistance compared to the control group (CK). The study also revealed that the target genes of Vvi-miR3633a regulate the expression of oxidase synthesis genes *VvSOD* and *VvCAT*, leading to reduced oxidase synthesis which may compromise the oxidation system. Additionally, the expression level of heat shock proteins in the transgenic lines was changed compared to the control (CK). Overall, this research provides valuable insights into understanding the molecular mechanisms involved in different crossing/breeding programs to produce heat-resistant grape varieties. Such varieties can be appropriate to propagate in warm climate areas with high temperature conditions.

Keywords: grape; microRNA; Vvi-miR3633a; heat stress

1. Introduction

The grapevine (*Vitis vinifera* L.), as a highly economically valuable fruit tree, is extensively cultivated worldwide [1,2]. In the 21st century, the impact of global warming on the development of the grape industry will be significant [3]. Various abiotic stresses have influenced grape growing and fruit quality [4]. When temperatures exceed 35 °C, changes in sugar and acid content in the grape's fruit occur, leading to a decrease in quality [5,6]. Therefore, enhancing high-temperature resistance in grapes has become an important focus for development and breeding.

This research discovered that the plant's reaction to high temperatures is a multifaceted process, encompassing phenotypic, physiological, and molecular regulatory changes [7,8]. In extreme high-temperature conditions, severe tissue damage and cell death can occur, leading to a reduction in photosynthetic rate and the accumulation of harmful substances related to reactive oxygen species [9,10]. Various methods for assessing heat-stress damage in plants were confirmed by the researchers, including Fv/Fm (maximum photosynthetic efficiency of photosystem II), SOD (superoxide dismutase) and CAT (catalase) activity, as well as MDA (malondialdehyde) content [11–13]. Due to the complex genetic control of high-temperature response involving phenotypic and physiological changes, traditional breeding techniques are not deemed suitable for producing high temperature-resistant varieties [6,14]. Consequently, it is essential to conduct comprehensive studies on heat resistance mechanisms. A crucial role in plant stress responses and growth development is played by miRNA, a type of non-coding RNA [15,16]; fundamental research into the function of miRNA could facilitate rapid understanding of novel genotype selection.

In the cell nucleus, genomic DNA is transcribed to produce long RNA molecules (up to 1000 nt) that are subsequently cleaved by the ribonuclease Drosha into hairpin structures of approximately 80 bases, which are known as precursor microRNA [16]. These structures undergo further processing by another ribonuclease called Dicer, resulting in the production of mature miRNA products ranging from 21 to 24 nt in size [17]. The biosynthetic pathway of miRNA involves various enzymes such as Dicer-like 1 (DCL-1), SERRATE (SE), Hua Enhancer 1 (HEN1), and Hyponastic Leaves 1 (HYL1) proteins [18]. Generally, miRNA regulates gene expression by cleaving mRNA and inhibiting translation [19,20]. Because some studies on the function of miRNA in plants are limited, further research is necessary.

Numerous research studies have demonstrated the pivotal role of miRNA in regulating high-temperature stress [21,22]. Upon exposure to heat, miR398 downregulates the expression of target genes, thereby enhancing plant heat resistance [23]. Concurrently, tocopherol production (vitamin E) positively influences the accumulation of miR398 under heat stress conditions, thus improving thermal stability [24]. Additionally, miR156, miR160 and miR166 also contribute to heat-resistant responses by modulating the expression of transcription factors [25,26]. In a previous study involving small RNA transcriptome analysis in grapevine leaves, 873 known miRNAs were identified with DEG-seq analysis revealing differential expression of miR3633a under high-temperature stress [27]. Despite being non-conserved in plants, the expression of Vvi-miR3633 was reported by Yin et al. during flower development in camellia [28]. Furthermore, it was observed that in grape fruit, Vvi-miR3633a can facilitate embryo abortion by mediating the synthesis of oxidase SOD and CAT in response to exogenous GA induction [29]. These findings suggest that miR3633a may regulate grape ROS stability in response to high temperatures.

In previous studies, we found differential expression of miR3633a in leaf miRNA libraries. Meanwhile, target gene GO analysis showed that Vvi-miR3633a was involved in grape development and stress resistance [27]. In order to further explore the function of Vvi-miR3633a, we cloned and obtained the sequence of Vvi-miR3633a. The effect of Vvi-miR3633a on callus was then studied experimentally. When Thompson seedless callus mass overexpressed Vvi-miR3633a, it exhibited reduced tolerance to high temperatures compared to the control group. These outcomes suggest that Vvi-miR3633a participates in plant heat resistance and can offer valuable insights for genetic breeding aimed at enhancing resistance in fruit trees.

2. Materials and Methods

2.1. Plant Growth Conditions and Treatments

The Shen yue [3] one-year cuttings were obtained from the Shanghai Academy of Agricultural Sciences and subsequently transplanted into the greenhouse at the College of Agriculture and Biology, Shanghai Jiao Tong University. The Thompson seedless grape plantlets underwent cultivation in a dedicated tissue culture room, with regular replacement every 35 days. The growth medium comprised MS, 30 g/L sucrose, 7 g/L Agar, and 0.25 mg/L IBA, maintaining a pH level around 5.8. The experimental details are illustrated in Supplementary Figure S1. Prior to incubation at a temperature of 25 °C, the grape cuttings were precultured for two days. Leaf samples from Shen yue grapes were collected at both 0 h (CK) and 4 h (HS, 45 °C), while Thompson seedless tissue culture samples were taken at 0 h (CK) and after exposure to HS for three hours (45 °C). All samples were promptly frozen using liquid nitrogen and stored at −80 °C prior to analysis. This treatment process was replicated across three separate biological groups.

The induction of embryonic callus in Thompson seedless explant medium was previously documented [30]. The methods for subculture and heat stress treatment of the callus mass were derived from earlier publications by our research team [31].

2.2. Measurement of Fv/Fm (The Maximal Photochemical Quantum Yield of PSII)

Three grape plantlets (cuttings) were randomly selected from each heat stress treatment. The Fv/Fm value was evaluated, and images were acquired using Imaging-PAM (WALZ, Munich, Germany) (chlorophyll fluorescence software).

2.3. Determination of Catalase and Superoxide Dismutase Activity

Catalase (CAT) and superoxide dismutase (SOD) activities were assessed using the Total Assay Kit (Sangon Biotech, Shanghai, China).

2.4. Extraction of RNA and cDNA Synthesis

The CTAB method [32] was utilized for total RNA extraction from grapevine leaves, while Trizol reagent from Sangon Biotech (Shanghai, China) was employed for total RNA extraction from grapevine mass callus. Initially, 0.2 g samples of grape leaves and callus were pulverized and placed into RNase-free 2 mL tubes. Subsequently, 800 µL preheated CTAB buffer at 65 °C and 50 µL β-mercaptoethanol were swiftly added to the tubes. The CTAB buffer consisted of pH 8.0, CTAB (2%), Tris-HCl (100 mM), EDTA (20 mM), NaCl (2 M). The mixture was thoroughly mixed before being incubated in an oven at 65 °C for 20 min, and shaken every 5 min. Lysate was extracted using vortexing with 850 uL of chloroform/isoamylol (24:1, *v*/*v*) followed by centrifugation at 12,000 rpm for 10 min at 4 °C. This chloroform/isoamylol extraction process was repeated twice. Next, a solution containing pre-cooled ethanol (98%) and NaAc (pH 5.2) was prepared and kept in an ice bath for 10 min. Afterward, chloroform/isoamylol alcohol was added, and the mixture was vortexed before centrifugation at 12,000 rpm for 10 min at 4 °C. Then 450 µL of the supernatant was collected and 150 µL LiCl (10 M) was added and mixed. The RNA was precipitated for 12 h at 4 °C and harvested by centrifugation at 12,000 rpm for 10 min at 4 °C. The supernatant was drawn using a pipette and washed with 75% pre-cooled alcohol at −20 °C. Finally, a pipette was used to add 50 mL DEPC ddH_2O for resuspension.

Total genomic DNA removal utilized DNase I (Vazyme, Nanjing, China). Subsequently, the total RNA was then transcribed into cDNA using the Fast King RT Kit (Vazyme, China) and random primers according to the manufacturer's instructions. The miRNA RT Enzyme Mix (Vazyme, China) was subsequently utilized to synthesize the cDNA.

2.5. Determination of RT-qPCR

The SYBR Green Super-mix enzyme from Vazyme in China was employed to prepare a 10 µL RT-qPCR (real-time quantitative PCR) reaction for the analysis of mRNA and miRNA expression. The gene expression level was determined using the $2^{-\Delta\Delta Ct}$ method [33]. All primers utilized in the reaction are listed in Supplementary Table S1.

2.6. Vector Construction

The sequence characteristics of Vvi-pri-miRNA were investigated using Thompson seedless and Shen yue varieties as experimental materials for PCR–Sanger sequencing analysis. Initially, fragments of the pri-Vvi-miR3633a sequence were inserted into linear pHB vectors BamH I and Xba I. The plasmid containing the pri-vvi-miR3633a sequence was then introduced into the GV3101 strain through Agrobacterium transformation. Subsequently, they were cultured on LB medium with 25 mg·L^{-1} rifampicin (Rif) and 50 mg·L^{-1} kanamycin (Kan) for 2–3 days at a temperature of 28 °C. Mono-clones were selected and incubated in LB liquid medium supplemented with 25 mg·L^{-1} Rif and 50 mg·L^{-1} Kan in a shaker for approximately 10 h at a speed of 200 rpm and a temperature of 28 °C.

2.7. *Transient Overexpression of Vvi-miR3633a in Thompson Seedless Leaves*

The transient overexpression of Vvi-miR3633a was investigated in 5-week-old Thompson seedless grape plantlets. Each experimental group consisted of 3 biological replicates, with 3–6 seedlings per replicate. The leaves were immersed in an Agrobacterium suspension (OD 600 = 0.85) and subjected to vacuum incubation at −0.1 MPa for 30 min. Subsequently, the leaves were removed and placed in a foam box. Co-cultivation was then conducted for 2 days, followed by washing with ddH_2O, drying, freezing with liquid nitrogen, and RNA extraction.

2.8. *Thompson Seedless Mass Callus Transformation*

Following previously documented procedures, callus formation was initiated from Thompson seedless flower explants. To facilitate callus subculturing, MSTP medium (comprising MS alkali salt, 20 $g{\cdot}L^{-1}$ sucrose, 1 $mg{\cdot}L^{-1}$ TDZ, 2.2 $mg{\cdot}L^{-1}$ picloram, pH5.8) was employed for growth under dark conditions at 26 °C [31]. The transformation process of the callus followed a previously published method. Transgenic callus cultures were maintained in a dark chamber for intervals of 20 days. The expression of miRNA in transgenic callus was validated using various primer pairs through PCR and RT-qPCR analysis (see Supplementary Table S1).

2.9. *Statistical Analysis*

All biological statistical analysis of the data was conducted with Microsoft Excel (2019) and SPSS (Statistical Package for the Social Sciences) (version, 22.0) software. The images were combined using Microsoft PowerPoint software. The chart was created with TB tools-II software (https://doi.org/10.1016/j.molp.2023.09.010).

3. Results

3.1. *Influence of High Temperature on Grape Development*

Exposure to high temperatures was observed to hinder grape growth and development. In our investigation of grape response under heat stress conditions, we conducted a comprehensive analysis of both morphological and physiological traits in annual Shen yue grapes during thermal treatments. Consistent with prior studies [6], we designated a temperature of 45 °C for a duration of 4 h as the experimental group while maintaining 0 h exposure as the control group, in order to mitigate potential confounding variables such as environmental factors or soil composition (Figure 1a). Drawing from previous experiments involving high-temperature treatments on Thompson seedless plantlets, we identified 0 h and 3 h intervals as representative time points for our heat treatments (Figure 1b). Post-treatment observations revealed wilted grape leaves and shoots accompanied by browning, a clear indication of suppressed growth resulting from prolonged exposure to elevated temperatures (Figure 1c). Correspondingly, key physiological markers exhibited a notable reduction ($p < 0.01$, **) in maximum photosynthetic efficiency (Fv/Fm) within Photosystem II (PSII), alongside heightened SOD and CAT enzyme activities (Figure 1e,f). The swift elevation in oxidase activity provided confirmation that plant regulatory mechanisms were activated under heat stress conditions, resulting in an expedited synthesis process converting reactive oxygen species (ROS) into benign compounds.

3.2. *The Expression Analyses of Vvi-miR3633a*

Leveraging prior research on the miRNA library datasets from Thompson seedless grapes exposed to elevated temperatures [27], we identified a specific miRNA (miR3633a) from among those differentially expressed ($p < 0.05$). To elucidate the involvement of miR3633a in grape function, we investigated its tissue-specific expression profile within the Shen yue variety. Our observations revealed prominent expression of Vvi-miR3633a in flowers, roots, and leaves while exhibiting lower levels in stems (Supplementary Figure S3a), indicating wide-ranging functionality for miR3633a. Next, we detected the expression level of the primary transcript of miR3633a, and the results showed that pri-miR3633a first increased, and then decreased, under heat stress, and that the expression level was the highest at 4 h. This suggests that the accumulation of miR3633a also has a changing process (Supplementary Figure S3b). High-throughput small RNA sequencing unveiled reduced TPM values for heat-treated Shen yue and Thompson seedless varieties with statistically significant disparities (** $p < 0.01$) (Figure 2a,b). Additionally, RT-qPCR results corroborated these findings by demonstrating substantial down-regulation of Vvi-miR3633a under high-temperature conditions (** $p < 0.01$) (Figure 2c,d), suggesting an important role for Vvi-miR3633a in response to heat stress.

Figure 1. Phenotypic and physiological indicators were determined. Phenotypes, fluorescence photographs, and oxidase (SOD and CAT) analysis: (**a**,**c**,**e**) high-temperature treatment (HS, 45 °C, 4 h) and control group (CK, 0 h) in annual Shen yue grape cuttings; (**b**,**d**,**f**) high-temperature treatment (HS, 45 °C, 1 h) and control group (CK, 0 h) in Thompson seedless plantlets. Three biological replicates were used for all data. All data were subjected to *t*-test analysis (*, $p < 0.05$; **, $p < 0.01$). The bar corresponds to a length of 1 cm.

3.3. Identification of Vvi-miR3633a and Its Target Genes

The pre-miR3633a sequence of Shen yue and Thompson seedless was obtained by PCR–Sanger sequencing, which was consistent with the sequence of the Pinot Noir reference genome and miRbase 22.0 database (Supplementary Figure S2). Agarose gel electrophoresis showed specific bands approximately 300 bp in size (Figure 3c). At the same time, the sequence of pre-Vvi-miR3633a could be folded to form a stable stem-loop structure (Figure 3b). The secondary structure was located at positions 5918078 to 5918232 of Chr17 (Figure 3a). The mature sequence and star sequence are the red marked fonts in Figure 2b, respectively, and the mature miRNA sequence is 5′-GGAAUGGAUGGUUAGGAGAG-3′.

To further explore the target of miR3633a under high-temperature conditions, we employed the psRNATarget online tool (https://www.zhaolab.org/psRNATarget/, access on 6 June 2023) for predicting the target gene of Vvi-miR3633a. Furthermore, through integration with mRNA transcriptome data from a previous study on heat-stressed grape leaves [34], potential target genes for Vvi-miR3633a were identified as follows: *VvGA3ox2* (VIT_219s0002g05300, *Gibberellin 3-β double oxygen synthase 2*), *VvGA2ox3* (VIT_209s0140g00140, *GIBBERELLIN 2-BETA-DIOXYGENASE 1*), *vv-GABA-T3* (VIT_204s0069g00300, *Gamma aminobutyrate transaminase 3*), *vv-Atg36* (VIT_217s0000g04770, *Autophagy-related protein 36*) and *vv-PIF4* (VIT_212s0028g01110, *phytochrome interacting factor 4*) (see Supplementary Table S2).

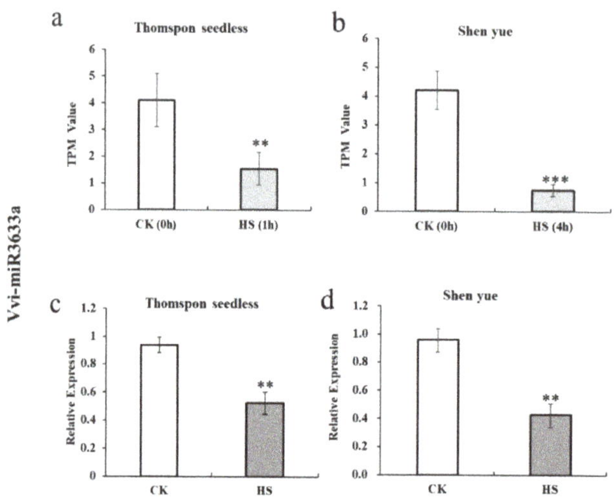

Figure 2. Expression level of miR3633a in Shen yue and Thompson seedless. (**a**,**b**) Based on TPM (transcripts per million) value of miR3633a in miRNAomes; (**c**,**d**) expression of miR3633a in RT-qPCR; (**a**,**c**) CK: 0 h, at 45 °C, HS: 1 h, at 45 °C; (**b**,**d**) CK: 0 h, at 45 °C, HS: 4 h, at 45 °C. Three biological replicates were used for all data. All data were subjected to *t*-test analysis (**, $p < 0.01$; ***, $p < 0.001$). The bar corresponds to a length of 1 cm.

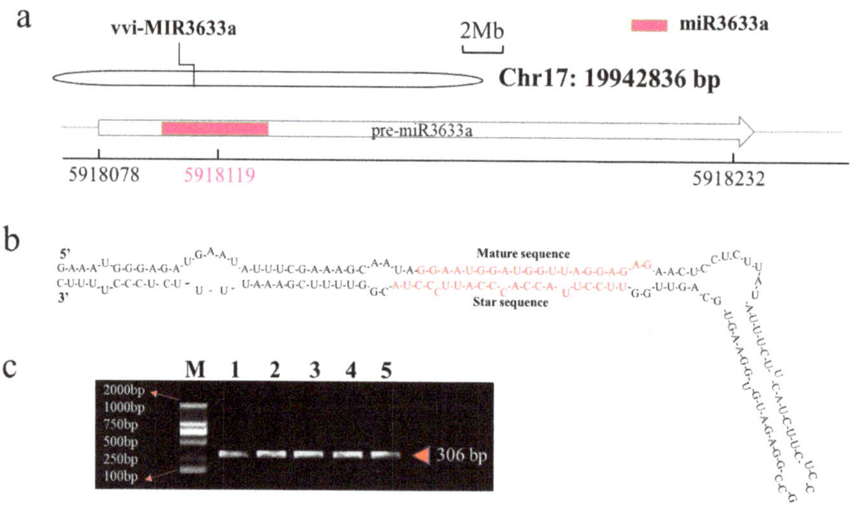

Figure 3. Sequence identification of miR3633a and pre-miR3633a in grapes. (**a**) Chromosome localization of Vvi-miR3633a and Vvi-pre-miR3633a; (**b**) stem-loop structure of Vvi-pre-miR3633a; (**c**) cloning of VvimiR3633a primary transcript (pri-miR3633a) sequence.

3.4. Expression Analyses of Target Genes of Vvi-miR3633a under Heat Stress

To investigate the target genes of Vvi-miR3633a in grape response to heat stress, we conducted a study using the leaves of Thompson seedless plantlets subjected to heat treatment as experimental materials. Our research focused on analyzing the expression patterns of marker genes related to heat stress. The findings indicated that *VvHsfA2*, *VvHSP70*, *VvSOD* and *VvCAT*, which are heat shock factor genes, were identified as markers for heat stress [34]. Their expression levels showed significant up-regulation (Figure 4a). The thermal response patterns of miR3633a potential target genes were different.

Vv-GA3ox2, *Vv-GA2ox3* and *Vv-PIF4* were significantly induced by heat treatment, while *Vv-Atg36* expression was down-regulated and *Vv-GABA-T3* expression was not significantly different (Figure 4b). The complementary expression pattern of miRNA–mRNA suggested that Vvi-miR3633a might be involved in the regulation of plant heat tolerance through *Vv-GA3ox2*, *Vv-GA2ox3* and *Vv-PIF4*.

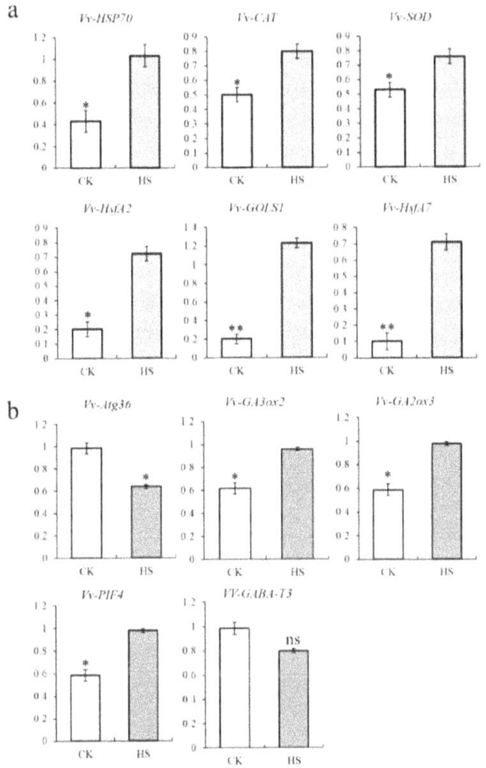

Figure 4. Identification of gene expression levels under heat stress. (**a**) RT-qPCR was used to identify key genes downstream of heat stress-related pathways; (**b**) high-temperature response analysis of miR3633a target gene. Three biological replicates were used for all data. All data were subjected to t-test analysis (*, $p < 0.05$; **, $p < 0.01$, ns, Not Significant).

3.5. Vvi-miR3633a Negatively Regulates Potential Target Genes Expression in Grapevine

To further identify potential target genes for Vvi-miR3633a, a transient expression experiment was conducted in leaves. Agrobacterium carrying the vector pHB-pri-miR3633a was introduced into Thompson seedless grape plantlet leaves, with the empty vector as control (Figure 5). The expression levels of Vvi-miR3633a and its target genes were assessed using RT-qPCR. Compared with empty vector (EV) control (CK), the relative expression of Vvi-miR3633a in the transformed leaves was significantly increased (** $p < 0.01$), while the relative expression of some predicted target genes was significantly decreased (** $p < 0.01$) (Figure 5b). For example, the expression of *VvAtg36* and *VvGA3ox2* was down-regulated in the transient leaves, while the other genes were significantly up-regulated and had no significant differences (Figure 5b). These findings suggest that miR3633a may regulate downstream gene expression by targeting the cleavage of *VvAtg36* and *VvGA3ox2*.

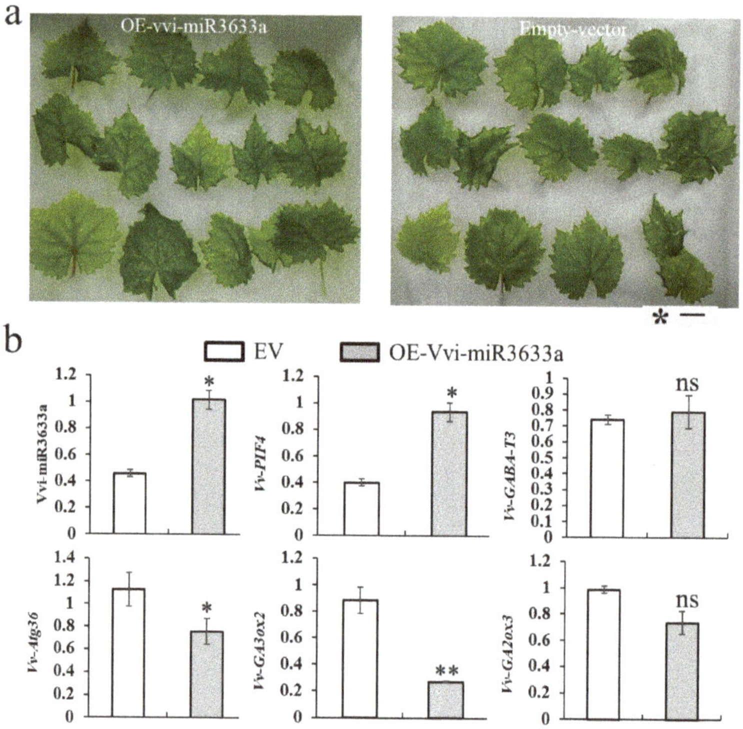

Figure 5. Vvi-miR3633a regulates the expression of target genes in grape leaves. (**a**) MiR3633a transient expression of Thompson seedless plantlet leaves, control group using empty vector (EV); (**b**) relative expression levels of miR3633a and its target genes based on RT-qPCR. The error bars represent the standard deviation of the three biological replicates. All data were subjected to *t*-test analysis (*, $p < 0.05$; **, $p < 0.01$, ns, Not Significant).

3.6. Overexpression of Vvi-miR3633a in Grape Callus

To decipher the function of Vvi-miR3633a in grapes, we generated a Vvi-pri-miR3633a overexpression vector (OE-miR3633a) and used pHB-GFP as a control (Figure 6a). The high cell division and differentiation capacity of callus is an ideal material for transgene manipulation. The overexpressed callus was validated using RT-qPCR to quantify the expression level of miR3633 in each OE mass callus. For the miR3633a-OE line, 10 mass calluses were randomly selected for assessing the extent of overexpression (Supplementary Figure S4). According to the differences in expression levels, we selected two lines with significant differences (* $p < 0.05$) and named them OE-Vvi-miR3633a-1 and OE-Vvi-miR3633a-2. To investigate the potential role of Vvi-miR3633a in response to high-temperature stress, we subjected transgenic OE callus to heat treatment at 45 °C. After 24 h, it was observed that the transgenic OE callus exhibited heightened susceptibility to high-temperature stress compared to the WT callus, which displayed reduced thermal damage (browning). Statistical analysis revealed a significantly lower thermal damage rate in WT compared to OE-Vvi-miR3633a-1 and OE-Vvi-miR3633a-2. To validate our findings, we replicated the high-temperature treatment experiment with consistent results (Supplementary Figure S6), indicating that overexpression of Vvi-miR3633a diminished heat resistance in the callus. Subsequent examination of *Vv-SOD* and *Vv-CAT* oxidase gene expression levels in the mass callus demonstrated decreased oxidase activity (Supplementary Figure S5) and reduced heat resistance in Vvi-miR3633a transgenic lines following heat treatment.

These findings suggest a negative regulatory effect of Vvi-miR3633a on grape under high-temperature stress.

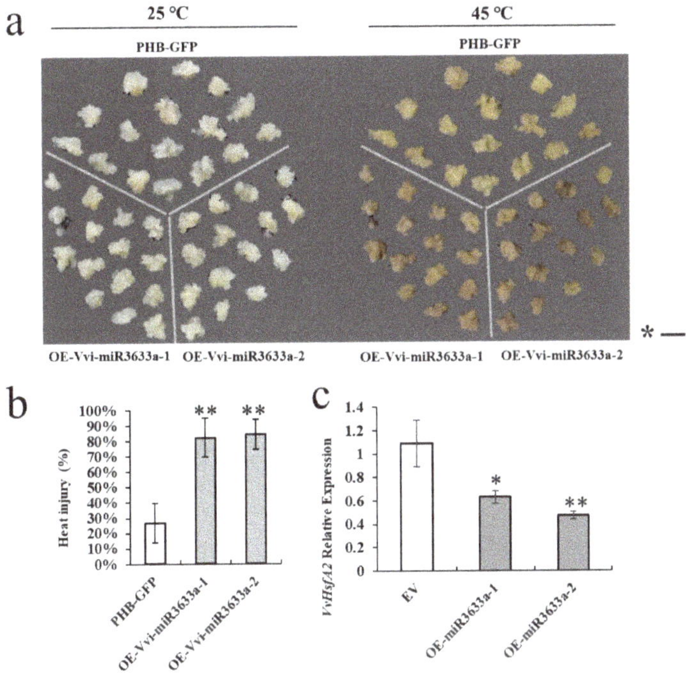

Figure 6. The phenotype of OE-Vvi-miR3633ain anther callus of grape under heat stress: (**a**) the phenotype of grape callus mass incubated under heat stress; (**b**) identification of heat injury rate of mass callus; (**c**) expression of heat response marker gene *VvHsfA2*. Line chart shows the SD with three biological replicates. Values represent means ± SE (n = 3). All data were subjected to *t*-test analysis (*, $p < 0.05$; **, $p < 0.01$). The bar corresponds to a length of 1 cm.

4. Discussion

As an abiotic stress factor, elevated temperature exerts a significant influence on the growth and fruit quality of grape plants [35,36]. Previous studies have demonstrated that heat stress results in decreased photosynthesis and oxidase activity in grapevine. Furthermore, we identified grape varieties resistant to high temperatures based on Fv/Fm values. To ensure the precision of our experimental treatments, we utilized Thompson seedless plantlets for high-temperature treatment to validate relevant physiological indicators. Under heat stress conditions, there was a reduction in Fv/Fm value and an elevation in CAT and SOD activities (Figure 1). These findings indicate that high-temperature environments can impact plant physiological responses, subsequently impeding growth and development.

MicroRNA, as a non-coding RNA, can regulate a variety of physiological responses [16]. Some miRNAs associated with plant development have also been shown to participate in abiotic stress responses [21,37]. For example, miR156 and the SPL module of the target gene not only regulate plant development [38], but also promote the stable expression of *HSPs* and *HsfA2* and enhance heat resistance [26]. MiR166 inhibited the expression of target gene *PHB* (*PHABULOSA*) during anthers development, regulated the spatial distribution of downstream genes *SPL/NZZ* and *WUS*, and participated in the flower development process [25]. At the same time, the miR165/miR166- *PHB* module has also been proved to regulate HSFA1 at the transcriptional and translation levels, enhancing plant response to heat stress [39]. With the development of next-generation sequencing (NGS), more miRNAs related to plant growth and development have been identified and their functions studied [40]. Based on the results of previous miRNA transcriptome sequencing studies, a series of differentially expressed miRNAs were identified [27,41,42]. Here, we reveal a non-conserved miRNA, known as miR3633a. RT-qPCR results confirmed that the expression level of Vvi-miR3633a was significantly down-regulated at high temperature (Figure 2). Meanwhile, we examined the grape seedlings at different seedling stages, and the results were consistent. Studies have shown that pre-miRNA sequences can affect the stability of secondary structures, thereby regulating miRNA expression [43]. Further, we amplified the sequence of pre-miR3633a in Shen yue and Thompson seedless varieties, and the results were consistent with those in the miRbase (Figure 3). This may explain the similarity of miR3633a expression trends in high-temperature environments.

In general, promoter cis-acting elements can influence gene expression. To decipher the possible function of *VvMIR3633a* in grape, we analyzed the promoter sequence. The first 2000 bp of the Vvi-pre-miR3633a sequence was analyzed based on the Thompson seedless genome and Plant Care online web page. A large number of light responsive elements were found, and others were found to respond to external environmental stress, such as the GC-motif, LTP, and ARE elements (Supplementary Table S3). These elements are easily bound by abiotic stress-related transcription factors to regulate miR3633a expression. Under heat stress, the expression of miR3633a is inhibited (Figure 2), while the expression level of its target gene is increased (Figure 4). Bai et al. found that the miR3633a-*GA3ox2* module can regulate the expression of downstream *vvSOD* and *vvCAT* genes and promote the synthesis of oxidase [29]. Under conditions of high temperature, the synthesis of oxidase can eliminate the accumulation of ROS species in plants. In phenotypic analysis, the heat damage rate of miR3633a-OE callus was significantly higher than that of WT (Figure 6 and Supplementary Figure S6). Therefore, in addition to its regulatory role, miR3633a may target key genes that inhibit oxidase synthesis and participate in the process of heat stress in plants.

Similarly, we previously constructed miRNA libraries of heat-induced grape leaves to analyze potential targets of Vvi-miR3633a. They were *VIT_219s0002g05300* (*VvGA3ox2*) [44], *VIT_209s0140g00140* (*VvGA2ox3*), *VIT_204s0069g00300* (*Vv-GABAT3*) [45], *VIT_217s0000g04770* (*Vv-Atg36*) [46] and *VIT_212s0028g01110* (*Vv-PIF4*) [47]. Most of the functions of these genes are related to plant hormones and light signaling. Under suitable temperature, *PHYTOCHROME-INTERACTING FACTOR4* (*PIF4*) changed the abundance of the flavonoid biosynthesis gene *YUCCA8* (*YUC8*) [48]. Overexpression of Vvi-miR3633a in grape leaves altered the expression of its target genes (Figure 5), indicating the regulatory role of miRNA and its involvement in grape growth and development. Meanwhile, *Vv-Atg36 and Vv-GA3ox2*, as potential target genes of Vvi-miR3633a, showed significant differences in expression levels under heat stress (Figure 5).

5. Conclusions

Grape thermal response is a complex biological process. In this study, we used morphological characteristics, physiological changes and miRNA–mRNA expression of high-temperature response to construct a model of grapevine interaction response mechanisms to regulate heat stress (Figure 7). Based on these findings, it was observed that elevated temperatures suppressed the manifestation of Vvi-miR3633a, resulting in an elevation of its potential target genes *Vv-Atg36* and *Vv-GA3ox2*. This led to a reduction in SOD and CAT oxidase activity and an increase in heat-induced damage, ultimately weakening the plant's resistance to high temperatures.

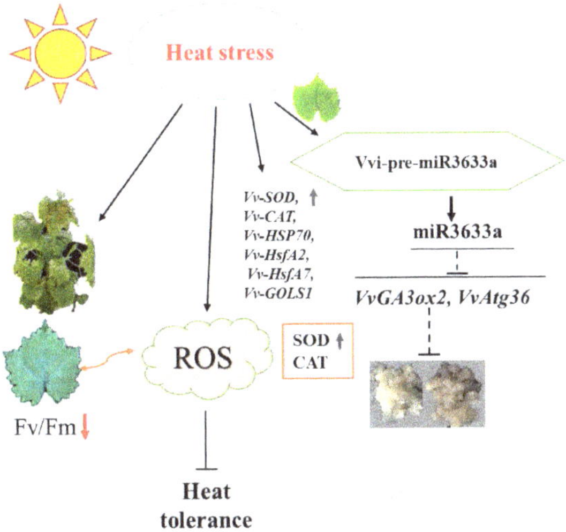

Figure 7. A hypothesized model for identification of the photosynthetic system, oxidase activity and related genes in grape response to high temperature. The upward arrow indicates positive regulation of heat stress, while the downward arrow indicates negative regulation. The dotted lines represent possible regulatory pathways.

Supplementary Materials: The following supporting information can be downloaded at: https://www.mdpi.com/article/10.3390/horticulturae10090983/s1, Supplementary Figure S1. The growth state of Shen yue annual grape cuttings and 5-week-old tissue culture plantlets of Thompson seedless. Supplementary Figure S2. Secondary structure sequence of Vvi-pre-miR3633a in grapes. Supplementary Figure S3. (a, b) Expression analysis levels of miR3633a in different tissues. Expression of pri-miR3633a under high temperature stress. Supplementary Figure S4. Expression levels of miR3633a detected by RT–qPCR in 10 mass callus tissue clusters. Supplementary Figure S5. The expressions of SOD and CAT genes were found in EV (empty vector) and OE (OE-miR3633a) callus under heat stress. Supplementary Figure S6. Phenotype of grape callus responding to high temperature stress. Supplementary Table S1. Sequence of primers used for the manuscripts. Supplementary Table S2. Target gene prediction of Vvi-miR3633a. Supplementary Table S3. Classification of cis-acting elements in promoter regions in Thompson Seedless grape.

Author Contributions: C.M. and H.L. developed the experiments. L.Z. performed the experiments and drafted the manuscript. Y.S. supplied study materials. Y.T. interpreted data. D.F. and S.S. analyzed the sequence data. Y.X., Z.Z. and L.W. collected phenotypes. J.L., J.H. and Y.R. revised the manuscript and vector construction. All authors have read and agreed to the published version of the manuscript.

Funding: This research received funding from the National Natural Science Foundation of China (Grant No. 32122076, 32341041), the Shanghai Agricultural Science and Technology Innovation Program (Grant No. 2023-02-08-00-12-F04607) and the earmarked fund for CARS-29.

Data Availability Statement: Experimental materials and data supporting the results of this study can be obtained by contacting the corresponding author.

Conflicts of Interest: The authors confirm that they have no competing interests to declare.

References

1. Chen, Q.J.; Deng, B.H.; Gao, J.; Zhao, Z.Y.; Chen, Z.L.; Song, S.R.; Wang, L.; Zhao, L.P.; Xu, W.P.; Zhang, C.X.; et al. A miRNA-Encoded Small Peptide, vvi-miPEP171d1, Regulates Adventitious Root Formation. *Plant Physiol.* **2020**, *183*, 656–670. [CrossRef] [PubMed]
2. Gao, Z.; Li, J.; Luo, M.; Li, H.; Chen, Q.; Wang, L.; Song, S.; Zhao, L.; Xu, W.; Zhang, C.; et al. Characterization and Cloning of Grape Circular RNAs Identified the Cold Resistance-Related *Vv-circATS1*. *Plant Physiol.* **2019**, *180*, 966–985. [CrossRef] [PubMed]

3. Zha, Q.; Xi, X.; He, Y.; Yin, X.; Jiang, A. Effect of Short-Time High-Temperature Treatment on the Photosynthetic Performance of Different Heat-Tolerant Grapevine Cultivars. *Photochem. Photobiol.* **2021**, *97*, 763–769. [CrossRef] [PubMed]
4. Zha, Q.; Xi, X.; Jiang, A.; Tian, Y. High Temperature Affects Photosynthetic and Molecular Processes in Field-Cultivated *Vitis vinifera* L. × *Vitis labrusca* L. *Photochem. Photobiol.* **2016**, *92*, 446–454. [CrossRef] [PubMed]
5. Zha, Q.; Xi, X.; He, Y.; Jiang, A. Transcriptomic analysis of the leaves of two grapevine cultivars under high-temperature stress. *Sci. Hortic.* **2020**, *265*, 109265. [CrossRef]
6. Zhang, L.; Song, Y.; Li, J.; Liu, J.; Zhang, Z.; Xu, Y.; Fan, D.; Liu, M.; Ren, Y.; Xi, X.; et al. Development, Identification and Validation of a Novel SSR Molecular Marker for Heat Resistance of Grapes Based on miRNA. *Horticulturae* **2023**, *9*, 931. [CrossRef]
7. Allakhverdiev, S.I.; Kreslavski, V.D.; Klimov, V.V.; Los, D.A.; Carpentier, R.; Mohanty, P. Heat stress: An overview of molecular responses in photosynthesis. *Photosynth. Res.* **2008**, *98*, 541–550. [CrossRef] [PubMed]
8. Song, P.; Jia, Q.; Xiao, X.; Tang, Y.; Liu, C.; Li, W.; Li, T.; Li, L.; Chen, H.; Zhang, W.; et al. HSP70-3 Interacts with Phospholipase Ddelta and Participates in Heat Stress Defense. *Plant Physiol.* **2021**, *185*, 1148–1165. [CrossRef] [PubMed]
9. Ashraf, M.; Harris, P.J.C. Photosynthesis under stressful environments: An overview. *Photosynthetica* **2013**, *51*, 163–190. [CrossRef]
10. Luo, H.B.; Ma, L.; Xi, H.F.; Duan, W.; Li, S.H.; Loescher, W.; Wang, J.F.; Wang, L.J. Photosynthetic responses to heat treatments at different temperatures and following recovery in grapevine (*Vitis amurensis* L.) leaves. *PLoS ONE* **2011**, *6*, e23033. [CrossRef]
11. Kalaji, H.M.; Jajoo, A.; Oukarroum, A.; Brestic, M.; Zivcak, M.; Samborska, I.A.; Cetner, M.D.; Łukasik, I.; Goltsev, V.; Ladle, R.J. Chlorophyll a fluorescence as a tool to monitor physiological status of plants under abiotic stress conditions. *Acta Physiol. Plant.* **2016**, *38*, 102. [CrossRef]
12. Mittler, R.; Vanderauwera, S.; Suzuki, N.; Miller, G.; Tognetti, V.B.; Vandepoele, K.; Gollery, M.; Shulaev, V.; Van Breusegem, F. ROS signaling: The new wave? *Trends Plant Sci.* **2011**, *16*, 300–309. [CrossRef] [PubMed]
13. Sharma, D.K.; Andersen, S.B.; Ottosen, C.O.; Rosenqvist, E. Wheat cultivars selected for high Fv / Fm under heat stress maintain high photosynthesis, total chlorophyll, stomatal conductance, transpiration and dry matter. *Physiol. Plant* **2015**, *153*, 284–298. [CrossRef] [PubMed]
14. Ye, Q.; Yu, J.; Zhang, Z.; Hou, L.; Liu, X. VvBAP1, a Grape C2 Domain Protein, Plays a Positive Regulatory Role Under Heat Stress. *Front. Plant Sci.* **2020**, *11*, 544374. [CrossRef]
15. Guan, Q.; Lu, X.; Zeng, H.; Zhang, Y.; Zhu, J. Heat stress induction of miR398 triggers a regulatory loop that is critical for thermotolerance in *Arabidopsis*. *Plant J.* **2013**, *74*, 840–851. [CrossRef]
16. Jones-Rhoades, M.W.; Bartel, D.P.; Bartel, B. MicroRNAs and their regulatory roles in plants. *Annu. Rev. Plant Biol.* **2006**, *57*, 19–53. [CrossRef]
17. He, M.; Kong, X.; Jiang, Y.; Qu, H.; Zhu, H. MicroRNAs: Emerging regulators in horticultural crops. *Trends Plant Sci.* **2022**, *27*, 936–951. [CrossRef] [PubMed]
18. Zhang, X.; Yazaki, J.; Sundaresan, A.; Cokus, S.; Chan, S.W.L.; Chen, H.; Henderson, I.R.; Shinn, P.; Pellegrini, M.; Jacobsen, S.E.; et al. Genome-wide High-Resolution Mapping and Functional Analysis of DNA Methylation in *Arabidopsis*. *Cell* **2006**, *126*, 1189–1201. [CrossRef] [PubMed]
19. Ma, C.; Burd, S.; Lers, A. miR408 is involved in abiotic stress responses in Arabidopsis. *Plant J.* **2015**, *84*, 169–187. [CrossRef] [PubMed]
20. Pei, M.; Liu, H.; Wei, T.; Jin, H.; Yu, Y.; Ma, M.; Song, X.; Dai, R.; Guo, D. Identification, characterization, and verification of miR399 target gene in grape. *Hortic. Plant J.* **2024**, *10*, 91–102. [CrossRef]
21. Ding, Y.; Huang, L.; Jiang, Q.; Zhu, C. MicroRNAs as Important Regulators of Heat Stress Responses in Plants. *J. Agric. Food Chem.* **2020**, *68*, 11320–11326. [CrossRef] [PubMed]
22. Zhu, C.; Ding, Y.; Liu, H. MiR398 and plant stress responses. *Physiol. Plant* **2011**, *143*, 1–9. [CrossRef] [PubMed]
23. Li, Y.; Li, X.; Yang, J.; He, Y. Natural antisense transcripts of *MIR398* genes suppress microR398 processing and attenuate plant thermotolerance. *Nat. Commun.* **2020**, *11*, 5351. [CrossRef] [PubMed]
24. Fang, X.; Zhao, G.; Zhang, S.; Li, Y.; Gu, H.; Li, Y.; Zhao, Q.; Qi, Y. Chloroplast-to-Nucleus Signaling Regulates MicroRNA Biogenesis in Arabidopsis. *Dev. Cell* **2019**, *48*, 371–382 e374. [CrossRef]
25. Clepet, C.; Devani, R.S.; Boumlik, R.; Hao, Y.; Morin, H.; Marcel, F.; Verdenaud, M.; Mania, B.; Brisou, G.; Citerne, S.; et al. The miR166-*SlHB15A* regulatory module controls ovule development and parthenocarpic fruit set under adverse temperatures in tomato. *Mol. Plant* **2021**, *14*, 1185–1198. [CrossRef]
26. Stief, A.; Altmann, S.; Hoffmann, K.; Pant, B.D.; Scheible, W.-R.; Bäurle, I. *Arabidopsis* miR156 regulates tolerance to recurring environmental stress through SPL transcription factors. *Plant Cell* **2014**, *26*, 1792–1807. [CrossRef]
27. Zhang, L.; Fan, D.; Li, H.; Chen, Q.; Zhang, Z.; Liu, M.; Liu, J.; Song, Y.; He, J.; Xu, W.; et al. Characterization and identification of grapevine heat stress-responsive microRNAs revealed the positive regulated function of vvi-miR167 in thermostability. *Plant Sci.* **2023**, *329*, 111623. [CrossRef]
28. Yin, Y.; An, W.; Zhao, J.-H.; Li, Y.-L.; Fan, Y.-F.; Chen, J.-H.; Cao, Y.-L.; Zhan, X.-Q. Constructing the wolfberry (*Lycium* spp.) genetic linkage map using AFLP and SSR markers. *J. Integr. Agric.* **2022**, *21*, 131–138. [CrossRef]
29. Bai, Y.; Zhang, X.; Xuan, X.; Sadeghnezhad, E.; Liu, F.; Dong, T.; Pei, D.; Fang, J.; Wang, C. miR3633a-*GA3ox2* Module Conducts Grape Seed-Embryo Abortion in Response to Gibberellin. *Int. J. Mol. Sci.* **2022**, *23*, 8767. [CrossRef]
30. Gambino, G.; Ruffa, P.; Vallania, R.; Gribaudo, I. Somatic embryogenesis from whole flowers, anthers and ovaries of grapevine (*Vitis* spp.). *Plant Cell Tissue Organ Cult.* **2007**, *90*, 79–83. [CrossRef]

31. Ren, Y.; Li, J.; Liu, J.; Zhang, Z.; Song, Y.; Fan, D.; Liu, M.; Zhang, L.; Xu, Y.; Guo, D.; et al. Transgenic Phenotypic Differences of Grapevine Circular RNA Between *Arabidopsis* and Grapevine Callus Imply the Species-dependent Functions of Vv-circPTCD1. *Front. Plant Sci.* **2023**, *12*, 2332. [CrossRef]
32. Li, M.; Li, X.; Han, Z.H.; Shu, H.; Li, T.Z. Molecular analysis of two Chinese pear (*Pyrus bretschneideri* Rehd.) spontaneous self-compatible mutants, Yan Zhuang and Jin Zhui. *Plant Biol.* **2009**, *11*, 774–783. [CrossRef] [PubMed]
33. Livak, K.J.; Schmittgen, T.D. Analysis of relative gene expression data using real-time quantitative PCR and the $2^{-\Delta\Delta CT}$ method. *Methods* **2001**, *25*, 402–408. [CrossRef] [PubMed]
34. Dou, F.; Phillip, F.O.; Liu, G.; Zhu, J.; Zhang, L.; Wang, Y.; Liu, H. Transcriptomic and physiological analyses reveal different grape varieties response to high temperature stress. *Front. Plant Sci.* **2024**, *15*, 1313832. [CrossRef]
35. Jiang, J.; Liu, X.; Liu, C.; Liu, G.; Li, S.; Wang, L. Integrating Omics and Alternative Splicing Reveals Insights into Grape Response to High Temperature. *Plant Physiol.* **2017**, *173*, 1502–1518. [CrossRef]
36. Zha, Q.; Xi, X.; Jiang, A.; Wang, S.; Tian, Y. Changes in the protective mechanism of photosystem II and molecular regulation in response to high temperature stress in grapevines. *Plant Physiol. Biochem.* **2016**, *101*, 43–53. [CrossRef]
37. Lin, J.S.; Kuo, C.C.; Yang, I.C.; Tsai, W.A.; Shen, Y.H.; Lin, C.C.; Liang, Y.C.; Li, Y.C.; Kuo, Y.W.; King, Y.C.; et al. MicroRNA160 Modulates Plant Development and Heat Shock Protein Gene Expression to Mediate Heat Tolerance in *Arabidopsis*. *Front. Plant Sci.* **2018**, *9*, 68. [CrossRef]
38. Ma, Y.; Xue, H.; Zhang, F.; Jiang, Q.; Yang, S.; Yue, P.; Wang, F.; Zhang, Y.; Li, L.; He, P.; et al. The miR156/SPL module regulates apple salt stress tolerance by activating MdWRKY100 expression. *Plant Biotechnol. J.* **2021**, *19*, 311–323. [CrossRef]
39. Li, J.; Cao, Y.; Zhang, J.; Zhu, C.; Tang, G.; Yan, J. The miR165/166-*PHABULOSA* module promotes thermotolerance by transcriptionally and post-translationally regulating HSFA1. *Plant Cell* **2023**, *35*, 2952–2971. [CrossRef]
40. Cao, C.; Long, R.; Zhang, T.; Kang, J.; Wang, Z.; Wang, P.; Sun, H.; Yu, J.; Yang, Q. Genome-Wide Identification of microRNAs in Response to Salt/Alkali Stress in Medicago truncatula through High-Throughput Sequencing. *Int. J. Mol. Sci.* **2018**, *19*, 4076. [CrossRef] [PubMed]
41. Chen, Q.; Deng, B.; Gao, J.; Zhao, Z.; Chen, Z.; Song, S.; Wang, L.; Zhao, L.; Xu, W.; Zhang, C.; et al. Comparative Analysis of miRNA Abundance Revealed the Function of Vvi-miR828 in Fruit Coloring in Root Restriction Cultivation Grapevine (*Vitis vinifera* L.). *Int. J. Mol. Sci.* **2019**, *20*, 4058. [CrossRef] [PubMed]
42. Zhang, L.; Chen, Q.; Liu, J.; Dou, F.; Wang, H.; Song, Y.; Ren, Y.; He, J.; Wang, L.; Zhang, C.; et al. Identification of grape miRNA revealed Vvi-miR164b involved in auxin induced root development. *Sci. Hortic.* **2022**, *295*, 110804. [CrossRef]
43. Yan, X.; Li, C.; Liu, K.; Zhang, T.; Xu, Q.; Li, X.; Zhu, J.; Wang, Z.; Yusuf, A.; Cao, S.; et al. Parallel degradome-seq and DMS-MaPseq substantially revise the miRNA biogenesis atlas in *Arabidopsis*. *Nat. Plants* **2024**, *10*, 1126–1143. [CrossRef] [PubMed]
44. Mitchum, M.G.; Yamaguchi, S.; Hanada, A.; Kuwahara, A.; Yoshioka, Y.; Kato, T.; Tabata, S.; Kamiya, Y.; Sun, T.P. Distinct and overlapping roles of two gibberellin 3-oxidases in *Arabidopsis* development. *Plant J.* **2006**, *45*, 804–818. [CrossRef] [PubMed]
45. Shimajiri, Y.; Oonishi, T.; Ozaki, K.; Kainou, K.; Akama, K. Genetic manipulation of the γ-aminobutyric acid (*GABA*) shunt in rice: Overexpression of truncated glutamate decarboxylase (*GAD2*) and knockdown of γ-aminobutyric acid transaminase (*GABA-T*) lead to sustained and high levels of GABA accumulation in rice kernels. *Plant Biotechnol. J.* **2013**, *11*, 594–604. [CrossRef] [PubMed]
46. Motley, A.M.; Nuttall, J.M.; Hettema, E.H. Pex3-anchored *Atg36* tags peroxisomes for degradation inSaccharomyces cerevisiae. *EMBO J.* **2012**, *31*, 2852–2868. [CrossRef] [PubMed]
47. Zhang, K.; Zheng, T.; Zhu, X.; Jiu, S.; Liu, Z.; Guan, L.; Jia, H.; Fang, J. Genome-Wide Identification of PIFs in Grapes (*Vitis vinifera* L.) and Their Transcriptional Analysis under Lighting/Shading Conditions. *Genes* **2018**, *9*, 451. [CrossRef]
48. Thines, B.C.; Youn, Y.; Duarte, M.I.; Harmon, F.G. The time of day effects of warm temperature on flowering time involve *PIF4* and *PIF5*. *J. Exp. Bot.* **2014**, *65*, 1141–1151. [CrossRef]

horticulturae

Article

Integrative Analysis of Transcriptome and Metabolome Reveals the Regulatory Network Governing Aroma Formation in Grape

Liping Huang, Yue Zhu, Min Wang, Zhili Xun, Xiaohe Ma * and Qifeng Zhao *

Pomology Institute, Shanxi Agricultural University, Jinzhong 030801, China; hlp-23@163.com (L.H.); zhy1012@sxau.edu.cn (Y.Z.); wangmin@163.com (M.W.); xzlgss@163.com (Z.X.)

* Correspondence: sxaugssmxh@163.com (X.M.); gssqfzhao@163.com (Q.Z.); Tel.: +86-0354-6215028 (Q.Z.)

Abstract: The aroma metabolites in grape berries have received attention in recent years, but a global analysis of gene-regulated metabolites is still lacking. In this study, three grape cultivars, "Kyoho", "Adenauer Rose", and "Mei Xiangbao", were used to determine the differential accumulation of metabolites and identify candidate genes related to grape berry aroma. A total of 27,228 genes were detected from the transcriptome, and 128 differentially accumulated metabolites (DAMs) were identified. Terpenoids and ester were the major substances in these three cultivars. KEGG enrichment showed that 12, 8, and 5 compounds were significantly enriched during the maturation process of these three grape cultivars, with most being terpenoids. A combined transcriptome and metabolome analysis found that the associated genes and metabolites were enriched in the following pathways: "Glycine, serine, and threonine metabolism", "Cysteine and methionine metabolism", "Tyrosine metabolism", "Phenylalanine metabolism", and "Phenylalanine, tyrosine, and tryptophan biosynthesis". Seven structural genes (*VvOMR1*, *VvGLYK*, *VvLPD2*, *VvAK2*, *VvSHM7*, *VvASP3*, and *VvASP1*) and four transcription factors (*Vv*ERF053, *Vv*ERF4, *Vv*MYB46, and *Vv*MYB340) related to grape berry aroma accumulation were discovered. Our findings provide new insights into grape aroma formation and regulatory mechanism research, and the results will be beneficial for grape aroma breeding in the future.

Keywords: aroma; metabolome; transcriptome; candidate genes; grape breeding; regulatory mechanism

1. Introduction

Grape berry aroma plays an essential role in consumers' choice of fruits and promotes their desire for re-consumption [1,2]; it is also a crucial criterion in grape breeding. Generally, the interaction among multiple metabolites contributes to the formation of aroma. Many studies have found that the volatile compounds in lipids and amino acids contribute to the enhancement of aroma in fruits [3]. Esters, as key odorants, have been found to influence the aroma quality of fruits such as apples and bananas [4,5], and polyunsaturated fatty acids have a considerable influence on the formation of esters in fruits. Ester formation in yellow-fleshed peaches has been related to the activity of alcohol acyltransferase (AAT) and gene expression [6].

Grape crops are a commercially significant fruit crop with a long history of cultivation, and their produce can be categorized as table grapes, raisins, juice, or wine [7]. Grape quality is determined by size, color, and aroma, among which aroma has received more attention in recent years. As a non-respiratory and non-climacteric fruit, grape berries have three development stages: the early stage of berry development, a short growth stagnation stage, and a maturation stage with rapid growth. The ripening of grape berries is accompanied by many physiological changes; for example, phenols (such as tannins) and aroma substances (such as pyrazines) accumulate in the early stage, while many aroma substances and their precursors accumulate in the maturation stage. The volatile aroma in grape berries usually exists in the form of a "free" or "bound" glycoside [8]. A large number of volatile aroma compounds in grape berries are found in the "bound" form and have no aroma; however, upon hydrolysis of the glycoside, the compounds can be volatilized. The volatile aroma in grape berries consists of hexacarbons, esters, terpenoids, alcohols, ketones, and aldehydes [9]. These compounds interact with each other and determine the aroma of grape berries. As direct precursors of fruit aromatic substances, amino acids can produce large amounts of aromatic substances, such as alcohols, aldehydes, esters, ketones, volatile terpenes, and carbonyls; these are the key components of grape aroma and can be used to classify neutral, strawberry, and muscat grape resources [10–15].

With the development of sequencing technology, combined RNA-Seq and GC/MS have been widely utilized to explore metabolites and related genes to reveal the coloration and quality formation in fruit crops [16–18]. Although aroma is an important characteristic of grapes, the molecular regulation mechanism is still unclear. In this study, three grape cultivars were used: "Adenauer Rose", which originated in France, belongs to *V. vinifera* L. with a strong muscat fragrance, and is a diploid cultivar. "Kyoho" belongs to *V. vinifera* × *V. labrusca* L. and has a strong strawberry aroma. "Mei Xiangbao" is a new tetraploid variety which possesses muscat and strawberry fragrances; and is crossbred from "Adenauer Rose" (female parent) and "Kyoho" (male parent). In this study, berries of these three grape cultivars in two developmental stages were explored using transcriptome and metabolome analyses, and then an integrative analysis of the transcriptome and metabolome was performed. A number of metabolites and genes associated with the formation of grape aroma were identified and quantified. The results will further our understanding of the molecular mechanism of the regulatory network governing aroma formation in grape berries and also provide us with a valuable reference for grape aroma breeding research.

2. Materials and Methods

2.1. Sample Collection

The grape cultivars (10-year-old seedlings) were planted in experimental arenaceous soil with a pH value of 7.8 at the Fruit Research Institute of Shanxi Agricultural University (39°58′ N and 116°13′ E) under a plastic cover and were trained into a two-wire vertical trellis system with a 2.5 m row space and a 1.2 m plant space. Organic fertilizer was applied in autumn, and phosphorus and potassium fertilizers were applied during the flowering and fruiting period. To avoid pests and fungal diseases, bagging was performed after pollination. Berry samples from three vines in 2017 were harvested at the development stages corresponding to the turning stage (when the berry begins to color and soften) and the maturation stage (when the berry reaches harvest ripeness). "Adenauer Rose" berries were collected on July 23 (58 days after full bloom) and August 10 (77 days after full bloom) at their turning and maturation stage; "Kyoho" berries were collected on July 25 (60 days after full bloom) and September 12 (110 days after full bloom) at their turning and maturation stage; and "Mei Xiangbao" berries were collected on July 23 (58 days after full bloom) and August 16 (83 days after full bloom) at their turning and maturation stage. Three biological replicates were used for each grape cultivar, and approximately 100 grape berries were randomly collected for each replicate. The whole berries were frozen in liquid nitrogen after removal of the skin and seeds and stored at −80 °C until needed.

2.2. mRNA Library Construction and Sequencing

The total RNA was isolated and purified using the TRIzol reagent (Invitrogen, Carlsbad, CA, USA), following the manufacturer's instructions. A NanoDrop ND-1000 (NanoDrop, Wilmington, DE, USA) and a Bioanalyzer 2100 (Agilent, Palo Alto, CA, USA) were used to determine the RNA concentration and integrity of each sample. Poly (A) RNA was achieved by using Dynabeads Oligo (dT) 25-61005 (Thermo Fisher, Waltham, MA, USA) and then fragmented into small pieces using the Magnesium RNA Fragmentation Module (New Englnd Biolabs (Beijing, China) LTD. cat. e6150, USA). After reverse-transcribing to cDNA, dUTP Solution (Thermo Fisher, cat.R0133, USA), *E. coli* DNA polymerase I (New Englnd Biolabs (Beijing) LTD. cat. m0209, USA), and RNase H (New Englnd Biolabs (Beijing) LTD. cat. m0297, USA) were used to synthesize U-labeled second-stranded DNAs, which were assembled in an adapter. After the heat-labile UDG enzyme (New Englnd Biolabs (Beijing) LTD., cat. m0280, USA) treatment of the U-labeled second-stranded DNAs, the ligated products were amplified with PCR.

Sequencing and analysis: Low-quality bases were removed using Cutadapt software (https://cutadapt.readthedocs.io/en/stable/, accessed on 28 October 2024, version:cutadapt-1.9) and then the clean reads were mapped to the reference genome and assembled using HISAT2 (https://daehwankimlab.github.io/hisat2/, accessed on 28 October 2024, version:hisat2-2.2.1) and StringTie software (http://ccb.jhu.edu/software/stringtie/, accessed on 28 October 2024, version:stringtie-2.1.6). After generating the final transcriptome, the transcript expression levels were then calculated and represented with FPKM (FPKM = [total_exon_fragments/mapped_reads (millions) $\times$ exon_length (kB)]). Genes with fold change >2 or fold change <0.5 and p-value < 0.05 were recognized as differentially expressed genes (DEGs), and then KEGG enrichment was applied to these DEGs.

2.3. Metabolite Extraction

The whole berries of the three cultivars at different development stages were ground to a powder in liquid nitrogen. Then, 500 mg (1 mL) of the powder was transferred immediately to a 20 mL headspace vial (Agilent, Palo Alto, CA, USA) containing a NaCl-saturated solution to inhibit any enzyme reaction. The vials were sealed using crimp-top caps with TFE-silicone headspace septa (Agilent). At the time of the SPME analysis, each vial was kept at 60 °C for 5 min, and then a 120 µm DVB/CWR/PDMS fiber (Agilent) was exposed to the headspace of the sample for 15 min at 60 °C.

2.4. GC-MS Analysis

After sampling, desorption of the VOCs from the fiber coating was carried out in the injection port of the GC apparatus (Model 8890; Agilent, CA, USA) at 250 °C for 5 min in the splitless mode. The identification and quantification of VOCs was carried out using an Agilent Model 8890 GC and a 7000D mass spectrometer (Agilent) equipped with a 30 m $\times$ 0.25 mm $\times$ 0.25 µm DB-5MS (5% phenyl-polymethylsiloxane) capillary column. Helium was used as the carrier gas at a linear velocity of 1.2 mL/min. The injector temperature was kept at 250 °C, and the detector at 280 °C. The oven temperature was programmed from 40 °C (3.5 min), increasing at 10 °C/min to 100 °C, at 7 °C/min to 180 °C, at 25 °C/min to 280 °C, and held for 5 min. Mass spectra were recorded in electron-impact (EI) ionization mode at 70 eV. The quadrupole mass detector, ion source, and transfer line temperatures were set, respectively, at 150, 230, and 280 °C. The MS was selected, and the ion monitoring (SIM) mode was used for the identification and quantification of the analytes.

2.5. Metabolomics Data Processing

Unsupervised PCA (principal component analysis) was performed using the statistics function within R (www.r-project.org/, accessed on 28 October 2024, 3.6). Before the unsupervised PCA was conducted, $\log_2$ transform and mean centering were used as preprocessing methods. The HCA (hierarchical cluster analysis) results for the samples and metabolites were presented as heatmaps with dendrograms, while Pearson correlation coefficients (PCCs) between the samples were calculated using the cor function in R and presented as heatmaps only. Both HCA and PCC were carried out using the R package ComplexHeatmap (http://bioconductor.org/packages/ComplexHeatmap/, accessed on 28 October 2024, version 2.8.0). For the HCA, the normalized signal intensities of the metabolites (unit variance scaling) were visualized as a color spectrum. For the two-group analysis, differential metabolites were determined using VIP (VIP > 1) and absolute Log2FC ($|$Log2FC$|$ $\geq$ 1.0). The VIP values were extracted from the OPLS-DA result, which also contained score plots and permutation plots, generated using the R package MetaboAnalystR (https://www.metaboanalyst.ca/docs/RTutorial.xhtml, accessed on 28 October 2024). The data were log-transformed (log_2) and mean-centered before the OPLS-DA. In order to avoid overfitting, a permutation test (200 permutations) was performed.

2.6. qRT-PCR Validation

Total RNA was extracted from the berries of the grape cultivars "Kyoho", "Adenauer Rose", and "Mei Xiangbao" using the Plant Total RNA Isolation Kit (SK8631; Sangon Biotech, Shanghai, China) according to the manufacturer's instructions. cDNA was obtained using PrimeScript™ RT-PCR Kit (RR047A; TaKaRa Bio, Kusatsu, Japan) according to the manufacturer's instructions. Quantitative real-time PCR (qRT-PCR) was conducted in the ABI QuantStudio™ 6 Flex System (Applied Biosystems, Waltham, MA, USA). The relative expression level of the selected genes was normalized to grapevine β-actin and calculated using the $2^{-\Delta\Delta CT}$ method. All reactions were performed using three biological replicates. The primers used in this study are listed in Table S1.

3. Results

3.1. Identification of Metabolites in Three Grape Cultivars

A total of 346 metabolites were identified at two developmental stages in three cultivars. The metabolites were divided into nucleosides, alkaloids and derivatives, benzenoids, lipids and lipid-like molecules, nucleotides and analogs, organic acids and derivatives, phenylpropanoids and polyketides, and other compounds (Table S2). Among them, lipids and lipid-like molecules contributed the most, followed by organic acids and derivatives and phenylpropanoids, which were the main components in these metabolites. Furthermore, hierarchal clustering of the metabolite profile during grape berry ripeness was performed, and the results are shown in the heatmap (Figure S1).

3.2. Differentially Accumulated Metabolites at Different Stages of Ripeness

For the analysis of differentially accumulated metabolites (DAMs), the two developmental stages of each cultivar and different cultivars were screened using the following criteria: (1) Metabolite accumulation levels with a fold change ≥2; (2) based on the OPLS-DA model analysis results, metabolites with VIP (variable importance in project) ≥ 1 [18]; and (3) with a |[log2 (fold change)]| > 1 and adjusted $p < 0.05$. A total of 128 DAMs were discovered in these three cultivars, and 94, 52, and 53 DAMs were screened by comparing A2 vs. A1, J2 vs. J1, and M2 vs. M1, respectively (Table S3). A total of 12 substances were found in the "Adenauer Rose", "Kyoho", and "Mei Xiangbao" cultivars (Figure 1a).

Terpenoids and esters were the major substances in these three cultivars. Most of the aromatic compound content in "Adenauer Rose" was significantly higher than in the other two cultivars (Table 1). Twelve compounds were discovered in A2 vs. A1, J2 vs. J1, and M2 vs. M1. Interestingly, three compounds were down-regulated in the grape cultivar "Kyoho" and up-regulated in "Adenauer Rose" and "Mei Xiangbao", including 6,6-dimethyl-Bicyclo [3.1.1]hept-2-ene-2-methanol, 1-methyl-4-(1-methyl ethylidene)-Cyclohexanol, and 2-methyl-5-(1-methylethenyl)-Cyclohexanol, which are terpenoids (Tables S4 and S5). However, 24 compounds were only discovered in A2 vs. A1 and M2 vs. M1, with most of these compounds being up-regulated in "Adenauer Rose" and "Mei Xiangbao". Three compounds, (E)-3-Hexen-1-ol-acetate, 2-methyl-5-(1-methylethyl)-Pyrazine, and (1R,2S,5S)-2-methyl-5-(1-methylethyl)-Bicyclo [3.1.0] hexan-2-ol, were up-regulated in "Adenauer Rose" and down-regulated in "Mei Xiangbao" (Tables S4 and S5). Twelve compounds were only discovered in J2 vs. J1 and M2 vs. M1; these compounds showed the same variation tendency in "Kyoho" and "Mei Xiangbao" (Tables S4 and S5). To explore the function of the metabolites, the differentially accumulated metabolites were then analyzed via KEGG enrichment (Figure 1b–d). In total, 12, 8, and 5 compounds were significantly enriched during the maturation process of these three grape cultivars, and most of them were terpenoids (Figure 1b–d, Table S6).

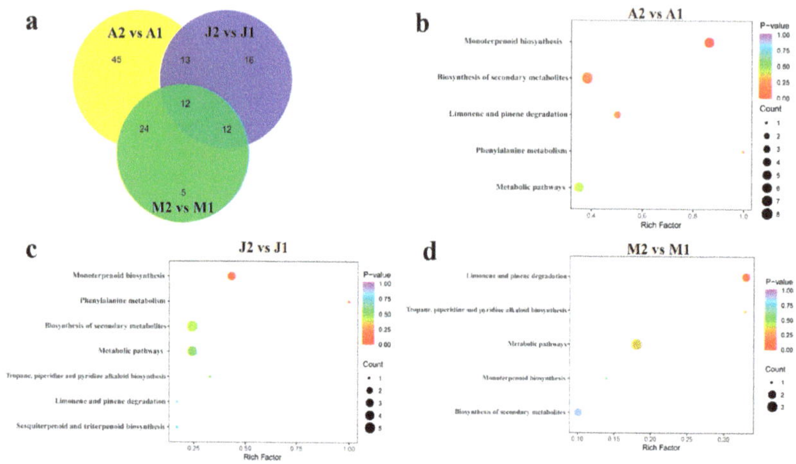

Figure 1. KEGG enrichment analysis of different accumulated metabolites for these three cultivars at different development stages. (**a**) Venn diagram for different groups. (**b**–**d**) Scatter plot of KEGG enrichment analysis for different development stages.

Table 1. Dynamic change in aromatic compounds in three cultivars at different development stages.

Class	A2 vs. A1			J2 vs. J1			M2 vs. M1		
	No.	Up	Down	No.	Up	Down	No.	Up	Down
Acids	1	0	1	2	1	1	1	1	0
Alcohols	7	7	0	6	5	1	6	6	0
Aldehydes	5	5	0	2	2	0	3	3	0
Amines	1	1	0	1	1	0	1	1	0
Aromatics	3	3	0	6	2	4	3	1	2
Esters	9	7	2	6	4	2	12	8	4
Heterocyclic compounds	17	16	1	6	3	3	8	5	2
Hydrocarbons	8	7	1	5	3	2	3	2	1
Ketones	6	6	0	4	3	1	3	3	1
Nitrogen compounds	2	2	0	0	0	0	0	0	0
Terpenoids	34	33	1	14	8	6	12	11	1
Other	1	0	1	1	0	1	1	0	1

3.3. *Expression Analysis of Key Metabolites Affecting Aroma Formation*

In order to identify key metabolites affecting aroma formation at the maturation stage of these three grape cultivars, we conducted a KEGG enrichment analysis for the comparison of the groups J2 vs. A2, J2 vs. M2, and A2 vs. M2. "Biosynthesis of secondary metabolites", "Monoterpenoid biosynthesis", "Limonene and pinene degradation", and "Metabolic pathways" were all discovered in the comparison of the groups J2 vs. A2 and J2 vs. M2. Besides these four KEGG pathways, "Tropane, piperidine and pyridine alkaloid biosynthesis" and "Phenylalanine metabolism" were discovered in the comparison of the groups A2 vs. M2 (Figure 2a–c, Table S6). In total, 14 metabolites were significantly enriched in these pathways, of which 10 were significantly enriched in the comparison of J2 vs. A2; 9 were significantly enriched in the comparison of J2 vs. M2; and 12 were significantly enriched in the comparison of A2 vs. M2 (Figure 3a, Table S6).

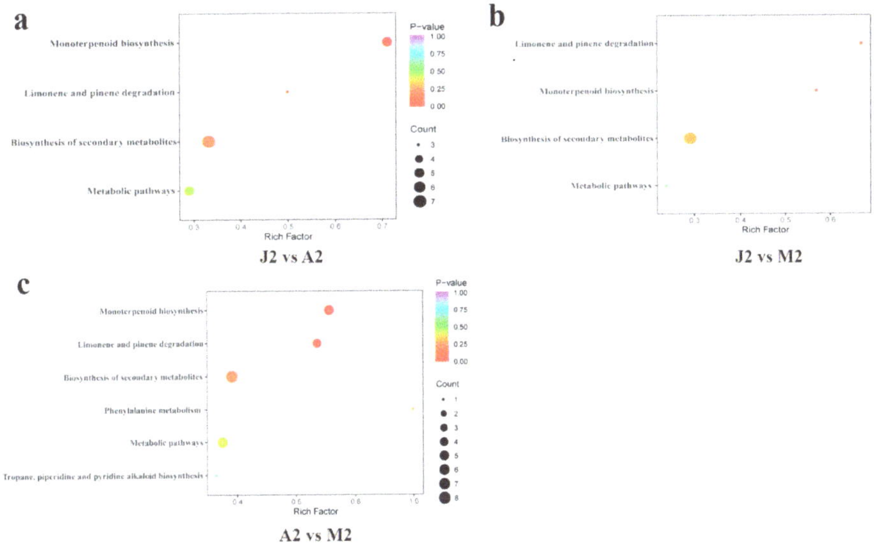

Figure 2. KEGG enrichment analysis of different accumulated metabolites for three cultivars at their maturation stage. (**a**) KEGG enrichment analysis of the groups J2 vs. A2, (**b**) KEGG enrichment analysis of the groups J2 vs. M2. (**c**) KEGG enrichment analysis of the groups A2 vs. M2.

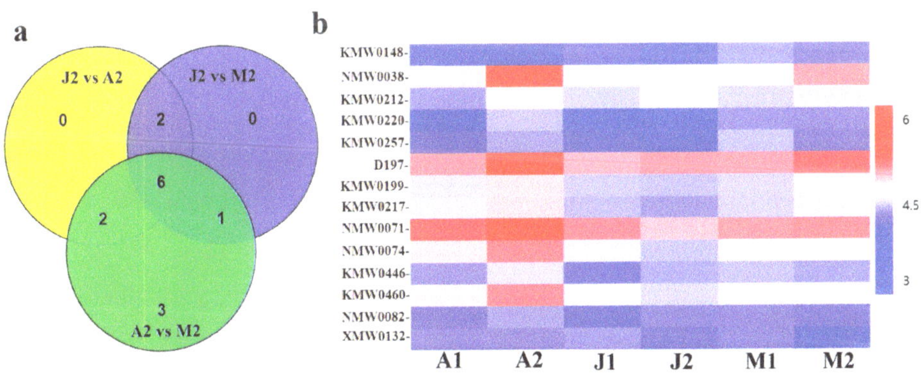

Figure 3. Statistical analysis of differentially accumulated metabolites for three cultivars at their maturation stage. (**a**) Venn diagram for different groups. (**b**) Heatmap of dynamic change for all significantly accumulated metabolites at different development stages.

The grape cultivar "Adenauer Rose" possesses a strong muscat fragrance and "Kyoho" possesses a strong strawberry fragrance. To further discover the key metabolites that could significantly affect the flavor formation of these two cultivars, six metabolites were selected according to the KEGG pathway enrichment analysis, i.e., 2,7,7-trimethyl-3-Oxatricyclo[4.1.1.0(2,4)]octane(D197), Myrcene(KMW0199), D-Limonene(KMW0217), 4-methyl-Benzenemethanol (NMW0038), L-alpha-Terpineol(NMW0071), and 6,6-dimethyl-Bicyclo[3.1.1]hept-2-ene-2-methanol (NMW0074) in J2 vs. A2, J2 vs. M2, and A2 vs. M2. Two metabolites, 3-Carene(KMW0220) and (1R,2S,5S)-2-methyl-5-(1-methylethyl)-Bicyclo [3.1.0] hexan-2-ol(KMW0257), were only selected according to J2 vs. A2 and J2 vs. M2. Two metabolites, Citral (KMW0446) and Geraniol (KMW0460) were only selected according to J2 vs. A2 and A2 vs. M2. One metabolite, α-Pinene (KMW0148), was only selected according to J2 vs. M2 and A2 vs. M2, and three metabolites, BenzeneacetAldehyde (KMW0212),

(-)-trans-Isopiperitenol (NMW0082), and 2,3,4,5-tetrahydro-Pyridine(XMW0132), were only selected according to A2 vs. M2 (Figure 3b, Table S6).

3.4. *RNA-Seq Analysis of Grape Berries at Different Stages of Ripeness*

RNA-Seq was performed to explore the gene expression in the different samples. The Q30 scores of all samples were >98%, and 4.96–8.03 GB of valid data were produced for each sample (Table S7). A total of 27,228 genes were identified from all transcriptome samples. The annotated numbers of GO and KEGG were 22697 and 4280, respectively. The PCA analysis demonstrated the good reproducibility of the transcriptome data (Figure S1). Then, the DEGs were identified by following the standard $|\log2\text{Fold Change}| \geq 1$ and FDR < 0.05, 3784, 7416, and 6159 DEGs were identified in J2 vs. J1, A2 vs. A1, and M2 vs. M1, respectively. Among them, there were 969, 2832, and 1765 up-regulated genes and 2815, 3327, and 5651 down-regulated genes in J2 vs. J1, A2 vs. A1, and M2 vs. M1, respectively (Figure 4). The numbers of DEGs in the different cultivars are exhibited in Figure 4.

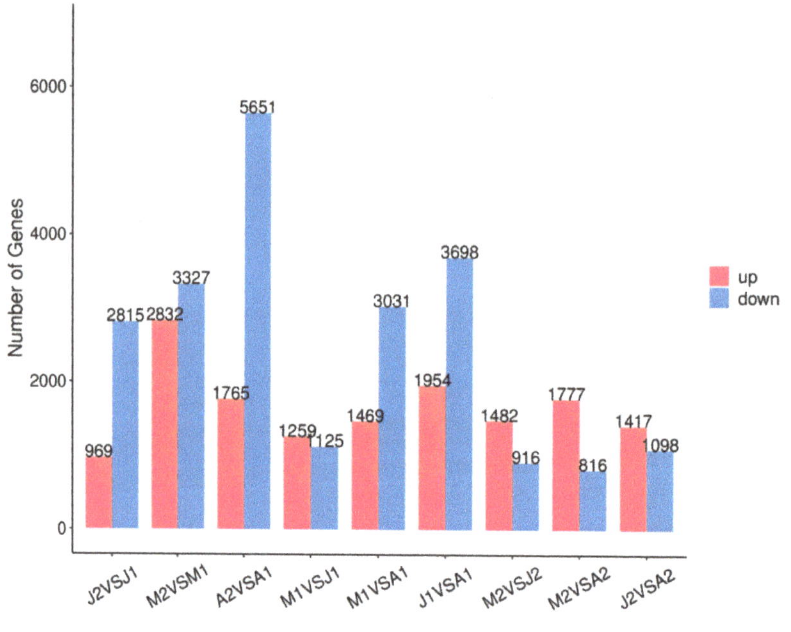

Figure 4. The number of DEGs (differentially expressed genes) in different groups. J: "Kyoho"; M: "Mei Xiangbao"; A: "Adenauer Rose"; 1: grape berries in the turning stage; 2: grape berries in the maturation stage.

3.5. *Figures, Tables, and Schemes*

In terms of identifying the relevant genes enriched in the metabolic or signaling pathways involved in grape aroma formation, the results showed that 793, 743, and 719 DEGs were enriched in the comparison of J2 vs. A2, M2 vs. J2, and M2 vs. A2 (Figure 5, Table S8). The aromatic substances in the fruits were mainly synthesized through fatty acid metabolism, amino acid metabolism, terpene metabolism, and monosaccharide and glycoside metabolism. In this study, "Glycine, serine and threonine metabolism", "Cysteine and methionine metabolism", "Glycine, serine and threonine metabolism", "Tyrosine metabolism", "Phenylalanine metabolism", and "Phenylalanine, tyrosine, and tryptophan biosynthesis" were selected according to the KEGG enrichment analysis of J2 vs. A2, M2 vs. J2, and M2 vs. A2. A total of 34, 18, and 12 DEGs were selected from the J2 vs. A2, M2 vs. J2, and M2 vs. A2 groups (Table S9). Five DEGs, i.e., *VvOMR1* (VIT_08s0007g04310), *VvGLYK* (VIT_09s0002g05200), *VvLPD2* (VIT_14s0060g01330), *VvAK2* (VIT_14s0068g01190), and *VvSHM7* (VIT_18s0001g -04340), were discovered in both J2 vs. A2 and M2 vs. J2 comparisons, while two DEGs, *VvASP3*(VIT_04s0008g03770) and *VvASP1*(VIT_08s0058g01000), were discovered in both J2 vs. A2 and M2 vs. A2 comparisons (Table S9). *VvOMR1* was significantly down-regulated in the "Kyoho" and "AR" and up-regulated in "MXB" from the turning stage to the maturation stage; *VvGLYK* was significantly down-regulated in "AR" and "MXB"; *VvLPD2* was significantly down-regulated in "Kyoho" and up-regulated in "AR" and "MXB"; *VvAK2* was significantly down-regulated in all three cultivars; *VvSHM7* and *VvASP3* were both up-regulated in "Kyoho" and "AR"; and *VvASP1* was significantly up-regulated in all three cultivars (Figure 6). Moreover, four common differentially expressed transcription factors, namely, *VvERF053*(VIT_12s0059g00280), *VvERF4*(VIT_19s0014g02240), *VvMYB46*(VIT_16s0039g01920), and *VvMYB340* (VIT_14s0066g01090), were discovered. qRT-PCR verification showed that *Vv*ERF053 was significantly down-regulated in "AR" and up-regulated in "MXB" from the turning stage to the maturation stage; *Vv*ERF4 was significantly down-regulated in "AR" and up-regulated in "Kyoho" and "MXB"; VvMYB46 was significantly up-regulated in all three cultivars; and VvMYB340 was significantly down-regulated in all three cultivars (Figure 7a).

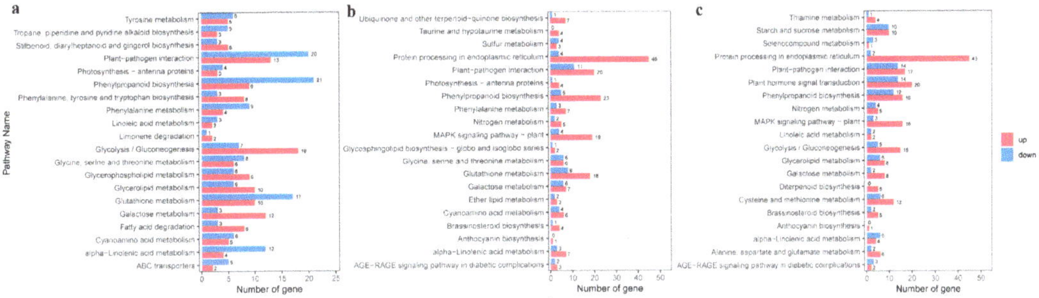

Figure 5. KEGG enrichment analysis of differentially expressed genes for three cultivars at their maturation stage.(**a**) DEGs of the groups J2 vs. A2, (**b**) DEGs of the groups J2 vs. M2. (**c**) DEGsof the groups A2 vs. M2.

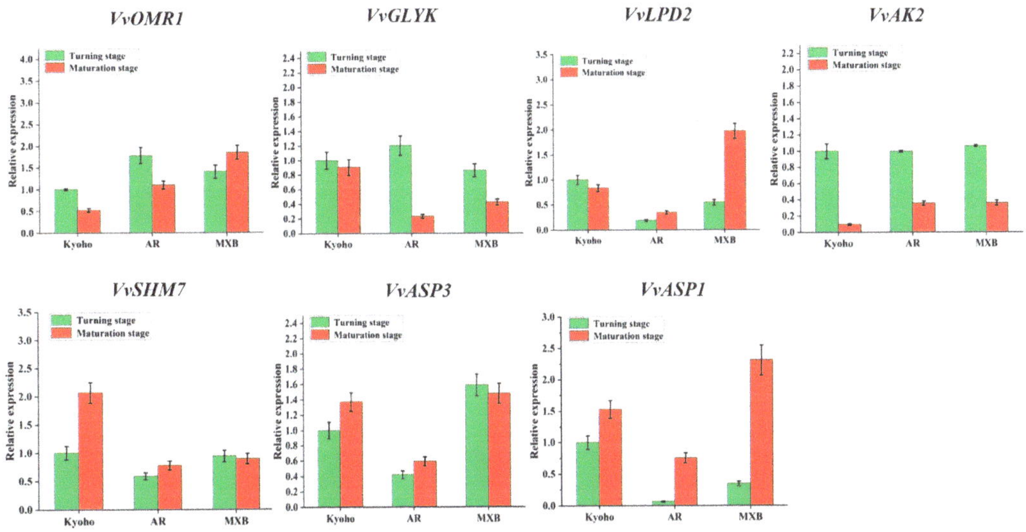

Figure 6. Candidate genes and qRT-PCR verification. Green bars represent grape turning stage and red bars represent grape maturation stage. Error bars represent the SD of three biological replicates.

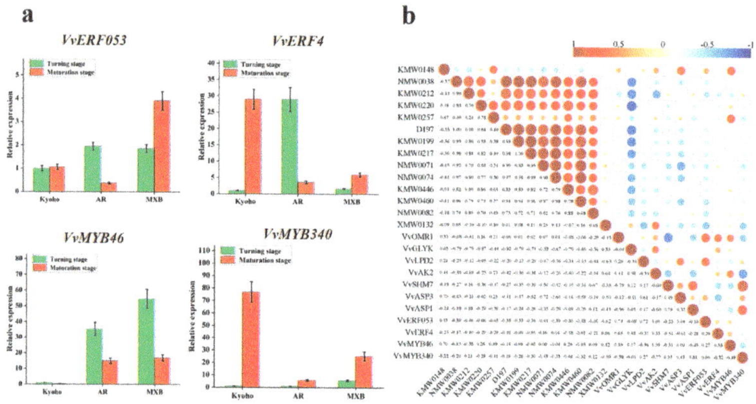

Figure 7. Correlation coefficients analysis for candidate genes and aromatic substances. (**a**) Transcription factor discovery and qRT-PCR verification. Green bars represent grape turning stage and red bars represent grape maturation stage. Error bars represent the SD of three biological replicates. (**b**) Pearson correlation coefficient analysis.

To discover the transcription factors significantly related to structural genes and key metabolites, we calculated the Pearson correlation coefficients ($p < 0.05$) for 11 candidate genes and 14 key metabolites (Figure 7b). The result showed that *VvGLYK* was negative with most of these metabolites; *VvAK2* was positive with XMW0132 and negative with KMW0212; *VvASP3* was positive with KMW0148 and *VvLPD2* and negative with NMW0071; *VvASP1* was positive with *VvSHM7* and negative with *VvOMR1* and *VvAK2*; *VvERF053* was significantly positive with *VvOMR1* and *VvLPD2*; *Vv*ERF4 was significantly positive with *VvOMR1* and negative with *VvASP3*; *Vv*MYB46 was significantly positive with KMW0148, KMW0257, *VvOMR1*, and *VvAK2*; and *Vv*MYB340 was significantly positive with *VvSHM7* and *VvASP1*.

4. Discussion

In grape aroma research, more than 2000 volatile components have been identified [19], and many studies have focused on the classification of aroma in different grapes. Terpenes, including sesquiterpenes and monoterpenes, are usually the main ingredient in muscat-scented cultivars; for instance, during the development of muscat-scented cultivars, linalool, and geraniol are the most important contributors to the aroma [13,20–22]. Recent research on the muscat grape cultivars "Shine Muscat", "Midnight Beauty", and "Centennial Seedless" showed that the contents of linalool, geraniol, geranic acid, and terpenes were higher than for other grape cultivars [23]. In our study, terpenoids were also enriched the most in the muscat cultivar, "Adenauer Rose", and linalool (KMW0291) was the highest aromatic substance, followed by 2,7,7-trimethyl- 3-Oxatricyclo [4.1.1.0(2,4)] octane (D197), fenchone (NMW0034), and geraniol (KMW0460) (Table S3). It was interesting that in the "Mei Xiangbao" grape cultivar, fenchone was the terpenoid with the highest content. Based on the KEGG enrichment analysis of "Adenauer Rose" and "Kyoho" at the maturation stage, 2,7,7-trimethyl- 3-Oxatricyclo [4.1.1.0(2,4)] octane (D197) was the most enriched aromatic substance in "Adenauer Rose".

Esters including ethyl acetate, ethyl butyrate, ethyl 2-butenoate, and ethyl 2-hexenoate are the main ingredients in strawberry-scented cultivars [12,24]. In our study, the esters in "Kyoho" were less enriched than in the other two cultivars. At the maturation stage, α-Terpineol (NMW0071), butanoic acid-ethyl ester (KMW0074), and benzeneacetic acid-ethyl ester (KMW0441) were the three highest aromatic substances in "Kyoho", while α-Terpineol was down-regulated from the turning stage to the maturation stage and butanoic acid-ethyl ester and benzeneacetic acid-ethyl ester were up-regulated. Esters and terpenoids were the aromatic substances with the highest content in the "Mei Xiangbao" cultivar. Differing from "Adenauer Rose" and "Kyoho", the ketone (methyl 2,2,3-trimethylcyclopentyl) (XMW1442) content in "Mei Xiangbao" was the highest, and it was also up-regulated from the turning stage to the maturation stage.

Amino acids and their biosynthetic intermediates play key roles as precursors for the biosynthesis of plant volatiles [25]. For instance, phenylalanine biosynthesis and metabolism pathways have been demonstrated to relate to plants' volatile diversity [26–29]. In our study, the two candidate genes *VvASP3*(VIT_04s0008g03770) and *VvASP1*(VIT_08 s0058g01000), which are involved in phenylalanine biosynthesis and the metabolism pathway, were first discovered based on the KEGG enrichment analysis. The Pearson correlation coefficients showed that these two structural genes were positively related to the content of α-Pinene (KMW0148), which is a terpenoid. Moreover, we also discovered a further four structural genes also belonging to amino acid metabolism pathways; until now, no studies have reported their role in grape berry aroma formation, and the function of these three genes still needs further investigation.

Ethylene response factors (ERFs), which belong to the AP2 family, play crucial roles in plant development and in the response to environmental stress. They can bind to the cis-acting element of the target gene promoter and promote the expression of ethylene-responsive genes [30–34]. Until now, little was known about ERFs' role in fruit aroma biosynthesis. In peach aroma biosynthesis, *PpERF5* and *PpERF7* bind together to form a protein complex that enhances the transcription of LOX4, which plays a key role in volatile biosynthesis during peach fruit ripening [35]. In the petals of sweet osmanthus, *OfERF61* can positively regulate the expression of *OfCCD4* and increase the production of β-ionone [36]. In our study, two ERFs were discovered: *VvERF053* and *VvERF4*. Based on the Pearson correlation coefficient analysis, they were positively related to the expression of the structural genes *VvOMR1* and *VvLPD2*, while *VvERF053* was found to be negatively related to most of the aromatic substances detected in this study.

MYB-related transcription factors also play crucial roles in plant development and the response to environmental stress through promoting their target gene expression [37–39]. Some reports have focused on the regulation effect of MYB transfactors in aromatic substance biosynthesis. In petunia, *PhMYB4* can suppress the production of benzenoid/phenylpropanoid biosynthesis by regulating the expression of their downstream genes [40]. In apples, *MdMYB85* directly interacts with the promoter region of *MdAAT1* and enhances ester aroma synthesis [23], while in sweet osmanthus, many MYB transfactors have been discovered, among which *OfMYB1R201* shows transcriptional activity in regulating the expression of genes related to floral volatile organic compounds [41]. *HcMYB* genes in *H. coronarium* can promote the key structural genes related to terpenoid and benzenoid biosynthesis [42]. In our study, two *MYBs*, *VvMYB46* and *VvMYB340*, were discovered. Based on the Pearson correlation coefficient analysis, *VvMYB46* was positively related to the expression of the structural genes *VvOMR1* and *VvAK2* and the production of α-Pinene (KMW0148) and 2-methyl-1-methylethyl-Bicyclo-hexan-2-ol(KMW0257). *VvMYB340* was positively related to the expression of the structural genes *VvSHM7* and *VvAPS1* and negatively related to the expression of the structural gene *VvAK2*.

5. Conclusions

In this study, we conducted RNA-seq and GC-MS analyses for the grape cultivars "Kyoho", "Adenauer Rose", and "Mei Xiangbao" to provide novel insights into the regulation of grape berry aroma formation. A total of 128 DAMs were identified; 12, 8, and 5 compounds were significantly enriched during the maturation process of these three grape cultivars, with most being terpenoids. Seven structural genes (*VvOMR1, VvGLYK, VvLPD2, VvAK2, VvSHM7, VvASP3*, and *VvASP1*) and four transcription factors (*VvERF053, VvERF4, VvMYB46*, and *VvMYB340*) involved in the amino acid metabolism pathway were discovered, although their function in grape berry aroma formation still requires further investigation. Our results will provide a reference for grape aroma breeding in the future.

Supplementary Materials: The following supporting information can be downloaded at: https://www.mdpi.com/article/10.3390/horticulturae10111159/s1, Figure S1: Hierarchal clustering of the metabolite profile during grape berry ripeness; Table S1: The primers used in this study; Table S2: Metabolite identification at two developmental stages in three cultivars; Table S3: Significantly differentially accumulated metabolites at different stages of ripeness; Table S4: Differential metabolic substance changes at different development stages; Table S5: Differential metabolic substance statistics between different groups; Table S6: Differentially accumulated metabolites analyzed via KEGG enrichment; Table S7: RNA-seq data statistics; Table S8: KEGG analysis of all DEGs between different groups; Table S9: KEGG enrichment analysis for all DEGs in different groups.

Author Contributions: Conceptualization, L.H. and Q.Z.; methodology, L.H.; software, X.M.; validation, L.H., Q.Z. and M.W.; formal analysis, Z.X.; investigation, Y.Z.; resources, Z.X.; data collection, L.H.; writing—original draft preparation, L.H.; writing—review and editing, L.H.; visualization, Y.Z.; supervision, X.M.; project administration, L.H.; funding acquisition, Q.Z. All authors have read and agreed to the published version of the manuscript.

Funding: This research was funded by the Shanxi Province Science Foundation for Youths, grant number 202103021223140, and the Open Bidding for Selecting the Best Candidates for Key Technology Research Projects in Shanxi Province, grant number 202201140601027-3.

Data Availability Statement: The article contains all raw data. For further inquiries, please contact the corresponding author.

Acknowledgments: We thank Tao Xu and Hangzhou Lianchuan Biotechnology Co., Ltd.

Conflicts of Interest: The authors declare no conflict of interest.

References

1. Barrett, D.M.; Beaulieu, J.C.; Shewfelt, R. Color, flavor, texture, and nutritional quality of fresh-cut fruits and vegetables: Desirable levels, instrumental and sensory measurement, and the effects of processing. *Crit. Rev. Food Sci. Nutr.* **2010**, *50*, 369–389. [CrossRef] [PubMed]
2. Qiu, W.; Su, W.; Cai, Z.; Dong, L.; Wu, Z. Combined analysis of transcriptome and metabolome reveals the potential mechanism of coloration and fruit quality in yellow and purple *Passiflora edulis* Sims. *J. Agric. Food Chem.* **2020**, *68*, 12096–12106. [CrossRef] [PubMed]
3. El Hadi, M.A.; Zhang, F.J.; Wu, F.F.; Zhou, C.H.; Tao, J. Advances in fruit aroma volatile research. *Molecules* **2013**, *18*, 8200–8229. [CrossRef] [PubMed]
4. Xu, Y.; Fan, W.; Qian, M.C. Characterization of aroma compounds in apple cider using solvent-assisted flavor evaporation and headspace solid-phase microextraction. *J. Agric. Food Chem.* **2007**, *55*, 3051–3057. [CrossRef] [PubMed]
5. Zhu, X.; Li, Q.; Li, J.; Luo, J.; Chen, W.; Li, X. Comparative study of volatile compounds in the fruit of two Banana cultivars at different ripening stages. *Molecules* **2018**, *23*, 2456. [CrossRef]
6. Li, C.; Xin, M.; Li, L.; He, X.; Yi, P.; Tang, Y. Characterization of the aromatic profile of purple passion fruit (*Passiflora edulis* Sims) during ripening by HS-SPME-GC/MS and RNA sequencing. *Food Chem.* **2021**, *355*, 129685. [CrossRef]
7. Mostafa, S.; Wang, Y.; Zeng, W.; Jin, B. Floral scents and fruit aromas: Functions, compositions, biosynthesis, and regulation. *Front. Plant Sci.* **2022**, *13*, 860157. [CrossRef]
8. Šikuten, I.; Štambuk, P.; Tomaz, I.; Marchal, C.; Kontić, J.K.; Lacombe, T. Discrimination of genetic and geographical groups of grape varieties (*Vitis vinifera* L.) based on their volatile organic compounds. *Front. Plant Sci.* **2022**, *13*, 942148. [CrossRef]
9. Lin, J.; Massonnet, M.; Cantu, D. The genetic basis of grape and wine aroma. *Hortic. Res.* **2019**, *6*, 81. [CrossRef]
10. Mario, N.C.; Tamames, E.L.; Jares, Y.C.M.G. Contribution to the study of the aromatic potential of three muscat *Vitis Vinifera* varieties: Identification of new compounds. *Food Sci. Technol. Int.* **1995**, *1*, 105–116. [CrossRef]
11. Escalona, J.M.; Flexas, J.; Schultz, H.R.; Medrano, H. Effect of moderate irrigation on aroma potential and other markers of grape quality. *Acta Hortic.* **1999**, *493*, 261–268. [CrossRef]
12. Buratti, S.R.A.; Benedetti, S.; Torreggiani, D. Electronic nose to detect strawberry aroma changes during osmotic dehydration. *J. Food Sci.* **2006**, *71*, 184–189. [CrossRef]
13. Fenoll, J.; Manso, A.; Hellín, P.; Ruiz, L.; Flores, P. Changes in the aromatic composition of the *Vitis vinifera* grape Muscat Hamburg during ripening. *Food Chem.* **2009**, *114*, 420–428. [CrossRef]
14. Yang, C.; Wang, Y.; Liang, Z.; Fan, P.; Li, S. Volatiles of grape berries evaluated at the germplasm level by head space SPME with GC–MS. *Food Chem.* **2009**, *114*, 1106–1114. [CrossRef]
15. Li, Y.; He, L.; Song, Y.; Zhang, P.; Chen, D.; Guan, L.; Liu, S. Comprehensive study of volatile compounds and transcriptome data providing genes for grape aroma. *BMC Plant Biol.* **2023**, *23*, 171. [CrossRef]
16. Lou, Q.; Liu, Y.; Qi, Y.; Jiao, S.; Tian, F.; Jiang, L.; Wang, Y. Transcriptome sequencing and metabolite analysis reveals the role of delphinidin metabolism in flower colour in grape hyacinth. *J. Exp. Bot.* **2014**, *65*, 3157–3164. [CrossRef]

17. Li, Y.; Fang, J.; Qi, X.; Lin, M.; Zhong, Y.; Sun, L.; Cui, W. Combined analysis of the fruit metabolome and transcriptome reveals candidate genes involved in flavonoid biosynthesis in *Actinidia arguta*. *Int. J. Mol. Sci.* **2018**, *19*, 1471. [CrossRef]
18. Zhang, Q.; Wang, L.; Liu, Z.; Zhao, Z.; Zhao, J.; Wang, Z. Transcriptome and metabolome profiling unveil the mechanisms of *Ziziphus jujuba* Mill. peel coloration. *Food Chem.* **2020**, *312*, 125903. [CrossRef]
19. Kourkoutas, D.; Elmore, J.S.; Mottram, D.S. Comparison of the volatile compositions and flavour properties of cantaloupe, Galia and honeydew muskmelons. *Food Chem.* **2006**, *97*, 95–102. [CrossRef]
20. Sun, L.; Zhu, B.; Zhang, X.; Zhang, G.; Yan, A.; Wang, H. Transcriptome profiles of three Muscat table grape cultivars to dissect the mechanism of terpene biosynthesis. *Sci. Data* **2019**, *6*, 89. [CrossRef]
21. Caffrey, A.; Ebeler, S.E. The occurrence of glycosylated aroma precursors in *Vitis vinifera* fruit and *Humulus lupulus* hop cones and their roles in wine and beer volatile aroma production. *Foods* **2021**, *10*, 935. [CrossRef]
22. Wang, W.; Feng, J.; Wei, L.; Khalil-Ur-Rehman, M.; Nieuwenhuizen, N.J.; Yang, L. Transcriptomics integrated with free and bound terpenoid aroma profiling during Shine Muscat (*Vitis labrusca* × *V. vinifera*) grape berry development reveals coordinate regulation of MEP pathway and terpene synthase gene expression. *J. Agric. Food Chem.* **2021**, *69*, 1413–1429. [CrossRef] [PubMed]
23. Li, L.X.; Fang, Y.; Li, D.; Zhu, Z.H.; Zhang, Y.; Tang, Z.Y. Transcription factors *MdMYC2* and *MdMYB85* interact with ester aroma synthesis gene *MdAAT1* in apple. *Plant Physiol.* **2023**, *193*, 2442–2458. [CrossRef] [PubMed]
24. Wu, Y.; Yu, W.W.; Zhao, W.J.; Song, L.P.; Xu, S.R.; Zhang, W.P.; Ma, C.X.; Wang, C.; Wang, L.; Shi, P. Study on the volatile composition of table grapes of three aroma types. *LWT-Food Sci. Technol.* **2019**, *115*, 108450. [CrossRef]
25. Maoz, I.; Lewinsohn, E.; Gonda, I. Amino acids metabolism as a source for aroma volatiles biosynthesis. *Curr. Opin. Plant Biol.* **2022**, *67*, 102221. [CrossRef] [PubMed]
26. Dixon, R.A.; Achnine, L.; Kota, P.; Liu, C.J.; Reddy, M.S.; Wang, L. The phenylpropanoid pathway and plant defence a genomics perspective. *Mol. Plant Pathol.* **2022**, *3*, 371–390. [CrossRef]
27. Boatright, J.; Negre, F.; Chen, X.; Kish, C.M.; Wood, B.; Peel, G. Understanding in vivo benzenoid metabolism in petunia petal tissue. *Plant Physiol.* **2004**, *135*, 1993–2011. [CrossRef]
28. Vogt, T. Phenylpropanoid biosynthesis. *Mol. Plant* **2010**, *3*, 2–20. [CrossRef]
29. Gonda, I.; Davidovich-Rikanati, R.; Bar, E.; Lev, S.; Jhirad, P.; Meshulam, Y. Differential metabolism of L-phenylalanine in the formation of aromatic volatiles in melon (*Cucumis melo* L.) fruit. *Phytochemistry* **2018**, *148*, 122–131. [CrossRef]
30. Cao, H.; Chen, J.; Yue, M.; Xu, C.; Jian, W.; Liu, Y. Tomato transcriptional repressor MYB70 directly regulates ethylene-dependent fruit ripening. *Plant J.* **2020**, *104*, 1568–1581. [CrossRef]
31. Huang, W.; Hu, N.; Xiao, Z.; Qiu, Y.; Yang, Y.; Yang, J. A molecular framework of ethylene-mediated fruit growth and ripening processes in tomato. *Plant Cell* **2022**, *34*, 3280–3300. [CrossRef] [PubMed]
32. Li, S.; Wu, P.; Yu, X.; Cao, J.; Chen, X.; Gao, L. Contrasting roles of ethylene response factors in pathogen response and ripening in fleshy fruit. *Cells* **2022**, *11*, 2484. [CrossRef] [PubMed]
33. Ma, X.; Zhao, F.; Su, K.; Lin, H.; Guo, Y. Discovery of cold-resistance genes in *Vitis amurensis* using bud based quantitative trait locus mapping and RNA-seq. *BMC Genom.* **2022**, *23*, 551. [CrossRef] [PubMed]
34. Zhao, Y.; Li, A.; Qi, S.; Su, K.; Guo, Y. Identification of candidate genes related to anthocyanin biosynthesis in red sarcocarp hawthorn (*Crataegus pinnatifida*). *Sci. Hortic.* **2022**, *298*, 110987. [CrossRef]
35. Wang, X.; Zhang, C.; Miao, Y.; Deng, L.; Zhang, B.; Meng, J. Interaction between *PpERF5* and *PpERF7* enhances peach fruit aroma by up regulating *PpLOX4* expression. *Plant Physiol. Biochem.* **2022**, *185*, 378–389. [CrossRef]
36. Han, Y.; Wang, H.; Wang, X.; Li, K.; Dong, M.; Li, Y. Mechanism of floral scent production in *Osmanthus* fragrans and the production and regulation of its key floral constituents, β-ionone and linalool. *Hortic. Res.* **2019**, *6*, 106. [CrossRef]
37. Jian, W.; Cao, H.; Yuan, S.; Liu, Y.; Lu, J.; Lu, W. *SlMYB75*, an MYB-type transcription factor, promotes anthocyanin accumulation and enhances volatile aroma production in tomato fruits. *Hortic. Res.* **2019**, *6*, 22. [CrossRef]
38. Li, W.F.; Ning, G.X.; Zuo, C.W.; Chu, M.Y.; Yang, S.J.; Ma, Z.H. MYB_SH[AL]QKY[RF] transcription factors *MdLUX* and *MdPCL-like* promote anthocyanin accumulation through DNA hypomethylation and *MdF3H* activation in apple. *Tree Physiol.* **2021**, *41*, 836–848. [CrossRef]
39. Su, K.; Xia, W.; Li, W.; Guo, Y.; Jiang, T.; Xiao, X. Discovery of candidate genes related to carotenoid accumulation based on yellow sarcocarp bud mutation peach resource. *Postharvest Biol. Technol.* **2023**, *206*, 112568. [CrossRef]
40. Yan, X.; Ding, W.; Wu, X.; Wang, L.; Yang, X.; Yue, Y. Insights into the MYB related transcription factors involved in regulating floral aroma synthesis in sweet *Osmanthus*. *Front. Plant Sci.* **2022**, *13*, 765213. [CrossRef]
41. Colquhoun, T.A.; Kim, J.Y.; Wedde, A.E.; Levin, L.A.; Schmitt, K.C.; Schuurink, R.C.; Clark, D.G. *PhMYB4* fine tunes the floral volatile signature of *Petunia x hybrida* through *PhC4H*. *J. Exp. Bot.* **2011**, *62*, 1133–1143. [CrossRef] [PubMed]
42. Abbas, F.; Ke, Y.; Zhou, Y.; Yu, Y.; Waseem, M.; Ashraf, U. Genome-Wide analysis reveals the potential role of MYB transcription factors in floral scent formation in *Hedychium coronarium*. *Front. Plant Sci.* **2021**, *12*, 623742. [CrossRef] [PubMed]

horticulturae

Article

Heterogeneity in Seed Samples from Vineyards and Natural Habitats Along the Eurasian *Vitis vinifera* Range: Implications for Domestication and Hybridization

Diego Rivera [1,*], Javier Valera [2], David Maghradze [3,4], Maia Kikvadze [3], Anna Nebish [5,6], Rafael Ocete [7], Carlos Álvar Ocete [7], Claire Arnold [8], Emilio Laguna [9], Francisco Alcaraz [1], Diego José Rivera-Obón [10], Gianni Lovicu [11], Massimino Farci [11] and Concepción Obón [12,*]

1 Department of Plant Biology, Faculty of Biology, University of Murcia, 30100 Murcia, Spain; falcaraz@um.es
2 Department of Prehistory, Archeology, and ancient History, University of Murcia, Campus de la Merced, C. Santo Cristo, 1, 30001 Murcia, Spain; javier.valera@um.es
3 Faculty of Viticulture and Winemaking, Caucasus International University, Chargali Street 73, 0141 Tbilisi, Georgia; david.maghradze@ciu.edu.ge (D.M.); maia.kikvadze@ciu.edu.ge (M.K.)
4 Faculty of Agrarian Sciences and Biosystems Engineering, Georgian Technical University, 0192, 17 David Guramishvili Ave., 0175 Tbilisi, Georgia
5 Institute of Vine and Wine Sciences ICVV (CSIC, UR, Government of La Rioja), 26007 Logroño, La Rioja, Spain; anna.nebish@icvv.es
6 Department of Genetics and Cytology, Yerevan State University, Charents str. 8, Yerevan 0025, Armenia
7 La Gualdia 9, 26211 Tirgo, La Rioja, Spain; ocete@us.es (R.O.); carlos_ocete@hotmail.com (C.Á.O.)
8 Office 331, Unicentre Building, University of Lausanne, 1015 Lausanne, Switzerland; claire.arnold@unil.ch
9 Generalitat Valenciana, Centre per a la Investigació i Experimentació Forestal, Avda. Comarques del País Valencià 114, 46930 Quart de Poblet, València, Spain; laguna_emi@gva.es
10 Jean Monnet Faculty, Paris-Saclay University, 54 Boulevard Desgranges, 92230 Sceaux, France; dieguitojos@gmail.com
11 Dipartimento per la Ricerca nell'Arboricoltura, Agris Sardegna, Via Mameli 126d, 09123 Cagliari, Italy; glovicu@agrisricerca.it (G.L.); mfarci@agrisricerca.it (M.F.)
12 Institute for Agri-Food and Agro-Environmental Research and Innovation (CIAGRO,) Orihuela Polytechnic School, Miguel Hernandez University, Ctra. Beniel, Km 3,2, 03312 Orihuela, Alicante, Spain
* Correspondence: drivera@um.es (D.R.); cobon@umh.es (C.O.); Tel.: +34-868884994 (D.R.)

Abstract: By exploring seed samples from vineyards and natural habitats across the Eurasian range of *Vitis vinifera*, our analysis revealed substantial morphological variation within populations. Through the analysis of domestication index values, probabilities, and entropy, we assessed seed diversity. Samples with high domestication probability values—predominantly from vineyards—exhibited low heterogeneity and entropy, with similar patterns observed in natural habitats, suggesting the presence of feral vines. In parallel, seeds with low domestication index values, found mainly in natural habitats, also displayed low entropy and are likely associated with *Vitis sylvestris* or other wild Vitaceae species. Intermediate domestication values pointed to hybrid swarms, highlighting the crucial role of hybridization in the development of modern grapevine cultivars. The study identified mixed populations across the Iberian Peninsula, Italy, and the South Caucasus, which act as significant gene reservoirs. A domestication gradient is evident, with higher domestication rates in the South Caucasus compared to Western Europe and East Asia. The results demonstrate the significance of these mixed populations as repositories of genetic diversity, underscoring their conservation value, particularly considering the negative impact of habitat alterations, especially in riparian forests due to major public works.

Keywords: feral; grapevine; *Vitis sylvestris*; introgression; heterogeneity estimate; Shannon index; hybridization; wild relatives; biodiversity

 https://doi.org/10.3390/horticulturae11010092

1. Introduction

Heterogeneity represents a fundamental organizing principle of biological systems, manifesting across multiple scales from the molecular to the ecosystem level. This inherent variability encompasses the morphological, functional, and interactive dimensions of organisms and their populations. In cultivated species such as *Vitis vinifera* L., population heterogeneity is particularly evident in the phenotypic and genetic variation among individuals within defined cultivar groups.

The application of entropy—a concept derived from information theory and thermodynamics—provides a robust quantitative framework for analyzing such biological diversity. In population biology, entropy metrics effectively quantify the distribution patterns of phenotypic traits, genetic markers, or physiological responses within populations [1]. Higher entropy values correspond to increased population heterogeneity, reflecting greater informational complexity and reduced predictability in population structure [2].

In viticultural systems, entropy values demonstrate significant variation between genetically diverse populations and clonally propagated monocultures. Higher entropy metrics in heterogeneous vineyards reflect increased informational complexity across multiple parameters, including genotypic diversity, pathogen-resistance profiles, and yield characteristics. This quantitative framework for assessing population heterogeneity has substantial implications for viticulture, informing germplasm selection, disease-resistance breeding programs, and the development of sustainable cultivation strategies. Moreover, understanding the relationship between population entropy and phenotypic plasticity provides valuable insights for enhancing vineyard resilience in the context of contemporary viticultural challenges.

Grapevine domestication is a complex process that has likely occurred over thousands of years, starting in the Near East and South Caucasus at least since the Neolithic period, about 11,000 years ago [3]. The wild grapevine species *Vitis sylvestris* C.C. Gmelin is widely regarded as the primary ancestor of the cultivated grapevine *Vitis vinifera* L., although the species rank for *Vitis sylvestris* is not universally accepted. The domestication process has resulted in the selective modification of key phenotypic traits, including sexual system evolution from dioecy to monoecy, increased berry and foliar dimensions, enhanced sugar accumulation, and altered seed morphology characterized by larger seeds with a reduced proportional size relative to the berry dimensions [4]. *Vitis sylvestris* is a dioecious species, meaning that male and female flowers are found on separate plants. In contrast, *Vitis vinifera* is predominantly hermaphroditic, with both male and female reproductive organs present in the same flower. The transition in reproductive strategy facilitated the domestication process through enabling a consistent fruit set and yield stability [4].

The domestication of *Vitis vinifera* from its wild progenitor *V. sylvestris* represents a pivotal transformation in agricultural development. This process, characterized by the selection of desirable traits, has had profound cultural and economic implications. The emergence of viticulture and oenology as specialized fields underscores the grapevine's importance in various ancient civilizations. The dissemination of grapevine cultivation facilitated intercultural exchange, encompassing technological innovations and diverse cultural practices across regions of the Old World [4]. This agricultural advancement has not only shaped ancient societies but continues to influence modern viticultural practices and the global wine industry.

Genetic studies have provided substantial insights into the complexities of grapevine domestication, with molecular evidence suggesting, though not conclusively, the possibility of multiple domestication events involving diverse *V. sylvestris* populations or related taxa. This hypothesis is corroborated by the high genetic diversity observed in modern cultivated grapevines, reflecting a rich evolutionary history. The genetic diversity preserved in *V.*

sylvestris populations, despite traits such as small bunches, poor-quality fruit, flowers that fail to ripen, or the lack of fruit production in male individuals, represents a valuable resource for contemporary grapevine breeding programs. This diversity harbors valuable traits, including disease resistance and tolerance to environmental stress, which could be introgressed into cultivated grapevines to enhance their resilience in a changing climate [4].

The range of wild grapevine (*V. sylvestris*) spans from the Hindu Kush mountain range to Western Europe and the Mediterranean region, covering latitudes from 30–31° N (Ourika River, Morocco) to 49–50° N (Rhine River, Ludwigshafen am Rhein, in Germany) [5,6]. Wild grapevines are primarily dioecious, meaning they have separate male and female individuals [7]. Female flowers exhibit morphological hermaphroditism, but their pollen is non-functional. On the other hand, male flowers carry functional pollen and a non-functional gynoecium, although there can be exceptions where male flowers exhibit hermaphroditic characteristics, referred to as androids by Levadoux [8].

In *Vitis* species, pollen grain morphology is closely associated with the plant's reproductive strategy. Male and hermaphroditic vines produce tricolporate pollen grains, characterized by three germinal furrows (colpi) and pores. These structures facilitate the release of male gametes. In contrast, female vines produce acolporate pollen grains, which lack these apertures, rendering them functionally sterile [9,10].

Hermaphroditism in grapevine populations emerged through mutations affecting male reproductive development. This evolutionary modification of sexual morphology enabled self-fertilization capacity, a characteristic that significantly enhanced cultivar stability during domestication [11]. Cultivated grapevine varieties (*Vitis vinifera*) exhibit high levels of heterozygosity and genetic diversity [11]. This genetic variability is a consequence of the species' evolutionary history, multiple domestication events, ongoing hybridization with wild relatives, and human-mediated selection processes.

The differentiation between wild grapevines (*Vitis sylvestris*) and their domesticated counterparts (*Vitis vinifera*) has been a subject of extensive research in ampelography and viticulture. While several morphological characteristics are commonly cited as distinguishing features, recent studies have highlighted the complexity and variability of these traits.

Wild grapevines are generally characterized by a greater environmental hardiness and distinct reproductive features. They typically produce small, black berries with a rounded shape and acidic flavor, arranged in loose, small to medium-sized clusters. The plants are dioecious, exhibiting sexual dimorphism in leaf morphology. Additionally, wild grapevines are often reported to exhibit an open U-shaped or very open petiolar sinus, though variations in this trait may occur under different stress conditions. Their seeds are generally described as small, rounded, and characterized by a short beak [12–14].

Turkovic [15] proposed that the open petiolar sinus, reminiscent of American *Vitis* rootstock genotypes or cultivars like *V. rupestris* Scheele "du Lot" or *V. riparia* Michx. "Gloire de Montpéllier" could serve as a key distinguishing feature. However, this characteristic is not exclusive to wild grapevines and may occur in some cultivars.

The purported enhanced disease resistance in wild grapevines, with the notable exception of susceptibility to the erineum strain of *Colomerus vitis*, has been historically cited as a distinguishing characteristic between wild and cultivated *Vitis* species. Recent investigations have challenged this generalization, revealing that heightened resistance is not a consistent trait across all wild grapevine populations. Instead, empirical evidence indicates that significantly elevated resistance levels are confined to a limited subset of wild accessions. This finding underscores the complexity of disease-resistance mechanisms in *Vitis* species and highlights the need for nuanced approaches in comparative studies of wild and cultivated grapevines [16].

Distinguishing between *V. vinifera*, female *V. sylvestris*, and their hybrids presents significant challenges. The primary differentiating factors include leaf size and trunk thickness, which are generally larger in *V. vinifera*. Sexual dimorphism in leaf morphology is observed in *V. sylvestris* populations, with geographical variations [17]. Subtle differences in rhytidome (bark) structure may also be present. Historical observations (19th century) by Rathay [18] in the upper Danube valley noted sexual dimorphism in wild grapevines, with females exhibiting nearly entire leaves and males possessing highly divided (three-lobed) leaves [19].

Levadoux [8] demonstrated the relativity of these distinguishing traits, observing that wild grapevines can exhibit characteristics typically associated with cultivated varieties, while some primitive cultivars may resemble wild grapevines. Certain traits, such as black berry color, small size, rounded shape, high acidity, and loose clustering, tend to dominate in sexually propagated cultivars. While cultivar characteristics are maintained through vegetative propagation, they can undergo significant modifications during sexual reproduction. For instance, seeds of cultivars like Pinot Noir or Petit Verdot may acquire traits resembling those of *V. sylvestris* within a few generations of sexual reproduction.

This complexity underscores the need for integrative approaches, combining morphological, genetic, and ecological data, to accurately distinguish between wild and domesticated grapevines. The intricate relationship between wild and cultivated grapevines continues to challenge researchers, highlighting the importance of comprehensive studies in grapevine biology and evolution.

We emphasize the significance of investigating seed heterogeneity in both vineyards and natural habitats to evaluate the phenotypic and genetic diversity within grapevines, and as an indicator of introgression events. The examination of grape seeds serves as a valuable tool for assessing the domestication levels of grapevines, both in contemporary populations and archaeological contexts [11,16–35].

This study builds on previous research to distinguish between indigenous wild grapevines, hybrids, and feral varieties. The term "wild" is commonly associated with organisms growing without human intervention in natural habitats. However, "wild" plants in the ecological sense constitute a very heterogeneous complex in terms of their origin and morphology [35], as Levadoux [8] clearly stated in the case of the grapevine (Figure 1). In this context, "wild" is used to denote any type of grapevine in natural settings, encompassing a broad range of origins and morphologies. The term "wild" or "autochthonous wild" refers to naturally occurring grapevines classified at the species level as *Vitis sylvestris*. In contrast, grapevines exhibiting clear domestication traits are classified as *Vitis vinifera*, where they are referred to as feral or sub-spontaneous when found in natural habitats and as cultivars when cultivated in vineyards. Hybrids between indigenous and cultivated wild individuals are termed hybrids, without a specific name. Asian and American species, along with their hybrids, are designated by their accepted scientific names as listed in POWO [36] or GRIN [37] except for *Vitis sylvestris*.

Our research questions and hypotheses are rooted in the work of Levadoux [8] that defines a wild grapevine, *lambrusque* in French, *lambrusca* in Italian, and *labrusca* in Spanish, as any grapevine seemingly growing in a wild state.

Various types of wild grapevines within the complex *V. vinifera–V. sylvestris* are identified based on their habitat and origin, and the major types are illustrated in Figure 1:

- Post-cultural wild or post-cultural lambruscae thrive in abandoned vineyards, closely resembling the ampelographic characteristics and seed parameters of the same cultivar in cultivated vineyards.
- Sub-spontaneous feral lambruscae grow in uncultivated soil from seeds originating in domesticated grapevine vineyards. While sharing most characteristics with the parent

grapevine cultivar, they exhibit greater differences from wild grapevines in the same natural habitat.

- Spontaneous wild *lambruscae* represents a natural autochthonous element of Western Eurasian flora. This category has multiple origins:
    (a) Colonial wild grapevines that arise from wild sub-spontaneous plants that have found favorable conditions for returning to the wild. They are naturalized but of cultivated origin.
    (b) Autochthonous wild (spontaneous native) grapevines descended from ancestors that were likely never cultivated; these grapevines are predominantly dioecious, wild, native, and autochthonous.
    (c) Wild mestizo grapevines resulting from the hybridization of native wild plants with post-cultural or sub-spontaneous lambruscae. They are naturalized but non-native due to exogenous parentage.

Figure 1. Relationships between cultivated and wild grapevine populations and different "wild" types. Image from D. Rivera based on Levadoux [8]. The yellow arrow represents a synthesis of widely accepted presumptions, scientific scenarios, and interpretations of findings that link *Vitis sylvestris* to the origins of *Vitis vinifera*. However, the potential contribution of other ancestors cannot be disregarded.

This implies that at least four different types of "wild" vines may occur and often overlap in natural habitats: wild sub-spontaneous lambruscae, wild colonial, wild mestizo, and wild autochthonous native. Their identification is a challenge that needs to be addressed.

Furthermore, various interspecific hybrids involving American *Vitis* species (*V. berlandieri* Planch. × *V. riparia*, *V. rupestris* × *V. berlandieri*, *V. solonis* (Planch.) Millardet × *V. riparia*) were developed in the 19th century to create phylloxera-resistant rootstocks for grapevine cultivation. These American species and their hybrids have escaped cultivation, entering natural habitats along European rivers, where they produce hybrids with *V. sylvestris* and even *V. vinifera*, as observed in Hungary. The local presence of American species and their hybrids has displaced preexisting *V. sylvestris* populations in certain areas of Hungary and France [38,39]. In Italy, numerous feral populations of American *Vitis* species have led to the discovery of interspecific hybrids, such as *V.* ×*baccoi* Ardenghi, Galasso, & Banfi (=*V. riparia* × *V. vinifera*), *V.* ×*goliath* Ardenghi, Galasso, & Banfi (=*V. riparia* × *V. rupestris* × *V. vinifera*), *V.* ×*instabilis* Ardenghi, Galasso, Banfi, & Lastrucci (=*V. riparia* × *V. rupestris*), *V.* ×*koberi* Ardenghi, Galasso, Banfi, & Lastrucci (=*V. berlandieri* × *V. riparia*), and *V.* ×*ruggerii* Ardenghi, Galasso, Banfi, & Lastrucci (=*V. berlandieri* × *V. rupestris*), growing in natural habitats. This complexity further complicates the study of wild grapevine populations [40]. Similarly, in Spain, hybrids have been named after natural populations, such as *V.* ×*gallica* F.M. Vázquez (=*V. berlandieri* × *V. vinifera*) and *V.* ×*hispanica* F.M. Vázquez & D. García (=*V. rupestris* × *V. vinifera*) [41]. In the Valencian hinterland, feral American grapevine rootstocks are known as *parrizos*, *parrizas*, or *vidueños* in Spanish.

The study aims to explore and differentiate the ampelographic characteristics and seed allometric relationships among the subtypes in the gradient from purely domesticated cultivars to native autochthonous wild grapevine (cultivars, postcultural vines in abandoned vineyards, wild sub-spontaneous *lambruscae*, wild colonial, wild mestizo, and wild autochthonous native), and in parallel determine their heterogeneity or entropy.

Our primary goal is to analyze and compare wild and domesticated Eurasian grapevine seed samples in terms of their heterogeneity, entropy, and the amount of information required to describe them. This analysis aims to identify links between heterogeneity and potential introgression or hybridization between wild and domesticated grapevines.

The fundamental objective is thus to develop a methodology for detecting introgressed or hybrid *Vitis* populations (wild mestizo) and to distinguish them from wild autochthonous native grapevines, through the study of seed morphology. Another goal is to identify feral populations of *Vitis vinifera* (wild sub-spontaneous lambruscae and wild colonial) in natural habitats based on the different degrees of domestication observed in their seeds.

To achieve this, we plan to utilize domestication index values and domestication probabilities, and the estimate of the heterogeneity within the seed ensemble of each single sample as indicators of introgressions and hybrid swarms, providing a quantitative estimate of variability within individuals and populations. Lastly, we seek to determine geographical patterns at different levels (countries and regions) and the connections between wild and domesticated grapevine diversity, shedding light on domestication events in these regions.

2. Materials and Methods

2.1. Geographic Locations of the Vineyards and Natural Habitats Sampled

A total of 4341 grape seeds from 816 samples were analyzed. But only 491 samples consisted of between 3 and 26 seeds, averaging eight seeds per sample. Of the 816 samples and 4341 seeds, 107 and 147, respectively, were fossils, 195 and 398 archaeobotanical, and 514 and 3796 modern. Modern samples included, in addition to *Vitis vinifera* and *V. sylvestris*, 5 samples of the genus *Ampelopsis*, 38 of American *Vitis* species, not only collected in America, but also escaped in Europe, and 8 Asian *Vitis* species, collected in India, China, and Japan.

From a geographical viewpoint, most samples were collected in Europe and the South Caucasus. In France, 20 samples were collected, primarily sourced from vineyards (19), with only 1 from a natural habitat. Georgia contributed 36 samples, with a balanced representation from both natural habitats (16) and vineyards (20). Armenia stands out with 73 samples, where 29 came from natural habitats and 44 from vineyards. Italy provided 50 samples, predominantly from natural habitats (38), while 12 were from vineyards. Spain exhibited the highest sample count at 183, comprising 80 from natural habitats and 103 from vineyards. Turkey and the USA, with 24 and 26 samples, respectively, showcased an inverse pattern. In the case of Turkey, 19 out of 24 samples came from vineyards, while in the USA, only 1 sample came from a vineyard, and 25 came from natural habitats. Further details can be found in Supplementary Tables S1 and S2.

2.2. Process of Seed Collection and Criteria for Inclusion in the Study

We aimed to achieve a comprehensive representation of variability within *Vitis*, encompassing both cultivated and wild populations, as well as archaeological and fossil materials. In the assessment of heterogeneity, we endeavored to sample wild grapevine populations extensively in Spain and Italy, supplemented by substantial samplings in Armenia and Georgia. In the case of cultivars, they were directly sampled from production vineyards or collections within various germplasm banks, notably in the Rioja's collection at Mendavia (Navarra, Spain), San Michele all'Adige (Trentino-Alto Adige, Italy), Agullent (Valencia, Spain), and Geilweilerhof—Institut für Rebenzüchtung (Siebeldingen, Germany).

Ripe grapes were manually collected in vineyards, natural habitats, and repositories by members or collaborators of this research team. Care was taken to collect each sample from one single individual, and when possible, from a couple of bunches of grapes. In natural populations, this was very difficult as many bunches had a very low number of grapes, so more were used.

The seeds were manually extracted from the grapes and then left to dry at room temperature and stored in glass test tubes, 100 × 16 mm, fitted with a screw cap, labeled inside and outside.

2.3. Methods Used for Calculating the Domestication Index and Probabilities of Domestication

Each of the 4341 seeds was individually described using 14 characters. Among these, 11 were quantitative, including the total seed length, maximum breadth and thickness, breadth of the stalk at the junction and seed base, length of the beak in the dorsal and ventral views, thickness of the beak at the seed base, total length of the chalaza scutellum, maximum breadth of the chalaza scutellum, and distance from the chalaza apex to the seed apex [27,35,42]. Additionally, there are three qualitative characteristics: contour type, with options like ovoid, quadrangular, triangular, rounded, and pentagonal; the arrangement of the fossettes, with choices such as parallel, furcate, convergent, and divergent; and the presence/absence of radial furrows. But, for the present work, only the 3970 seeds of the samples with three or more seeds were considered.

The quantitative and qualitative traits were assessed by analyzing digital images of seeds. These seeds were photographed in the dorsal, ventral, and lateral views. Measurements were conducted utilizing the open-source Fiji software 2.9.0, released on 14 September 2022 [43]. Additionally, scaled images of fossilized and archaeological seeds, sourced from specialized literature, were incorporated for reference measurements. All recorded traits were documented in an Excel spreadsheet, and allometric relationships were automatically computed through dedicated algorithms.

The first step was to calculate six classical domestication indices for each individual seed: Stummer's [12], Facsar and Perret's [20–23,26], and Mangafa and Kotsakis' 1 to 4 indices [25]. The next step was to combine these indices into a single index, which we call the domestication index (DI), and its complementary wildness (WI) index [35,42].

We used logistic regression models [35,44,45] and in parallel the random forest technique [35,46–49] to evaluate, using morphometric data and comparison collections, the probability of individual seeds coming from a domesticated grapevine, or a wild native autochthonous one, or presenting intermediate characteristics that prevent a clear allocation to one of the two categories. In the present study, "randomForest" in R version 4.4.1 (14 June 2024) and R Studio [50,51] assigns a probability value of being "domesticated" to each single seed (PDIrF, probability of domestication estimated with randomForest), which ranges from 0 to 1. And Logit proceeds similarly in R [35] (PDILO, the probability of domestication estimated with Logit). These together with the DI index served to identify the "domestication syndrome" in grapevine seeds [35].

Details of the different calculation methods are given in the references cited and accompany this paper, with Supplementary References [52–54], as Supplementary Materials File S1: Methods.

2.4. Methods Used for Calculating the Heterogeneity and Entropy Levels

The heterogeneity within a sample tends to increase when there are a greater number of distinct categories and a more even distribution of abundance or representation across those categories. To quantify the degree of sample heterogeneity, various statistical estimators can be employed, based on the behavior of the domestication values observed for the individual seeds within each sample. Indicators reflecting the dispersion or spread of values within each sample, such as the standard deviation or the range between the minimum and maximum values, can effectively represent this sample-level heterogeneity. Such heterogeneity may potentially be associated with the hybrid or admixed origin of the sample under investigation.

The introgression index (*HI*), denoted by the standard deviation of domestication index values (σ (*DI*)), serves as an indicator of heterogeneity within the sample.

The proportion of wild seeds (*PW*) within each sample refers to the fraction of seeds below the domestication threshold ($DI \leq 0.2$), denoted as "wild," ranging from 0 to 1. The term "wild" here signifies phenotypic characteristics, not merely habitat, and its interpretation considers values across the entire sample and other contextual data, aiding in distinguishing native wild plants, hybrids or mestizo, ferals (sub-spontaneous or colonial), or complex populations with different types cohabiting.

The wild sum/2 (*WS*/2) for each sample is calculated using Equation (1).

$$\frac{WS}{2} = \frac{mean\ (1 - DI) + PW}{2} \tag{1}$$

This index integrates factors such as the wildness index (1 − domestication index) and proportion of wild seeds, contributing to a comprehensive assessment of the sample. The wild sum index is calculated by dividing the sum by 2 to ensure that the range of values remains between 0 and 1. This approach maintains comparability with other indexes, where 0 represents fully domesticated and 1 represents fully wild. The mean value of $1 - DI$ provides an indication of the wildness degree of the sample (with values > 0.8 meaning wild autochthonous native), and when combined with *PW*, the proportion of wild seeds in the sample, it allows for a preliminary evaluation of the heterogeneity within the sample. A purely wild sample will yield a value of 1 due to $DI = 0$ and $PW = 1$, while conversely, a purely domesticated sample will yield 0. Intermediate values suggest introgression.

The maximum and minimum value of the domestication index (max (*DI*) and min (*DI*), respectively) within the sample allow us to calculate the range $\rho(DI)$ (max (*DI*) − min (*DI*)), which also serves as a distinctive parameter, facilitating the differentiation of major types.

2.5. Methods Used for Calculating the Information and Entropy Levels

Theoretically, addressing the heterogeneity of a seed sample can involve calculating information using the Shannon index. While this approach has found success in the case of the *Phoenix* palms' genus [2], its direct application to seed observations may lack informativeness. However, when calculated based on individual values of *DI*, *PDILO* (the probability of domestication estimated with Logit), and *PDIrF* (the probability of domestication estimated with random forest), the Shannon index proves to be extremely useful.

To compute the Shannon index for grapevine seed sample heterogeneity (*Ht*), we start with the premise that *DI* (domestication index) may have seven distinct values (0, 0.17, 0.33, 0.5, 0.66, 0.83, 1), while *PDILO* (the probability of domestication estimated with Logit) and *PDIrF* (the probability of domestication estimated with random forest), ranging continuously from 0 to 1, are discretized into ten intervals of 0.1 each. Probabilities are computed based on the frequency distribution of possible values within each sample. The index is then calculated using Equation (2).

$$H_t = -\sum_{i=1}^{7} p(DI_i) \times log_2\,(p(DI_i) - \sum_{i=1}^{10} p(PDILO_i) \times log_2\,(p(PDILO_i) - \sum_{i=1}^{10} p(PDIrF_i) \times log_2\,(p(PDIrF_i) \quad (2)$$

The magnitude of the total heterogeneity index (H_t) serves as a quantitative indicator of phenotypic diversity within seed assemblages. Higher H_t values denote greater morphological heterogeneity within a given seed sample, whereas lower H_t values indicate reduced phenotypic variation. When H_t = 0, this signifies complete homogeneity, representing a monomorphic sample containing solely one domestication phenotype from the possible morphological spectrum. This metric thus provides insights into the degree of domestication-related diversity present within modern and archaeobotanical assemblages.

We employed the Shannon uniformity (or equitability) index (E_h) to quantify the relative abundance distribution of morphological variants within each seed assemblage. This metric enables the assessment of phenotypic evenness, thereby elucidating the distributional patterns of domestication-related traits and providing insights into the taxonomic origin and cultivation status of the source population. The *Eh* index functions as a reliable discriminatory metric for differentiating among domesticated cultivars, their wild progenitors, and the intermediate hybrid populations that constitute the domestication gradient. This differentiation is achieved through the quantitative assessment of seed morphotype uniformity patterns, which reflect the varying degrees of phenotypic stabilization characteristic of populations at distinct stages along the domestication continuum. The presence of hybrid swarms, representing admixed populations with intermediate phenotypes, can be effectively identified through their distinctive *Eh* values (Equation (3)).

$$E_h = \sum_{i=1}^{3} \frac{H_t}{log_2(N_{types}) \times 3} = \frac{H_t(DI)}{log_2\,(7) \times 3} + \frac{H_t(PDILO)}{log_2\,(10) \times 3} + \frac{H_t(PDIrF)}{log_2\,(10) \times 3} \quad (3)$$

The parameter E_h varies between 0 and 1, where E_h = 0 indicates minimal evenness, suggesting pure samples of one single type, while the rest are missing. On the other hand, maximum evenness, with E_h = 1, is observed in samples where all potential types are present and in equal proportions. The values of equitability are approximately 1/10 of the Shannon index scores.

3. Results

3.1. Phenotypic Seed Patterns of Variability Within Samples

The samples analyzed exhibit a broad spectrum of variation, ranging from phenotypically purely wild to fully domesticated (Figure 2A,B). Notably, there are several transitional degrees in between, some of which may correspond to ongoing introgression processes and represent the full range from cultivars (PW = 0, 1 − DI < 0.2) (Figure 2A) to postcultural vines in abandoned vineyards, wild sub-spontaneous lambruscae, wild colonial, wild mestizo, and wild autochthonous native plants (PW > 0.6, 1 − DI > 0.8) (Figure 2B).

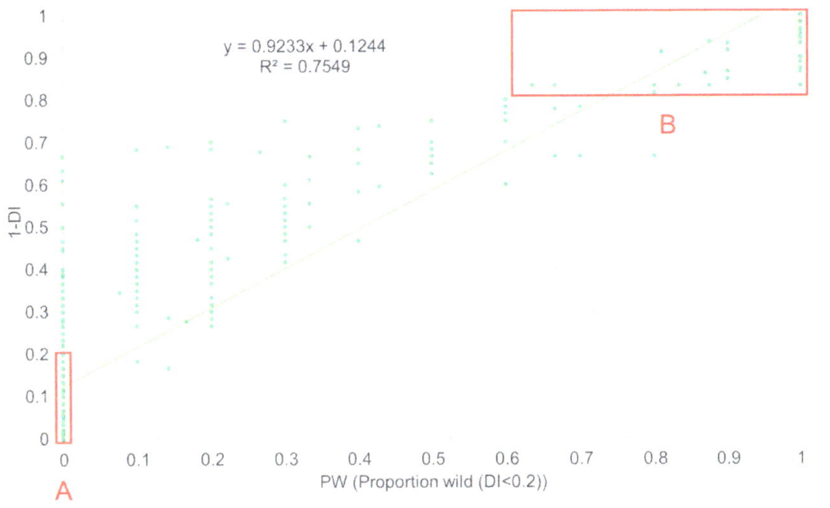

Figure 2. Relationships between the *PW* index (based on the proportion of wild seeds) and average $1 - DI$. Wild index ($1 - DI$) = 1 minus the average of the domestication index values of the seeds in each sample. (**A**) cultivars, purely domesticated; (**B**) purely wild autochthonous native plants. "Proportion wild", *PW*, = proportion of seeds below the wild threshold (DI < 0.2). Green dots, data from seed samples; Dotted line of best fit, that models the data's trend.

An inverse relationship is observed between the proportion of seeds in the sample exhibiting the wild phenotype and the mean values of the domestication index. More specifically, this relationship is directly correlated with the square of the mean wildness degree within the sample (Figure 2). However, the significant dispersion of points for a given *PW* value suggests the high heterogeneity of *DI* values across a substantial portion of the samples. This method shows the variability and characterizes the cultivars well but does not allow us to accurately characterize the intermediate types of mestizos or wild colonial or sub-spontaneous lambruscae, and only to partially characterize the native autochthonous wild types.

Morphological variability in seeds is depicted in Figure 3, where the mean values of width and stalk length parameters, in mm, per seed sample are compared for different grapevine types. The graph illustrates that the minimum values of stalk length are characteristic of wild grapevines, regardless of their modern or fossil nature or their continental origin, thus including Eurasian and American wild *Vitis* species. In the case of *Vitis vinifera*, variability is notably high, but stalks generally tend to be longer than in the rest of the *Vitis* species. In intermediate positions, we find samples interpreted as hybrids, which we will explore later as they may be attributed to introgression processes from wild vines into domesticated populations—a phenomenon that is not uncommon.

In addition to the above, Figure 3 allows us to distinguish the seeds of wild vines from the South Caucasus, called *V. caucasica*, due to their greater width and longer stalk length compared to *Vitis sylvestris* from central and western Europe. Differentiation of Asian wild vines, American wild vines, and fossil seeds is not possible from *Vitis sylvestris* in Figure 3, although some of the former have extremely short stalks. "Wild" *Vitis vinifera* includes cultivars that in general show "primitive" traits that would link them to *Vitis sylvestris*.

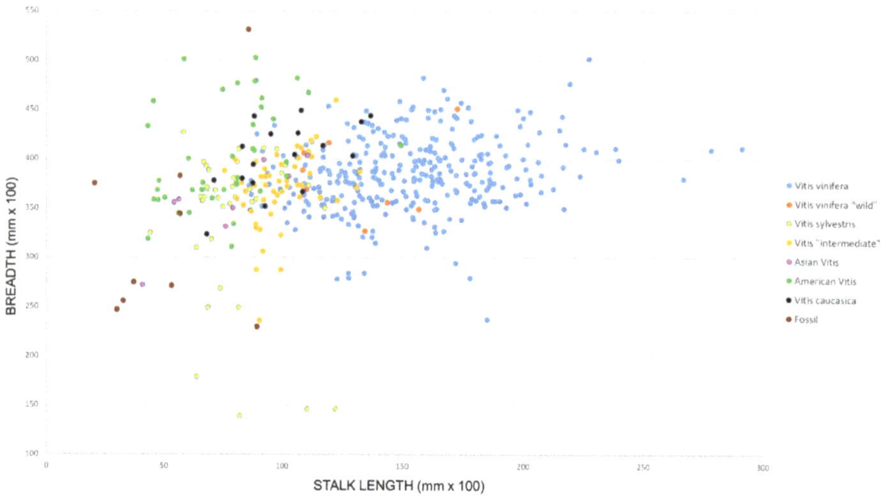

Figure 3. Seed breadth and stalk length as an example of phenotypic differences between wild, intermediate, and domesticated *Vitis*. Note: Asian *Vitis* here includes *Vitis romanetii* Rom. Caill., *V. piasezkii* Maxim. var. *pagnuccii* (Rom. Caill. ex Planch.) Rehder, *V. heyneana* Schult., *V. flexuosa* Thunb., *V. ficifolia* Bunge, and *V. amurensis* Rupr., from China, India, and Afghanistan; and American *Vitis* includes *Vitis vulpina* L., *V. shuttleworthii* House, *V. rupestris* Scheele, *V. rotundifolia* Michx., *V. riparia* Michx., *V. popenoei* J. H. Fennel, *V. peninsularis* M. E. Jones, *V. palmata* Vahl, *V. nesbittiana* Comeaux, *V. mustangensis* Buckley, *V. monticola* Buckley, *V. labrusca* L., *V. cinerea* var. *helleri* (L. H. Bailey) M. O. Moore, *V. cinerea* (Engelm.) Millardet, *V. californica* Benth., *V. bryoniifolia* Bunge, *V. bloodworthiana* Comeaux, *V. blancoi* Munson, *V. biformis* Rose, *V. arizonica* Engelm., *V. aestivalis* var. *linsecomii* (Buckley) L. H. Bailey, *V. aestivalis* Michx. var. *aestivalis*, *V. acerifolia* Raf., *V.* ×*doaniana* Munson ex Viala, and *V. girdiana* Munson, from the USA, Mexico, and natural habitats of Europe (where some are invader species).

3.2. The Place and Role of Intermediate Samples: Markers of Introgression

The detection of intermediate or mixed heterogeneous phenotypes (Figure 3) reveals introgression in varying levels among distinct groups within *Vitis*. Among those initially labeled as "wild" due to their natural habitat, approximately 20% of samples exhibit intermediate phenotypes that suggest introgression, although not only introgression, because wild colonial plants partially reverting to or converging with wild phenotypes could present similar values to those of truly wild mixed varieties and thus would be different to those of wild autochthonous native varieties. This proportion significantly rises to 40–55% for samples categorized as intermediate based on overall seed morphology. Cultivated samples exhibit the lowest introgression, ranging only from 7 to 10%, likely those domesticated phenotypes with wild traits ("wild" *Vitis vinifera* in Figure 3). Interestingly, this supports the hypothesis of Levadoux and other authors regarding the hybrid origin of certain cultivars (Figure 3).

We demonstrate that employing the mean domestication index for the sample μ (DI) and its variation level ρ (DI) can also aid in detecting potential introgression or hybridization cases (Figure 4).

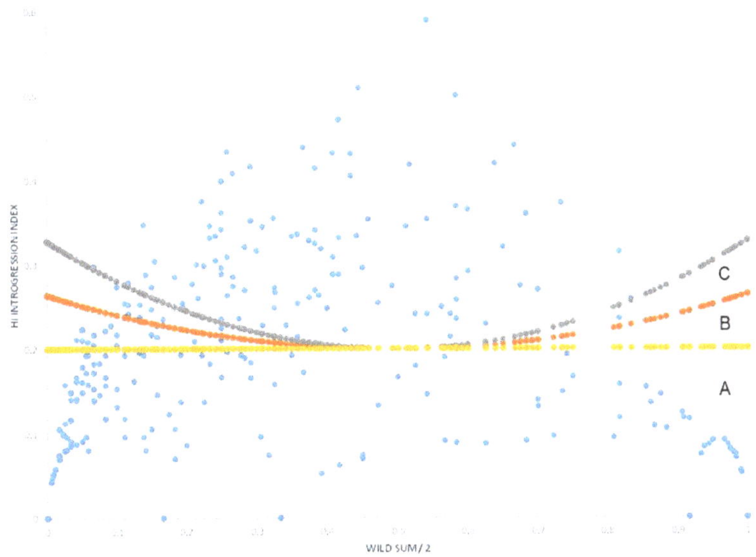

Figure 4. Thresholds for recognizing introgression along the WS/2, wild sum index. The highest values of the introgression index are found in samples that present an intermediate position with respect to the domestication syndrome, expressed in terms of WS/2 (Figure 4). Blue dots represent the 491 samples with 3 or more seeds. (**A**), yellow dots, constant IH = 0.2 threshold, (**B**), orange dots, catenary threshold (Formula (3)), and (**C**), grey dots, hyperbolic cosine threshold (Formula (4)).

Contrary to a constant threshold value (TH(x) = 0.2) (Figure 4A), the introgression index threshold appears to consist of values defined by hyperbolic curves. These curves could take the form of a catenary (Figure 4B) and Equation (4),

$$TH\ (x) = 0.5\cosh\left(x\ /0.5\right) - 0.3, 0 \leq x \leq 1 \tag{4}$$

Or, more likely, a hyperbolic cosine (Figure 4C) and Equation (5),

$$TH\ (x) = \cosh\ (x - 0.5) - 0.8, 0 \leq x \leq 1 \tag{5}$$

This depends on the sample's domestication level. Notably, higher threshold values are observed towards the lower and upper extremes of the wild sum index (Figure 4).

Most samples identified as hybrid grapevines typically feature small seeds, displaying mean dimensions more akin to those found in wild populations but with higher introgression index (*HI*) values. Interestingly, this occurrence of elevated *HI* values, suggesting introgression or hybridization, is not observed among samples with larger seeds, commonly associated with cultivars.

Furthermore, the introgression index (*HI*) exhibits elevated values in other *Vitis* species, such as *V. girdiana* Munson and *V. nesbittiana* Comeaux from Mexico, or *V. piasezkii* Maxim. var. *pagnuccii* (Rom. Caill. ex Planch.) Rehder and *V. romanetii* Rom. Caill. from China. This could be attributed to the inherent characteristics of each species, implying a notable degree of heterogeneity, or possibly to hybridization events in the history of analyzed accessions. Given that our introgression index is specifically designed to discern the boundaries between *V. sylvestris* and *V. vinifera*, its application to other species is not straightforward. Therefore, it should be considered as an indicator deserving further exploration and testing.

The behavior of the domestication index (*DI*) can facilitate the identification of seeds based on their putative origin. In the context of vineyards, when the mean of *DI* ($\mu(DI)$) is less than or equal to 0.5, the minimum *DI* (min(*DI*)) is less than or equal to 0.5, and the standard deviation of *DI* ($\rho(DI)$) is less than 0.5, these conditions are indicative of phenotypically wild cultivars (Table 1), which may be recognized as potentially primitive. When comparable parameter values of the domestication index (*DI*, $\mu(DI, min(DI))$ are observed in samples originating from natural habitats across Western Eurasia, these findings may be interpreted as indicative of the presence of the wild grape species *Vitis sylvestris* within those samples. The similarities in *DI* profiles likely reflect the shared phylogenetic affinities and evolutionary histories between the wild *V. sylvestris* populations and the putatively primitive cultivated grapevine accessions exhibiting analogous *DI* characteristics. Similar *DI* parameter values are prevalent in nearly all analyzed ancient plant fossils and approximately half of the Eurasian and American wild *Vitis* species (Table 1 and Figure 5).

Table 1. Joint distribution of seed domestication parameters and major grapevine types.

Parameters			Types							
μ (DI)	**min (DI)**	**ρ (DI)**	**Int.**	**Feral**	**Cultivars**	**Wild**	**Others**	**Fossils**	**Primitive**	**S**
			Vineyards							
≤0.5	≤0.5	0	-	-	-	-	-	-	2 **	2
		0–0.4	-	-	-	-	-	-	2 **	2
		0.6–1	2 *	-	-	-	-	-	-	2
>0.6	≤0.5	0–0.4	-	-	2	-	-	-	-	2
		0.4–0.6	-	-	12	-	-	-	-	12
		0.6–1	2 *	-	0	-	-	-	-	2
	>0.5	0	-	-	185	-	-	-	-	185
		0–0.4	-	-	79	-	-	-	-	79
			Natural habitats							
≤0.5		0	-	-	-	7	12	84		103
		0–0.4	1 °	-	-	8	17	9	-	35
		0.4–0.6	1 °	-	-	12	8	-	-	21
		0.6–1	7	-	-	2 a	3	-	-	12
>0.6	≤0.5	0–0.4	-	8	-	-	2	-	-	10
		0.4–0.6	1	12	-	-	1	-	-	14
		0.6–1	20	-	-	-	5	-	-	25
	>0.5	-	-	7	-	-	2	17	-	26
		0–0.4	-	28	-	-	0	0	-	28
			34	55	278	29	50	110	4	

Abbreviations: μ (*DI*), Mean domestication index value; *min* (*DI*), minimum domestication index value; ρ (*DI*), domestication index range; Int., intermediate, introgressed, or hybrid; S, marginal frequencies. Notes: (*) Forcallat, Albariño, Chardonnay Blanc, Chitistvala_Meskhuri, Brancellao, Garnacha Blanca, Merlot, (**) Pardillo, Merseguera, Planta Nova, Rozsaszolo; (°) Guadiana River, Aragon River Dam, Samebis seri (Georgia), ([a]) Kvetari_04, presents one single seed with *DI* = 0.83 and the rest are clearly wild; similarly, in Huelva, one single seed presents *DI* = 0.67. The colors used in the image are as follows: Pink (used for "Int."), Light gray (used for "Feral with heterogeneous domestication traits"), Dark gray (used for "Feral with homogeneous domestication traits"), Bright blue (used for "Cultivars"), Light blue (used for Fossils), Yellow (used for "Primitive"), Bright green (used for "Wild"), Beige/Orange (used for "Others").

Conversely, in vineyards, the majority of cultivated grapevine accessions exhibit $\mu(DI)$ greater than 0.6, with min(*DI*) exceeding 0.5 and $\rho(DI)$ ranging from 0 to 0.4. An analogous pattern is observed in samples from natural habitats, which we interpret as indicative of feral grapevine populations (Table 1 and Figure 5). Interestingly, some American wild *Vitis* species and ancient plant fossils also exhibit this cultivar-like *DI* parameter profile (Table 1). This suggests that the domestication syndrome detectable in seeds, while unambiguous, may in some cases replicate preexisting patterns present within wild *Vitis* populations.

Putative hybrid specimens, both from natural habitats and vineyards, are characterized by a high $\rho(DI)$ value, typically ranging from 0.6 to 1.0 (Table 1). The elevated variability of *DI* values within a single sample may imply a process of introgression with domesticated *V. vinifera*, whether of ancient or more recent origin.

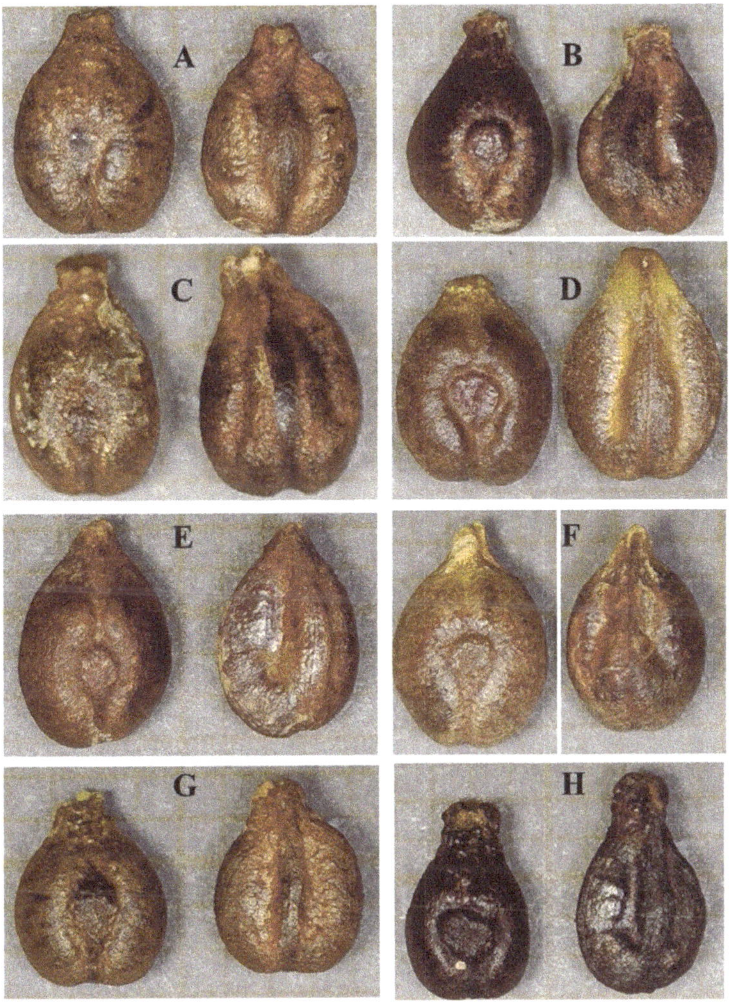

Figure 5. Different seed types recorded from natural habitats in Italy and Spain. Digital microscopy images: background, 1 mm orange graph paper. Samples, their types, and hybridization and domestication index values: (**A**) WILD, 246 Antchatea-Plentzia, *HI* = 0.2, *DI* = 0.27; (**B**) HYBRID, 245 Armintza-Arroyo Amorrada, *HI* = 0.22, *DI* = 0.54; (**C**) INTROGREDED? 244 Balmaseda-Cadegua, *HI* = 0.15, *DI* = 0.65; (**D**) INTROGREDED? 582 Estena River Gorge, *HI* = 0.27, *DI* = 0.7; (**E**) WILD, 249 Urkuleta 1, *HI* = 0.12, *DI* = 0.08; (**F**) FERAL ROOTSTOCK, 260 *V. berlandieri* × *V. riparia* 5BB, *HI* = 0.19, *DI* = 0.13; (**G**) INTROGREDED? 568 Rio Irati (Navarra), *HI* = 0.24, *DI* = 0.43; (**H**) INTROGREDED? 570 Piemonte (Italy), HI = 0.14, DI = 0.33.

Intermediate characteristics are not limited to wild populations colonizing natural habitats; we have identified primitive traits in cultivars such as Albariño, Brancellao, Cainho, Tempranillo (p.p.), Chardonnay Blanc, Forcallat, Garnacha Blanca, Merlot, and Tempranillo Blanco. Samples of varieties like Chardonnay Blanc, Forcallat, and Garnacha Blanca fall within the range of wild-type characteristics. Additionally, varieties such as Rozsaszolo, Planta Nova, Merseguera, or Pardillo display distinctly wild-type seeds and low introgression index (HI) values (0.09–0.15).

3.3. Domestication Probability Variability Within Samples and Entropy Estimates

Obón et al. [35] recently showed that estimating the probability of domestication for each seed using methods such as logistic regression and machine learning, specifically "randomForest", with the use of training sets, allows for a more comprehensive understanding of the overall domestication degree of the samples. When combined with the analysis of heterogeneity, entropy, or information within each sample, we can achieve a more precise insight. In the case of the dataset, the results are summarized in Table 2, where a comparison is drawn between samples obtained from vineyards and those collected from natural habitats (Figure 5).

Table 2. Comparative analysis of heterogeneity and mean domestication probability in samples from vineyards and natural habitats [1].

					Vineyards						
*Het.**PD*	**0–0.1**	**0.1–0.2**	**0.2–0.3**	**0.3–0.4**	**0.4–0.5**	**0.5–0.6**	**0.6–0.7**	**0.7–0.8**	**0.8–0.9**	**0.9–1**	***MH***
0 ≤ Het. < 1	0.0000	0.0000	0.0000	0.0000	0.0000	0.0000	0.0000	0.0000	0.0000	**0.6747**	**0.6747**
1 ≤ Het. < 2	0.0000	0.0000	0.0000	0.0000	0.0000	0.0000	0.0000	0.0000	0.0000	0.0683	0.0683
2 ≤ Het. < 3	0.0000	0.0000	0.0040	0.0000	0.0000	0.0000	0.0000	0.0000	0.0321	0.0643	0.1004
3 ≤ Het. < 4	0.0000	0.0080	0.0000	0.0000	0.0000	0.0000	0.0000	0.0000	0.0281	0.0120	0.0482
4 ≤ Het. < 5	0.0000	0.0000	0.0000	0.0120	0.0000	0.0000	0.0040	0.0000	0.0241	0.0000	0.0402
5 ≤ Het. < 6	0.0000	0.0000	0.0000	0.0040	0.0040	0.0040	0.0000	0.0120	0.0201	0.0000	0.0442
6 ≤ Het. < 7	0.0000	0.0000	0.0000	0.0000	0.0000	0.0080	0.0000	0.0000	0.0000	0.0000	0.0080
7 ≤ Het. < 8	0.0000	0.0000	0.0000	0.0000	0.0120	0.0040	0.0000	0.0000	0.0000	0.0000	0.0161
MP	0.0000	0.0080	0.0040	0.0161	0.0161	0.0161	0.0040	0.0120	0.1044	**0.8193**	1.0000
					Natural Habitats						
Het.\\PD	**0–0.1**	**0.1–0.2**	**0.2–0.3**	**0.3–0.4**	**0.4–0.5**	**0.5–0.6**	**0.6–0.7**	**0.7–0.8**	**0.8–0.9**	**0.9–1**	**MH**
0 ≤ Het. < 1	0.1033	0.0083	0.0041	0.0000	0.0041	0.0000	0.0000	0.0000	0.0000	0.0331	0.1529
1 ≤ Het. < 2	0.0455	0.0207	0.0000	0.0000	0.0000	0.0041	0.0041	0.0000	0.0000	0.0413	0.1157
2 ≤ Het. < 3	0.0165	0.0248	0.0041	0.0083	0.0000	0.0000	0.0000	0.0041	0.0165	0.0331	0.1074
3 ≤ Het. < 4	0.0041	0.0207	0.0207	0.0083	0.0000	0.0124	0.0041	0.0083	0.0289	0.0124	0.1198
4 ≤ Het. < 5	0.0000	0.0165	0.0372	0.0083	0.0124	0.0041	0.0124	0.0207	0.0496	0.0083	0.1694
5 ≤ Het. < 6	0.0000	0.0000	0.0165	0.0331	0.0083	0.0124	0.0124	0.0331	0.0083	0.0000	0.1240
6 ≤ Het. < 7	0.0000	0.0000	0.0083	0.0331	0.0537	0.0207	0.0000	0.0165	0.0000	0.0000	0.1322
7 ≤ Het. < 8	0.0000	0.0000	0.0000	0.0041	0.0165	0.0413	0.0165	0.0000	0.0000	0.0000	0.0785
MP	0.1694	0.0909	0.0909	0.0950	0.0950	0.0950	0.0496	0.0826	0.1033	0.1281	1.0000

[1] Data are presented as proportions of the total, summing up to 1. Abbreviations: *Het.*, heterogeneity estimate using Shannon index. *PD*: mean domestication probability for the entire sample considering the *DI* (domestication index), *PDILO* (the probability of domestication estimated with Logit), and *PDIrF* (the probability of domestication estimated with randomForest) values. *MH*: marginal sums for each heterogeneity range. *MP*: marginal sums for each probability range. The colors used in the image are as follows: Yellow (used for "probability values of samples from cultivars above 0"), Bright green (used for "probability values of samples from natural habitats above 0").

As expected, vineyards contributed highly homogeneous samples, characterized by an extreme degree of domestication (Table 2). However, occasional instances from vineyards revealed highly heterogeneous varieties with markedly different seeds, which collectively yielded average domestication values. Some varieties even exhibited predominantly wild traits.

Natural spaces harbor significantly higher heterogeneity, encompassing a spectrum of domestication degrees as deduced from the seed phenotype analysis, with intermediate values predominantly associated with elevated heterogeneity (Table 2). Samples exhibiting a maximum domestication probability and low heterogeneity are interpreted as naturalized or semi-spontaneous, while those with minimal domestication probability and low heterogeneity are identified as originating from native wild vines or naturalized American vines, which are indistinguishable using domestication estimates based on seed morphology. Finally, most samples display intermediate domestication values and maximum heterogeneity, indicating the existence of persistent introgression phenomena from cultivated vine populations to wild vines (Figure 6).

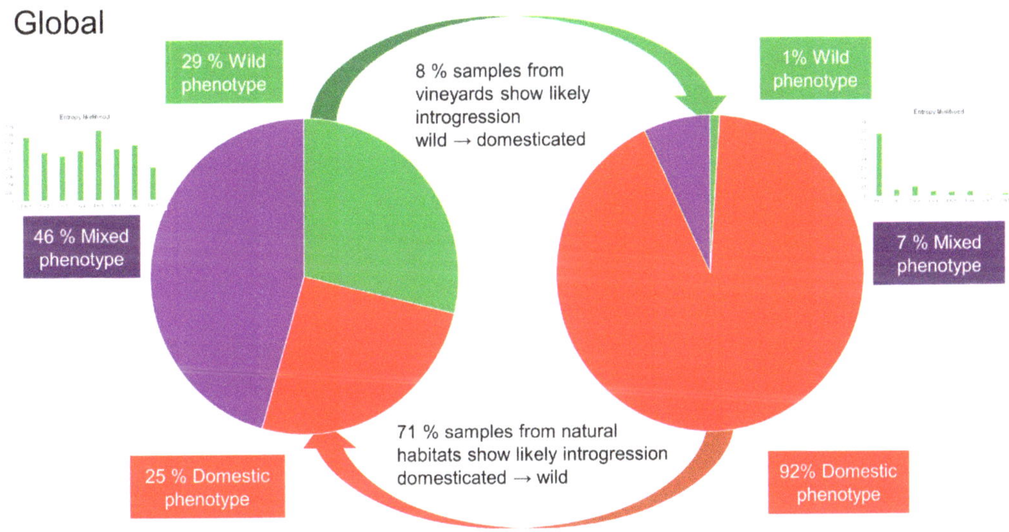

Figure 6. Differences in heterogeneity and the proportion of domesticated, wild, and intermediate types in vineyards and natural habitats, along with an estimation of introgression flows. Color codes: Red, domestic phenotype. Green, wild phenotype. Purple, intermediate or mixed phenotypes.

4. Discussion

4.1. Interpreting the Results in the Context of Domestication and Hybridization

In general, we observe a significant influence of domesticated grapevines on wild populations, which is to be expected given the antiquity of cultivation and domestication processes of Eurasian grapevines, often dating back several thousand years. However, it is crucial to consider the size of both wild and domesticated populations, as this governs the direction of genetic flow. During the initial phases of cultivation, regardless of the local or exotic nature of primitive cultivars, the potential for introgression from wild populations into cultivated ones could have been substantial. Once cultivation is established, the disproportion heavily favors domesticated varieties.

A striking example is found in Castilla–La Mancha, Spain, where more than 150,000 hectares are dedicated to vine cultivation in La Mancha Protected Designation of Origin (PDO) [55], 25,000 in Valdepeñas PDO, 12,500 in Manchuela DO [56], 9000 in

Almansa PDO [57], and 23,000 in Jumilla PDO [58], with an average planting density of about 3000 to 4000 vines per hectare, thus reaching 650,000,000 to 900,000,000 vines. On the other hand, wild vines in the Cuenca Alta del Guadiana, one of the rivers that crosses this re-gion, barely exceed a hundred and, in most stretches, are practically non-existent as also occurs in other rivers of the region such as Tajo, Jucar, Cabriel, or Segura. The dispropor-tion is overwhelming.

In Italy, we have detected a prevalence of domesticated and intermediate vine phenotypes in natural habitats, where it seems that wild vines are reduced to a vestigial pres-ence (Figure 7). In Tuscan populations, D'Onofrio [59] identified redundant genotypic profiles among wild grapevine specimens, indicating clonal propagation in natural conditions. Analysis of specimens from Piedmont, Italy, demonstrated that certain popula-tions initially classified as *V. sylvestris* were identified as naturalized *V. vinifera* (Figure 7). Most identified sylvestris genotypes possessed chlorotype A, while others had chlorotype D, akin to the predominant chlorotype in *V. vinifera* cultivated in Italy. Genotypic analysis of Tuscan populations identified *V. sylvestris* specimens exhibiting hybridization with cul-tivated varieties, demonstrating asymmetric introgression favoring gene flow from *V. vi-nifera* to wild populations [59]. Additionally, certain genotypes seemed to be crosses within *V. sylvestris*, indicating a significant level of sexual exchange between *V. sylvestris* and *V. sylvestris* × *V. vinifera* genotypes.

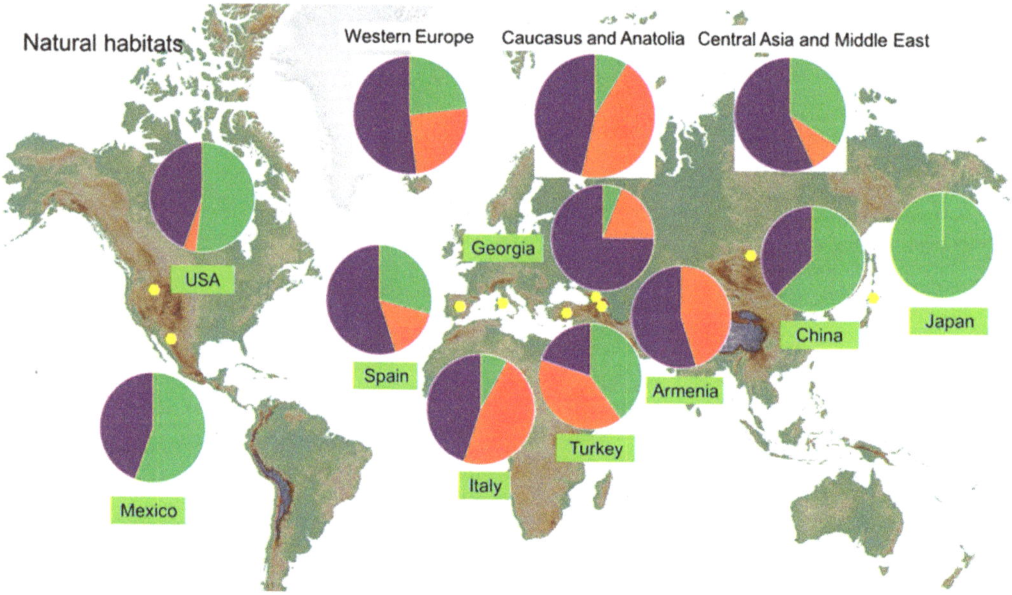

Figure 7. Proportion of domesticated, wild, and intermediate grapevine seed types within natural habitats, across selected countries and regions. Color codes: Red, domestic phenotype. Green, wild phenotype. Purple, intermediate or mixed phenotypes. The yellow dots mark the centroid corresponding to the nearest pie charts, which appear displaced for reasons of space.

The dispersal of pollen from cultivated vineyards in proximity may exert a substantial influence on the evolution of adjacent wild grapevine populations [60]. On the one hand, the low level of pollen-mediated gene flow from cultivated to wild vines could contribute to a risk of extinction for the wild component (i.e., the totality of wild individuals). Conversely, pollen dispersal within diminished wild populations could induce inbreeding depression among wild vines.

We must highlight the trade-off between domestication traits such as higher yield and larger berries, which often come at the expense of reduced resistance to environmental stresses [61].

4.2. Discussing the Importance of Finding Similar Samples in Both Vineyards and Natural Habitats

In general, it is apparent that the likelihood of encountering fruits of wild grapevine species, such as *Vitis sylvestris* and other taxa with similar seed characteristics, is significantly higher in natural habitats compared to vineyards (Figures 6–8), as indicated by an odds ratio of nearly 26 (26% in natural habitats versus 1% in vineyards). Conversely, the probability of encountering domesticated grapevines, specifically *Vitis vinifera*, is not so notably higher, with an odds ratio of 4 (92% in vineyards versus 23% in natural habitats) within vineyard settings (Figures 6 and 8), a relationship that is not intuitively expected, given that 92% of vineyard samples present a phenotype that is clearly domesticated. However, the probability of encountering domesticated grapevines in natural habitats, albeit lower, is still notable, at approximately 0.23.

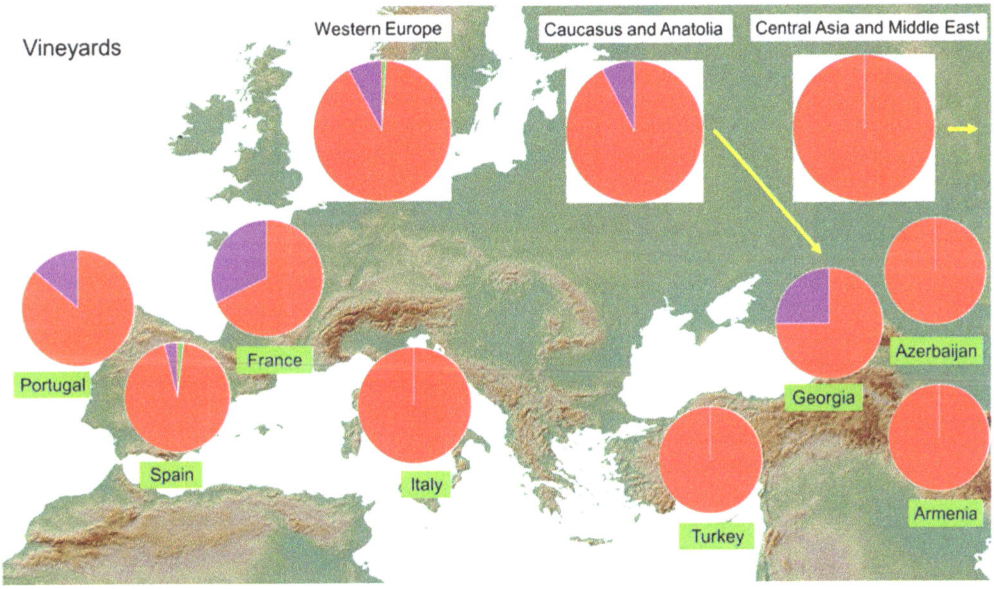

Figure 8. Proportion of domesticated, wild, and intermediate grapevine seed types within vineyards, across selected countries and regions. Color codes: Red, domestic phenotype. Green, wild phenotype. Purple, intermediate or mixed phenotypes. The yellow arrows mark the centroid corresponding to the nearest pie charts, which appear displaced for reasons of space.

Consequently, feral grapevines currently constitute a significant component of grapevine populations inhabiting natural habitats, while phenotypes resembling wild grapevines in their seed morphology are extremely rare in vineyards. Intermediate phenotypes appear in both vineyards and natural habitats, being predominant in the latter (Figures 6 and 7).

The presence of wild traits in vineyards is primarily associated with a select few "primitive" cultivars characterized by low domestication values, particularly evident in their seed morphology (Table 3). Conversely, domesticated traits within wild grapevine populations manifest in a diverse array of origins and types, aligning with Levadoux's [5] classification system. This spectrum spans from postcultural vines, which thrive in abandoned vineyards reclaimed by natural vegetation from nearby habitats, to wild sub-spontaneous lambruscae newly established in natural environments. Additionally, there are wild colonial vines, direct descendants of cultivars, which may have gradually lost some domestication traits over successive generations. Hybrid swarms of *Vitis vinifera*, exhibiting varying levels of introgression, further contribute to this complexity, mixing with preexisting wild autochthonous native *Vitis sylvestris*.

Table 3. Summary of descriptive parameters and indexes for various grapevine types: mean values and standard deviation [1].

Types According to Levadoux	Simplified	1 − DI	PW	WS/2	ME	SH	L (mm)	B (mm)	Stalk (mm)	Stummer Index
Wild autochthonous native	*Vitis sylvestris* modern	0.8 ± 0.2	0.7 ± 0.3	0.7 ± 0.2	0.2 ± 0.2	2.1 ± 1.6	4.8 ± 0.4	3.6 ± 0.4	0.8 ± 0.2	76.1 ± 4.7
Asian *Vitis* spp.	Asian *Vitis*	0.7 ± 0.3	0.6 ± 0.4	0.6 ± 0.4	0.4 ± 0.3	3.5 ± 3.1	4.7 ± 1.0	3.6 ± 4.2	0.8 ± 0.3	77.5 ± 8.5
American *Vitis* spp.	American *Vitis*	0.7 ± 0.2	0.5 ± 0.4	0.6 ± 0.3	0.3 ± 0.2	3.1 ± 2.3	5.3 ± 0.8	4.0 ± 0.5	0.8 ± 0.2	77.1 ± 8.0
"Primitive" cultivars	*Vitis vinifera* "wild"	0.6 ± 0.2	0.3 ± 0.3	0.5 ± 0.2	0.3 ± 0.3	2.7 ± 2.4	5.1 ± 0.3	3.7 ± 0.3	1.1 ± 0.2	73.9 ± 3.8
Vitis caucasica	*Vitis caucasica*	0.5 ± 0.2	0.3 ± 0.2	0.4 ± 0.2	0.6 ± 0.2	5.4 ± 1.9	5.2 ± 0.6	4.0 ± 0.4	1.0 ± 0.2	77.1 ± 4.3
Wild mestizo (hybrids, different levels of introgression)	*Vitis* intermediate modern	0.4 ± 0.1	0.2 ± 0.2	0.3 ± 0.1	0.5 ± 0.3	4.5 ± 2.6	5.2 ± 0.4	3.8.01±	1.0 ± 0.2	72.2 ± 4.0
Wild colonial	Feral	0.3 ± 0.1	0.0 ± 00	0.1 ± 00	0.3 ± 0.3	2.9 ± 2.5	5.4 ± 3.7	3.7 ± 0.2	1 ± 0.1	69.5 ± 4.9
Cultivars	*Vitis vinifera* modern	0.1 ± 0.1	0.0 ± 00	0.0 ± 0.1	0.1 ± 0.2	1.0 ± 1.6	6.2 ± 0.7	3.9 ± 0.4	1.6 ± 0.3	63.2 ± 5.5
Postcultural vines in abandoned vineyards	Postcultural *Vitis vinifera*	0.1 ± 0.1	0.0 ± 00	0.0 ± 00	0.1 ± 0.1	0.8 ± 1.3	5.6 ± 0.3	3.8 ± 0.3	1.3 ± 0.1	68.0 ± 4.4
Wild sub-spontaneous lambruscae	Sub-spontaneous *Vitis vinifera*	0.1 ± 0.1	0.0 ± 00	0.1 ± 00	0.2 ± 0.3	2.2 ± 2.4	5.8 ± 4.4	3.9 ± 0.3	1.3 ± 0.2	67.3 ± 4.1

[1] Note: Descriptive parameters and indexes, including heterogeneity, are presented as mean values ± standard deviation for different types of grapevines, including wild, domesticated, and sub-spontaneous varieties. Abbreviations: *1 − DI*, wild index. PW, proportion wild, considering *DI* < 0.2. *WS/2*, wild sum/2. ME, mean equitability. SH, Shannon index estimates of heterogeneity within samples. L (mm), seed length in mm. B (mm), seed breadth in mm. Stalk (mm), seed-stalk length in mm.

The diversity within natural habitats is further compounded by distinctions between wild autochthonous native grapevines in Western Europe and those in the Caucasus (Table 2). Furthermore, the increasing presence of escaped American *Vitis* grapevines adds to this complexity, as they demonstrate seed morphology values akin to those of *Vitis sylvestris* (Table 3).

This phenomenon is interpreted as indicative of a substantial seed dispersal flux from cultivated fields to riparian forests and other natural habitats, where cultivars subsequently establish themselves as sub-spontaneous or feral populations. Repeatedly sowing a cultivar does not necessarily lead to convergence with *V. sylvestris*. Some recessive but homozygous traits of *V. vinifera* remain conserved. In cases of heterozygous cultivars, or when one of the ancestors possesses dominant traits like those of *V. sylvestris*, and in instances of multifactorial inheritance, descendants of one or more cultivars may, after a few generations, exhibit an increasing number of forms closely resembling the wild ones (Figure 1) [8].

Considering the observations, the prevalence of intermediate seed types, which are the primary group within natural habitats, reaching 41% (Figure 6), may be attributed to several factors. Firstly, it could reflect a process of partial regression through successive generations of feral grapevines. Additionally, it may result from continued introgression with preexisting native autochthonous wild grapevines. It is noteworthy to mention that samples exhibiting intermediate domestication values typically demonstrate significantly higher levels of heterogeneity or entropy compared to those representing pure wild or common domesticated categories (Table 2).

Throughout ampelographic history, various wild grapevine forms have been documented, often representing hybrid swarms with domesticated populations in different re-gions. In Andalusia, Simón de Roxas Clemente [62–64] identified six wild grapevine types, including Mollar negro bravío and Garabatona. Bronner [65,66], in the mid-19th century, distinguished 36 wild grapevine varieties in Alsace, some with oval seeds and a sweet taste, indicating a mix of true wild, feral, and possibly introgressed individuals. In Italy, authors like Mendola, Negri, and Franchino recognized wild grapevines under names such as Uzelina and Zampina. Uzelina is rustic with small black berries, while Zampina exhibits dioecious traits and foliar sexual dimorphism. Italian populations show considerable variability, with hybrid swarms and mixed populations due to the intentional grafting of cultivars on wild grapevines in natural habitats [67]. Scienza et al. [68,69], Anzani et al. [70,71], and Failla et al. [72] identified considerable variability in wild Italian grapevine populations. Scienza et al. [69] evidenced the presence of feral varieties. Eight plants from natural habitats with ampelographic traits of cultivated grapevine were discovered in Georgia by Kikvadze et al. 2024 [73].

Molecular analyses unveiled the potential introgression of *V. sylvestris* into certain cultivars in Italy and France [61], which could explain our detection of cultivars with prim-itive seed traits associated with low domestication markers (Table 3).

Biblical texts provide early documented evidence of vineyard heterogeneity and deterioration, notably in prophetic passages addressing the Israelites. This phenomenon may be attributed to propagation through seedlings rather than clonal reproduction. Two significant references illustrate this viticultural challenge: In Isaiah 5, the text employs the metaphor of a carefully tended vineyard that produces inferior fruit, stating, "He waited for the vineyard to yield good grapes, but the fruit it produced was sour." Similarly, Jeremiah 2 extends this metaphor, lamenting, "Yet I planted you as an elect vineyard, with only true seed. Then how have you been turned away from me, toward that which is depraved, O strange vineyard?" These passages not only serve as religious allegories but also potentially document early observations of genetic drift in grapevine cultivation. Such deterioration may have manifested particularly in female raisin and table grape cultivars exposed to cross-pollination from nearby wild grapevine populations, resulting in offspring with undesirable characteristics.

While *V. vinifera* cultivars cultivated in Germany, propagated solely vegetatively, ex-hibit hermaphroditic traits, both female and male forms emerge in their crosses with di-oecious wild grapevines (*V. sylvestris*). Consequently, feral hybridogenic populations may manifest dioecy, as highlighted by Wagner in 1960 [74]. Avramov et al. [75] initially pro-posed a two-allele model for controlling flower type in *V. vinifera* L. grape varieties, while Antcliff's model introduced a single genetic determinant with three alleles for sex deter-mination, and subsequent research identified two markers linked to *Vitis* species flower sex loci on chromosome 2 [76–80].

Regardless of the explanatory model adopted to interpret dioecy control in *Vitis*, it is evident that dioecy, while a prerequisite for considering the possibility of a specific vine belonging to the species *V. sylvestris*, is not a sufficient characteristic. In natural habitats across Eurasia, dioecious vines can arise from the introgression of wild vines with cultivated varieties from nearby vineyards. Additionally, they may originate from American or Asian vines that have escaped cultivation or from direct dioecious hybrid producers.

Consequently, the occurrence of dioecious individuals across natural habitats cannot be exclusively attributed to *V. sylvestris* populations. These observations demonstrate the intricate interactions between natural evolutionary processes and anthropogenic selection pressures in determining genetic structure and population dynamics across diverse geographical distributions. Further research is warranted to elucidate the mechanisms underlying dioecy expression and its ecological implications within wild *Vitis* populations.

4.3. *Cultivar Origins Unveiled: Exploring the Influence of Hybridization on the Evolution of Modern Grapevine Varieties*

The analysis by Riaz et al. [77] of genetic relationships among selected grapevine genotypes provided evidence of genetic relationships between wild and cultivated samples from the Mediterranean basin to Central Asia. The genetic structure indicated a con-siderable amount of gene flow, which limited the differentiation between *Vitis vinifera* and *V. sylvestris*. Georgian wild grapevine accessions showed genetic clustering not only with local cultivars (classified under proles Pontica according to Negrul's system) but also with Western European varieties (proles Occidentalis), lending further support to Georgia's status as an ancient center of grapevine domestication. This may provide an explanation for the high degree of admixture detected in our analyses of samples from natural habitats in Georgia and, to a lesser extent, Armenia, where the presence of feral vines in those habitats is extremely high (Figure 7). Cluster analysis revealed that Western European wild grapevines grouped with cultivated varieties from the same region, suggesting that the local proles Occidentalis played a more significant role in Western European viticulture development than the introduction of wild or proto-domesticated vines from Eastern Europe [77]. The results also indicated that grapes with mixed ancestry occur in regions where wild vines were domesticated. This may be at the origin of the high degree of mix-ing and heterogeneity detected in the samples of the cultivars analyzed from Georgia, but also from France and Portugal, and to a lesser extent Spain (Figure 8).

Cunha et al. [81] in their study on local grapevine varieties grown in Portugal showed that part of the diversity recorded was mostly local in some cases, as demonstrated by the proximity of several varieties (Vinhos Verdes) to the wild cluster in different analyses. All these findings, together with the known pairing between the wild maternal lineage of the Iberian Peninsula and an important number of Portuguese grapevine varieties (with chlorotype A), point to the fact that some of these varieties derive, directly or indirectly, from originally local wild populations, supporting the possible occurrence of an introgression process from wild to cultivated grapevine. The predominantly local genetic diversity pat-terns and the east-to-west decline in diversity could alternatively suggest that local pop-ulations likely developed distinct sub-gene pools through time, primarily due to gene flow from domesticated to wild populations, rather than supporting wild-to-cultivated introgression.

4.4. *Unraveling the Geographical Gradient of Phenotypic Characteristics in Western Europe*

The spatial gradient from western to eastern regions of Spain exemplifies, on a reduced scale, the broader phenomenon of a higher prevalence of mixed populations or predominance of feral cultivars in the eastern portions of the species' range. Initial field sampling efforts aimed to collect morphologically identifiable fruits of the wild grape *Vitis sylvestris*, while attempting to exclude those from feral, rootstock, and recognizable hybrid individuals. However, subsequent analysis revealed notable differences when the data were organized by major river basins, providing overwhelming evidence of intermediate, mixed, or domesticated trait expressions (Figure 9).

The Guadiana River basin stands out, with a significant proportion of samples exhibiting wild characteristics (Cabañeros, Badajoz, and Chanza in Figure 9) and a relatively low degree of introgression. Conversely, the Ebro River basin shows a troubling situation, where nearly all samples are feral, domesticated, or intermediate, suggesting wide introgression processes from vineyards to natural habitats, seemingly replacing wild populations, except for one sample of Erro River with typical wild traits. However, other samples from the same river Erro, which is a tributary of the Irati, showed intermediate or clearly domesticated values. In all the samples from the banks of the Erro, a relatively high heterogeneity was observed, ranging between 3 and 5, when in pure populations of wild vines or in the fields of cultivated vines the heterogeneity varies between 0 and 2.

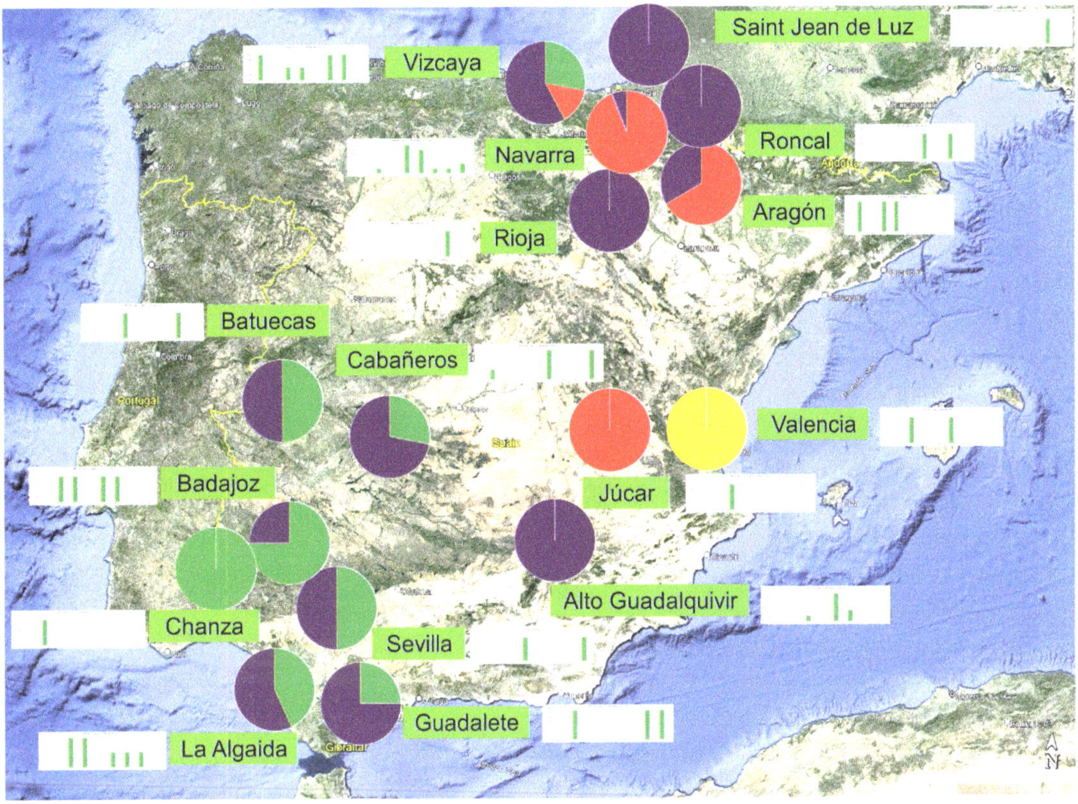

Figure 9. Analysis of the proportions of domesticated, wild, and intermediate seed types in selected Spanish populations from natural habitats and the heterogeneity within samples. Color codes: Red, domestic phenotype *V. vinifera*. Green, wild phenotype *V. sylvestris*. Purple, intermediate or mixed phenotypes. Yellow, American rootstocks. Note: the number of sampled localities in each river basin, ranges 1 to 17. Bar charts represent the proportion of heterogeneity within zones, where x is the heterogeneity class, ranging from 0 to 8 and y is the likelihood of each class. Image map: Diego Rivera. Background: Google Earth.

The small basins draining into the Cantabrian Sea primarily display pure wild or mixed characteristics, with only one population classified as feral with domestication traits (at Monchué). The mixed condition of populations in Urdaibai (Vizcaya, Spain) was analyzed by Albizuri et al. [82], where they recorded native wild Eurasian grapevines co-existing with feral *V. vinifera* varieties, feral rootstocks resembling *Vitis rupestris* or *V. ri-paria*, and direct producer hybrid grapevines. Our sample from Urdaibai has wild charac-teristics with heterogeneity values less than 1; thus, it is attributed to *V. sylvestris*.

In the few sampled localities along the western Tagus River, only intermediate or wild vines were found (Figure 8). In the Jucar and Turia basins, feral *V. vinifera* cultivars and American rootstock grapevines were identified, as previously reported by Laguna on an ampelographic basis [83–85].

In the study by De-Andrés et al. [86] on a molecular basis, in the structure model, depending on the number of clusters identified by adjusting the parameter K of Evano, individuals from wild populations of Spanish grapevines increasingly differentiate from cultivated ones as the number of clusters grows. However, traces of introgression persist in both cases, while a notable difference between wild populations from the northern and southern regions of the Peninsula is distinguished. A portion of these populations has also been studied in the present work, where the north–south gradient is detected but is partly masked by a notable east–west difference (Figure 9).

Regarding heterogeneity within each sample (Figure 9), we observe the lowest values in Chanza, associated with wild vines, and the highest values in La Rioja and along the banks of the Bidasoa River in Saint Jean de Luz, associated in this case with mixed or intermediate populations, with a notable presence of "hybrids" or "mestizos" in the Levadoux sense [8]. In the rest of the areas, we find significant variability among samples, ranging from low to very high in Cabañeros, Biscay, and Guadalete, intermediate in Aragón, Valencia, Jucar, La Algaida, Upper Guadalquivir, Badajoz, or Batuecas, and from moderate to very high in Navarra and El Roncal.

The analyzed samples collected in natural habitats from France and Switzerland are limited in number, hindering a comprehensive understanding of the issue. In the case of Saint Jean de Luz (France), there appears to be an intermediate grapevine population with a significant feral influence. Ocete et al. [87] described a wide range of Stummer's index values for the Saint Jean de Luz population, varying between 0.5 and 0.8, with a mode below 0.75, supporting the notion of a wide introgression with a substantial feral influ-ence. Our results coincide, showing an intermediate mean domestication probability of 0.56 associated with a high heterogeneity above 6 which is characteristic of mixed indi-viduals and populations, which may imply the presence in the population of sub-sponta-neous wild vines, colonial or feral, and of crossbreeds (mestizos in the sense of Levadoux) with varying degrees of introgression between the native wild autochthonous and the re-maining. Conversely, the population in the Rhône Valley, Mont d'Autan, Valais, in Swit-zerland, with a mean domestication probability of only 0.1 and heterogeneity index below 2, seems unequivocally native wild (*V. sylvestris*).

In Sardinia, the situation is even more complex (Figure 10), since only one locality, Montresta, shows clearly wild characteristics, with a domestication probability of only 0.07, although with heterogeneity values above 2. In Santadi, in the Pula area, one sample was identified as wild, with a very low probability of domestication, 0.12, and a heterogeneity below 2. But other samples from the area were clearly intermediate (with a domestication probability of 0.67, but extremely high heterogeneity above 7) or domesticated (with a domestication probability of 0.97, and a heterogeneity index value of only 1.3). A third sample, among the 31 analyzed from Sardinia, could eventually be identified as wild, collected at Flumini Maggiore–Malacalzetta, but with a very high heterogeneity within the sample, close to 6, and a probability of domestication higher than 0.2; we should consider it a result of successive introgressions with domesticated vines. Regarding the heterogeneity within each sample, it is highest in the localities of Padria and Talana, and lowest in Tiana. Meanwhile, intermediate values are observed in Montresta, Santu Lussurgiu, Domusnovas, Nuoro, and partly in Urzulei. There are populations with samples showing varied levels of heterogeneity, typically ranging from intermediate to high, such as Malacalzetta, Uta, Pula, Villanova Tulo, Mamoiada, and Orgosolo (Figure 10).

Using molecular markers, phenomena of introgression were detected by Zecca et al. [88] in the wild grapevine populations of Sardinia, originating from nearby vineyards. Contrary to our findings, the authors emphasize the predominantly wild nature of the sam-pled individuals. We propose that the molecular markers currently utilized to differ-entiate between wild and domesticated grapevines should be validated through compre-hensive phenotypic analyses of voucher specimens.

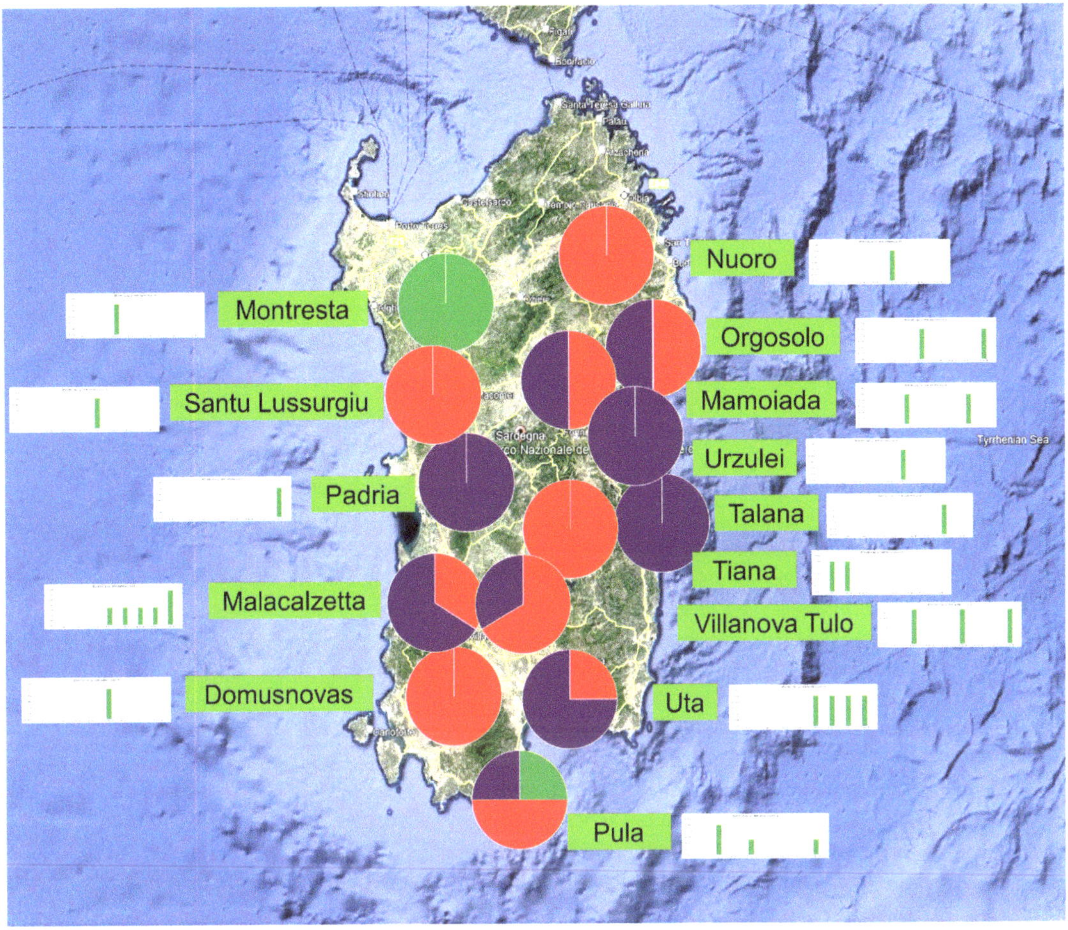

Figure 10. Analysis of the proportions of domesticated, wild, and intermediate seed types in selected Sardinian populations (Italy) from natural habitats and their heterogeneity within samples. Color codes: Red, domestic phenotype *V. vinifera*. Green, wild phenotype *V. sylvestris*. Purple, intermediate or mixed phenotypes. Bar charts represent the proportion of heterogeneity within zones, where x is the heterogeneity class, ranging from 0 to 8 and y is the likelihood of each class. Image map: Diego Rivera. Background: Google Earth.

Our findings suggest an alternative interpretation to the proposed secondary center of grapevine domestication in Sardinia by Grassi et al. in 2003 [89]. While the genetic proximity between wild grapevine specimens from Nuoro and local Sardinian cultivars (Bo-vale Murru and Bovale Muristellu) has been interpreted as evidence of local domestication, our data support an alternative hypothesis. The presence of exclusively domesticated-type seeds in our Nuoro wild grapevine samples suggests these populations may represent naturalized or feral vines rather than truly wild specimens, thus challenging the local domestication theory (Figure 10).

In summary, the examination of wild grapevine populations in Western Europe, particularly in Spain and Sardinia (Italy), reveals a substantial degree of introgression or hy-bridization, with seeds exhibiting domesticated traits that can be interpreted as either re-peated introgressions or feral occurrences. This phenomenon is not confined to Western Europe, as similar patterns are observed in the Caucasus, where the degree of domestication in "wild" grapevines is higher [90] (Figure 7). Riaz et al. [61] analyzed the relation-ships between 1378 wild and cultivated grapevine genotypes as represented by the first two principal coordinates of a PCoA using allelic profiles from 20 SSR molecular markers. The figures clearly indicate the existence of two well-differentiated groups within *Vitis sylvestris:* one centered in the Caucasus (Armenia, Azerbaijan, and Georgia) and another in Western Europe (Spain, Croatia, France, and Italy), separated by the core of cultivars originating from France, Spain, Georgia, Italy, Pakistan, and Turkmenistan. This provides us with an alternative perspective on the anomalies detected in the phenotypes of wild vines from the South Caucasus with a higher degree of domestication.

4.5. Highlighting Conservation Significance: Recognizing Mixed Grapevine Populations as Crucial Repositories of Genetic Diversity

The relatively high stability of natural riparian habitats over the centuries until the mid-20th century made them an optimal refuge for escaped cultivars, whose seeds were transported from the vineyards by fruit-eating birds and mammals. In addition, there is evidence of the occasional but intentional practice of grafting cultivars onto wild grape-vine individuals in riparian forests, thus increasing the presence of domesticated grape-vines among wild plants. Terpó [21,22,91,92] described examples from a Hungarian beech forest of grapevine populations with 80% male individuals, including feral forms of *V. riparia* and escaped *V. vinifera* cultivars.

This continuous process of introgression into wild populations is not merely recent, since it was noted by authors in the nineteenth and twentieth centuries before and after the introduction of phylloxera and other grapevine pests, which led to the abandonment of European vineyards, followed by their replantation using American grapevines and their hybrids as rootstocks. It underscores the gene reservoir role of populations in natural habitats. This reservoir extends beyond wild grapevines to include cultivated grapevines, offering the potential discovery of ancient cultivars lost from vineyards, as previously noted by De Andrés et al. [86].

Sometimes no relationships between wild populations and nearby vineyards are detected, which may be due to recent and profound changes in the repertoire of cultivars growing in the vineyards. Molecular marker studies suggest no genetic contribution from the autochthonous vines of the Iregua River, a tributary of the Ebro, to the current local varieties, and vice versa [93]. Notably, despite an *HI* value of 0.31 and *DI* value of 0.62 (Supplementary Table S1, Sample 254), the seeds studied suggest introgression, possibly with varieties that have not been grown in the area for a century.

The adverse effects of anthropogenic habitat alteration on riparian forests extend beyond the mere destruction of tree cover. The introduction of invasive species and pathogens further exacerbates these impacts, potentially posing significant consequences for biodiversity conservation and overall ecosystem survival.

The invasion of phylloxera, powdery mildew, and downy mildew in 1850 caused the disappearance, in only half a century, of most vineyards and numerous wild grapevines, in France and other countries of Europe [8]. This resulted in significant transformations in the structure of grapevine populations, both within vineyards and in natural habitats.

Given the great diversity of wild populations of *Vitis* found in riparian forests and other natural habitats, from the Iberian Peninsula to Central Asia, which harbor not only wild vines but also modern and ancient feral cultivars, as well as a complex network of introgressions, the systematic "cleaning and rearrangement" of riverbanks with the elimination of existing vegetation means the destruction and loss of this diversity of great interest for agriculture and food.

Another aspect to consider is the growing presence of feral American vines in the natural habitats of Eurasia because of the massive use of these species and their hybrids as rootstocks, of which we have an example in Valencia (Spain) [83–85]. This phenomenon leads to the substitution of the hybrid swarm *Vitis vinifera–Vitis sylvestris* by American vines with lesser aptitudes to produce grapes of interest for table consumption or wine-making. The remarkable presence of feral rootstocks along both banks of the Danube River between Esztergom and Budapest stands out. In this area, true wild specimens of *V. sylvestris* are only preserved on certain islets, as documented by Popescu et al. in 2013 [94]. Feral individuals of American rootstocks and their hybrids have also been identified in Italy and Spain [40,41,95,96]. American rootstock genotypes now constitute the predomi-nant wild *Vitis* germplasm in several European viticultural regions. This is particularly evident in areas such as Campania, Italy, where they have largely displaced native wild *Vitis* populations—if these populations had not already been lost due to phylloxera [97].

Recent morphometric approaches to grapevine seed analysis have yielded valuable insights. Karasik et al. [98] demonstrated clear varietal differentiation using Fourier coefficient analysis, though they did not explore intra-sample heterogeneity or wild–cultivated relationships. More comprehensively, Bouvy et al. [99] analyzed over 19,000 archaeological grape pips, revealing a transition from wild to domestic varieties in France around 600–500 BCE, coincident with Mediterranean influences and eastern table cultivar introduction. Their discovery of abundant wild-like morphologies suggests both early-stage domestication and significant gene flow between the introduced and wild popula-tions, supporting the concept of continuous genetic exchange during domestication.

Riaz et al. [61] emphasized the importance of exploring wild relatives of crops to identify genetic factors associated with stress resistance. Specifically, they mention the presence of salt-tolerant grape accessions in certain *V. sylvestris* populations and the recent discovery of wild and cultivated accessions from Germany, Iran, and Georgia with tolerance to mildew diseases. This stresses the potential of wild ancestors as genetic resources for breeding disease-resistant varieties.

In summary, we must underscore the significance of identifying, preserving, and characterizing wild grapevine germplasm (including true wild, feral, and intermediate), for their potential contributions to disease and stress resistance, as well as berry quality traits, in the wine and grape industry's future.

Further research is needed to assess the extent to which American rootstock genotypes have displaced both naturalized *Vitis vinifera* populations and wild *Vitis sylvestris* communities in natural and semi-natural habitats. Additionally, it is crucial to evaluate the conservation status and potential threats to remnant Eurasian grapevine populations in their native ecosystems.

5. Conclusions

The coexistence of mixed populations, comprising dioecious autochthonous wild subpopulations, feral hermaphroditic cultivars, and hybrids, presents opportunities for genetic diversification in both cultivated crops and wild populations. Moreover, these mixed populations serve as potential repositories for ancient varieties that are currently absent from cultivation. However, the drastic intervention in natural habitats, particularly riparian forests, poses a significant threat. Activities such as riverbank channeling, improper clearing of waterways, and large-scale public infrastructure projects have resulted in the burial of numerous pure or hybrid wild grapevine populations.

Following such interventions, secondary colonization by modern cultivars from nearby vineyards or by rootstocks that produce seeds in abandoned vineyards often occurs. In this process, birds play a crucial role in facilitating the recolonization of altered habitats by seedlings from cultivated or abandoned vineyards.

It is imperative to raise awareness about the fragility of grapevine natural habitats, the ongoing reduction in their area in Europe, and the anticipated impact of climate change. These factors underscore the necessity to protect and preserve these habitats as repositories of significant diversity legally and practically. By doing so, we can safeguard the genetic resources essential for the sustainability and resilience of grapevine cultivation in the face of environmental challenges.

Potential directions for future research should encompass aspects such as the following:

Investigating the specific mechanisms driving secondary colonization by modern cultivars and rootstocks in altered habitats could provide insight into the dynamics of grapevine population shifts. Assessing the effectiveness of different conservation strategies would be crucial for preserving wild grapevine populations and their genetic diversity. Exploring the potential for utilizing mixed populations as sources of genetic material could enable breeding programs aimed at improving crop resilience and adaptability. Additionally, examining the role of other wildlife, in addition to birds, in facilitating the dispersal and colonization of grapevines in altered habitats may reveal important ecological interactions. Conducting parallel studies to quantify phenotypic and genetic diversity in mixed populations could shed light on their potential for enhancing grapevine resilience to environmental stressors.

Supplementary Materials: The following supporting information can be downloaded at https://www.mdpi.com/article/10.3390/horticulturae11010092/s1, Table S1: Samples analyzed organized by decreasing heterogeneity and increasing proportion of seeds with wild traits; Table S2: Samples analyzed organized by hybridization index, domestication index values, and geographical origins; Methods S1: Standard methods.

Author Contributions: Conceptualization, D.R. and C.O.; Data curation, J.V., D.M., M.K., A.N., E.L., F.A., D.J.R.-O., G.L. and M.F.; Formal analysis, D.J.R.-O.; Investigation, D.R., J.V., M.K., A.N., R.O. and M.F.; Methodology, D.R., F.A. and D.J.R.-O.; Resources, D.R., D.M., A.N., R.O., C.Á.O., C.A., E.L. and G.L.; Software, F.A. and D.J.R.-O.; Validation, R.O. and C.O.; Visualization, D.R.; Writing—original draft, D.R.; Writing—review and editing, D.R., D.M., E.L. and C.O. All authors have read and agreed to the published version of the manuscript.

Funding: This research received no external funding.

Data Availability Statement: The original contributions presented in the study are included in the article/Supplementary Materials, and further inquiries can be directed to the corresponding author/s.

Acknowledgments: We extend our gratitude to Erika Maul from The Institute for Grapevine Breeding Geilweilerhof, Germany, for generously providing samples of grapes from contemporary Near Eastern varieties. Additionally, we express our appreciation to Gayane Melyan from the Center of Agrobiotechnology of the Armenian National Agrarian University in Etchmiadzin, Armenia, for sending wild grape samples from Armenia. We would also like to acknowledge the contributions of several institutions and individuals. Specifically, we thank the U.S. National Plant Germplasm System, Rancho Santa Ana, Istituto ed Orto Botanico di Palermo, Botanisches Garten Johannes Gutenberg—Universität Mainz, Smithsonian Institution, and Giardino Botanico di Padova for supplying us with seeds of American and Asian Vitis species and related genera. Furthermore, our gratitude extends to the curators of grapevine collections, including those from the Rioja region (Spain) in Mendavia, La Casa de las Vides nurseries in Agullent (Valencia, Spain), Rojas Clemente grapevines from the Royal Botanical Garden of Madrid (Spain), and the Istituto Agrario di San Michele all'Adige (Trentino, Italy), for granting us access to leaves, shoots, and grapes for our research. Wild populations were sampled by the authors and Encarna Carreño.

Conflicts of Interest: The authors declare no conflicts of interest.

References

1. Magurran, A.E. *Ecological Diversity and Its Measurement*; Springer Science & Business Media: Secaucus, Spain, 2013.
2. Rivera, D.; Alcaraz, F.; Rivera-Obón, D.J.; Obón, C. Phenotypic Diversity in Wild and Cultivated Date Palm (*Phoenix*, Arecaceae): Quantitative Analysis Using Information Theory. *Horticulturae* **2022**, *8*, 287. [CrossRef]
3. Dong, Y.; Duan, S.; Xia, Q.; Liang, Z.; Dong, X.; Margaryan, K.; Musayev, M.; Goryslavets, S.; Zdunić, G.; Bert, P.F.; et al. Dual domestications and origin of traits in grapevine evolution. *Science* **2023**, *379*, 892–901. [CrossRef] [PubMed]
4. Grassi, F.; De Lorenzis, G. Back to the origins: Background and perspectives of grapevine domestication. *Int. J. Mol. Sci.* **2021**, *22*, 4518. [CrossRef] [PubMed]
5. Arnold, C. Ecologie de la vigne sauvage, *Vitis vinifera* L. ssp. *sylvestris* (Gmelin) Hegi, dans les forêts alluviales et colluviales d'Europe. *Geobot. Helv.* **2002**, *6*, 1–256.
6. Arnold, C.; Gillet, F.; Gobat, J.M. Occurrence of the wild grapevine *Vitis vinifera* ssp. *sylvestris* in Europe. *Vitis* **1998**, *37*, 159–170.
7. Martínez de Toda, F. *Biología de la Vid*; Mundi-Prensa: Madrid, Spain, 1991; pp. 1–346.
8. Levadoux, L. Les populations sauvages et cultivées de *Vitis vinifera* L. *Ann. Amélior. Plant.* **1956**, *1*, 59–118.
9. Mercuri, A.M.; Torri, P.; Florenzano, A.; Clò, E.; Mariotti Lippi, M.; Sgarbi, E.; Bignami, C. Sharing the Agrarian Knowledge with Archaeology: First Evidence of the Dimorphism of *Vitis* Pollen from the Middle Bronze Age of N Italy (Terramara Santa Rosa di Poviglio). *Sustainability* **2021**, *13*, 2287. [CrossRef]
10. Gallardo, A.; Ocete, R.; López Martínez, M.A.; Lara, M.; Rivera, D. Assesment of pollen dimorphism in populations of *Vitis vinifera* L. subsp. *sylvestris*(Gmelin) Hegi in Spain. *Vitis* **2009**, *48*, 59–62.
11. This, P.; Lacombe, T.; Thomas, M.R. Historical origins and genetic diversity of wine grapes. *Trends Genet.* **2006**, *22*, 511–519. [CrossRef]
12. Stummer, A. Zur Urgeschichte der Rebe und des Weinbaumes. *Mitt. Anthrop. Ges. Wien* **1911**, *41*, 283–296.
13. Ocete, C.A.; Martínez-Zapater, J.; Ocete, R.; Lara, M.; Cantos, M.; Arroyo, R.; Morales, R.; Iriarte-Chiapusso, M.; Hidalgo, J.; Valle, J.M.; et al. La vid silvestre euroasiática; un recurso fitogenético amenazado ligado a la historia de la humanidad. *Enoviticultura* **2018**, *50*, 1–16.
14. Ocete, R.; Arnold, C.; Failla, O.; Lovicu, G.; Biagini, B.; Imazio, S.; Lara, M.; Maghradze, D.; Lopez, M.A. Considerations on the European wild grapevine (*Vitis vinifera* L. ssp. *sylvestris* (Gmelin) Hegi) and Phylloxera infestation. *Vitis* **2011**, *50*, 97–98.
15. Turkovic, Z. Untersuchungs ergebnisse über *Vitis sylvestris* Gmelin im Jahre 1954. *Weinb. Keller* **1955**, *2*, 74–79.
16. Bitsadze, N.; Kikilashvili, S.; Chipashvili, R.; Mamasakhlisashvili, L.; Maghradze, T.; Kikvadze, M.; Toffolatti, S.L.; De Lorenzis, G.; Failla, O.; Ocete, R.; et al. Resistance to downy mildew in wildly growing Eurasian *Vitis vinifera* L. grapevines. *J. Plant Pathol.* **2024**, *106*, 1759–1771. [CrossRef]
17. Campostrini, F.; Anzani, R.; Failla, O.; Iacono, F.; Scienza, A.; de Micheli, L. Application of the phyllometric analysis to the geographic classification of Italian population of wild grapevine (*Vitis vinifera* L. ssp. *sylvestris* Hegi). *J. Int. Sci. Vigne Vin* **1993**, *27*, 255–262.
18. Rathay, E. *Die Geschlechtsverhältnisse der Reben und ihre Bedeutung für den Weinbau*; K.K. Hofbuchhandlung Wilhem Frick: Wien, Austria, 1889; Volume 2.
19. Zdunić, G.; Maul, E.; Eiras Dias, J.; Muñoz-Organero, G.; Carka, F.; Maletić, E.; Savvides, S.; Jahnke, G.; Nagy, Z.; Nikolić, D.; et al. Guiding principles for identification, evaluation and conservation of *Vitis vinifera* L. subsp. *sylvestris*. *Vitis* **2017**, *56*, 127–131. [CrossRef]
20. Facsar, G. Összehasonlitö morfolögiai vizsgälatok kerti szölöfajtäk magjain I (Vergleichende morphologische Unter-suchungen der Samen von Gartenrebensorten. I.). *Bot. Közlemenyek* **1970**, *57*, 221–231.
21. Terpó, A. The carpological examination of wild growing grapevine species of Hungary. I. Qualitative and quantitative characteristics of grapevine seeds. *Acta Bot. Hung.* **1976**, *22*, 209–247.
22. Terpó, A. The carpological examination of wild growing grapevine species of Hungary. II. Qualitative and quantitative characteristics of grapevine seeds. *Acta Bot. Hung.* **1977**, *23*, 247–273.
23. Facsar, G.; Jerem, E. Zum urgeschichtlichen Weinbau in Mitteleuropa. Rebkernfunde von *Vitis vinifera* L. aus der urnenfelder-, hallstatt-und latenezeitlichen Siedlung Sopron-Krautacker. *Wiss. Arb. Burgenland* **1985**, *71*, 121–144.
24. Rivera, D.; Walker, M. A review of paleobotanical findings of early *Vitis* in the Mediterranean and of the origins of cultivated grape-vines, with special reference to prehistoric exploitation in the Western Mediterranean. *Paleobot. Palynol.* **1989**, *61*, 205–237. [CrossRef]
25. Mangafa, M.; Kotsakis, K. A new method for the identification of wild and cultivated charred grape seeds. *J. Archaeol. Sci.* **1996**, *23*, 409–418. [CrossRef]
26. Perret, M. *Caractérisation et Evaluation du Polymorphisme des Génotypes Sauvages et Cultivés de Vitis vinifera L. à L'aide de Marqueurs RAPD et de Certains Traits Morphologiques*; Travail de Diplôme, Université de Neuchâtel: Neuchâtel, Switzerland, 1997.
27. Rivera, D.; Miralles, B.; Obón, C.; Carreño, E.; Palazón, J.A. Multivariate analysis of *Vitis* subgenus *Vitis* seed morphology. *Vitis* **2007**, *46*, 158–167.

28. Obón, C.; Rivera, D.; Carreño, E.; Alcaraz, F.; Palazón, J.A. Seed Morphology of *Vitis vinifera* and Its Relationship to Ecogeographical Groups and Chlorotypes. *Acta Hortic.* **2008**, *799*, 51–69. [CrossRef]
29. Orrù, M.; Grillo, O.; Lovicu, G.; Venora, G.; Bacchetta, G. Morphological characterisation of *Vitis vinifera* L. seeds by image analysis and comparison with archaeological remains. *Veg. Hist. Archaeobot.* **2012**, *22*, 231–242. [CrossRef]
30. Orrù, M.; Grillo, O.; Venora, G.; Bacchetta, G. Computer vision as a method complementary to molecular analysis: Grapevine cultivar seeds case study. *C. R. Biol.* **2012**, *335*, 602–615. [CrossRef]
31. Orrù, M.; Grillo, O.; Venora, G.; Bacchetta, G. Seed morpho-colotimetric análisis by computer vision: A helpful tool to identify grapevine (*Vitis vinifera* L.) cultivars. *Aust. J. Grape Wine Res.* **2015**, *21*, 508–519. [CrossRef]
32. Ucchesu, M.; Orrù, M.; Grillo, O.; Venora, G.; Usai, A.; Serreli, P.F.; Bacchetta, G. Earliest evidence of a primitive cultivar of *Vitis vinifera* L. during the Bronze Age in Sardinia (Italy). *Veg. Hist. Archaeobot.* **2015**, *24*, 587–600. [CrossRef]
33. Ucchesu, M.; Orrò, M.; Grillo, O.; Venora, G.; Pagietti, G.; Ardu, A.; Bacchetta, G. Predictive method for correct identification of archaeological charred grape sedes: Support for advances in knowledge of grape domestication process. *PLoS ONE* **2016**, *11*, e0149814. [CrossRef]
34. Bonhomme, V.; Ivorra, S.; Lacombe, T.; Evin, A.; Figueiral, I.; Maghradze, D.; Marchal, C.; Pagnoux, C.; Pastor, T.; Pomarèdes, H.; et al. Pip shape echoes grapevine domestication history. *Sci. Rep.* **2021**, *11*, 21381. [CrossRef]
35. Obón, C.; Rivera-Obón, D.J.; Valera, J.; Matilla, G.; Alcaraz, F.; Maghradze, D.; Kikvadze, M.; Ocete, C.A.; Ocete, R.; Nebish, A.; et al. Is there a domestication syndrome in *Vitis* (Vitaceae) seed morphology? *Genet. Resour. Crop Evol.* **2024**, 1–25. [CrossRef]
36. POWO. Plants of the World Online. Available online: https://powo.science.kew.org/ (accessed on 15 February 2024).
37. GRIN. Search Taxonomy Data in GRIN-Global. Available online: https://npgsweb.ars-grin.gov/gringlobal/taxon/taxonomysearch (accessed on 15 February 2024).
38. Facsar, G.; Udvardy, L. Adventive grapevine species (*Vitis*-hybrids). In *The Most Important Invasive Plants in Hungary*; Botta-Dukát, Z., Balogh, L., Eds.; Institute of Ecology and Botany, Hungarian Academy of Sciences: Vácrátót, Hungary, 2008; pp. 47–54.
39. Schnitzler, A.; Arnold, C.; Guibal, F.; Walter, J.M. Histoire de la vigne sauvage, *Vitis vinifera* ssp. *sylvestris*, en Camargue / Wild grapevine *Vitis vinifera* ssp. *sylvestris* in Camargue, southern France. *Ecol. Mediterr.* **2018**, *44*, 53–66. [CrossRef]
40. Ardenghi, N.M.; Galasso, G.; Banfi, E.; Zoccola, A.; Foggi, B.; Lastrucci, L. A taxonomic survey of the genus *Vitis* L. (Vitaceae) in Italy, with special reference to Elba Island (Tuscan Archipelago). *Phytotaxa* **2014**, *166*, 163–198. [CrossRef]
41. Vázquez, F.M.; García Alonso, D. Aproximación al conocimiento de los táxones del género *Vitis* L. (Vitaceae) que viven silvestres en Extremadura (España). *Folia Bot. Extrem.* **2017**, *11*, 5–37.
42. Valera, J.; Matilla-Seiquer, G.; Obón, C.; Alcaraz, F.; Rivera, D. Grapevine in the ancient Upper Euphrates: Horticultural implications of a Bayesian morphometric study of archaeological seeds. *Horticulturae* **2023**, *9*, 803. [CrossRef]
43. Schindelin, J.; Arganda-Carreras, I.; Frise, E.; Kaynig, V.; Longair, M.; Pietzsch, T.; Preibisch, S.; Rueden, C.; Saalfeld, S.; Schmid, B.; et al. Fiji: An Open-Source Platform for Biological-Image Analysis. *Nat. Methods* **2012**, *9*, 676–682. [CrossRef]
44. Hosmer, D.W., Jr.; Lemeshow, S.; Sturdivant, R.X. *Applied Logistic Regression*; John Wiley & Son: Hoboken, NJ, USA, 2013.
45. Couronné, R.; Probst, P.; Boulesteix, A.L. Random forest versus logistic regression: A large-scale benchmark experiment. *BMC Bioinform.* **2018**, *19*, 1–14. [CrossRef]
46. randomForest. randomForest: Breiman and Cutler's Random Forests for Classification and Regression. Available online: https://cran.r-project.org/web/packages/randomForest/index.html (accessed on 18 December 2023).
47. Breiman, L. Random Forests. *Mach. Learn.* **2001**, *45*, 5–32. [CrossRef]
48. Breiman, L.; Cutler, A.; Liaw, A.; Wiener, M. Package 'randomForest' 2022. Available online: https://cran.r-project.org/web/packages/randomForest/randomForest.pdf (accessed on 18 December 2023).
49. Rossiter, D.G. Random Forests Two Ways. Available online: https://www.css.cornell.edu/faculty/dgr2/_static/files/R_PDF/CompareRandomForestPackages.pdf (accessed on 18 December 2023).
50. R Project. The R Project for Statistical Computing. Available online: https://www.r-project.org/ (accessed on 18 December 2023).
51. RStudio. RStudio Desktop. Available online: https://posit.co/download/rstudio-desktop/ (accessed on 18 December 2023).
52. Liu, Y.; Wang, Y.; Zhang, J. New Machine Learning Algorithm: Random Forest. In *ICICA 2012, LNCS 7473*; Liu, B., Ma, M., Chang, J., Eds.; Springer: Berlin/Heidelberg, Germany, 2012; pp. 246–252.
53. R-Bloggers. Random Forest in R. 2023. Available online: https://www.r-bloggers.com/2021/04/random-forest-in-r/ (accessed on 18 December 2023).
54. Negrul, A.M. Origin of the cultivated grapevine and its classification. In *Ampelografiia SSSR [Ampelography of the USSR]*; Varanov, A., Ed.; Pischepromizdat: Moscow, Russia, 1946; Volume 1, pp. 133–216. (In Russian)
55. La Mancha Denominación de Origen. Available online: https://lamanchawines.com/denominacion-de-origen/mapa-y-estadisticas/ (accessed on 3 September 2024).
56. Manchuela Denominación de Origen. Available online: https://www.manchuela.wine/cifras-y-anadas/ (accessed on 3 September 2024).

57. Almansa Denominación de Origen. Available online: https://denominacion-origen-almansa.com/zona-de-produccion/ (accessed on 3 September 2024).
58. Jumilla Denominación de Origen. Available online: https://jumilla.wine/municipios-de-la-dop/ (accessed on 3 September 2024).
59. D'Onofrio, C. Introgression among cultivated and wild grapevine in Tuscany. *Front. Plant Sci.* **2020**, *11*, 202. [CrossRef]
60. Di Vecchi-Staraz, M.; Laucou, V.; Bruno, G.; Lacombe, T.; Gerber, S.; Bourse, T.; Boselli, M.; This, P. Low level of pollen-mediated gene flow from cultivated to wild grapevine: Consequences for the evolution of the endangered subspecies *Vitis vinifera* L. subsp. *silvestris*. *J. Hered.* **2009**, *100*, 66–75. [CrossRef]
61. Riaz, S.; De Lorenzis, G.; Velasco, D.; Koehmstedt, A.; Maghradze, D.; Bobokashvili, Z.; Musayev, M.; Zdunic, G.; Laucou, V.; Andrew Walker, M.; et al. Genetic diversity analysis of cultivated and wild grapevine (*Vitis vinifera* L.) accessions around the Mediterranean basin and Central Asia. *BMC Plant Biol.* **2018**, *18*, 137. [CrossRef] [PubMed]
62. Clemente, S. *Ensayos Sobre las Variedades de vid que Vegetan en Andalucia*; Villalpando: Madrid, Spain, 1807.
63. Clemente, S. *Essai sur les Variétés de la Vigne qui Végétent en Andalousie*; Poulet: Paris, France, 1814.
64. Clemente, S. *Ensayo Sobre las Variedades de la vid Común que Vegetan en Andalucía*; Estereotipia Perojo: Madrid, Spain, 1879.
65. Bronner, J.P. *Die Wilden Trauben des Rheinthales*; Buchdruckerei von Georg Mohr: Heidelberg, Germany, 1857.
66. Bronner, J.P. Description des vignes sauvages de la vallée du Rhin. *La Bourgogne* **1859**, *1*, 97–110.
67. Scossiroli, R.E. Origine ed evoluzione della vite. *Atti Ist. Bot. e Lab. Critt.* **1988**, *7*, 35–55.
68. Scienza, A.; Failla, O.; Anzani, R. Wild grapevine (Vitis vinifera L. *sylvestris* C.C.Gmel.) in Italy: Diffusion, characteristics and germplasm preservation. In Proceedings of the Germplasm Preservation of Horticultural Crops: Wild and·Cultivated Plants, Bejing, China, 1–3 September 1988.
69. Scienza, A.; Protii, A.; Conca, E.; Romano, F. Diffusione e caratteristiche della *Vitis vinifera sylvestris* Gmelin in ltalia. In Proceedings of the Compte Rendu IV Symposium Lntern. de Génétique de la Vigne, Verona, Italy, 13–18 April 1986; pp. 86–95.
70. Anzani, R.; Failla, O.; Scienza, A.; Campostrini, E. Wild grapevine (Vitis vinifera ssp. *sylvestris*) in Italy: Distribution characteristics and germplasm preservation. In Proceedings of the 5th International Symposium on Grape Breeding, St. Martin/Pfalz, Germany, 12–16 September 1989; pp. 97–113.
71. Anzani, R.; Failla, O.; Scienza, A.; De Micheli, L. Individuazione e conservazione del germoplasma di vite selvatica (*Vitis vinifera sylvestris*) in Italia. *Vignevini* **1993**, *6*, 51–60.
72. Failla, O.; Anzani, R.; Scienza, A. La vite selvatica in Italia: Diffusione, caratteristiche e conservazione del germoplasma. *Vignevini* **1992**, *19*, 37–46.
73. Kikvadze, M.; Kikilashvili, S.; Bitsadze, N.; Maghradze, T.; De Lorenzis, G.; Rubio, R.O.; Rivera, D.; Bacilieri, R.; Failla, O.; Maghradze, D. Wildly growing Eurasian grapevine (*Vitis vinifera* L.) in Georgia: Composition, research and efforts for the preservation. *Acta Hortic. ISHS* **2024**, *1385*, 19–24. [CrossRef]
74. Wagner, E. Die Feststellung des Geschlechts bei Rebensämlingen. *Vitis* **1960**, *2*, 190–197.
75. Avramov, L.; Jelenkovi, G.; Jovanovi, M.; Rodic, Z. Inheritance of flower type in some grape varieties (*Vitis vinifera* L.). *Vitis* **1967**, *6*, 129–135.
76. Dalbo', M.A.; Ye, G.N.; Weeden, N.G.; Steinkellner, H.; Sefc, K.M.; Reisch, B.I. A gene controlling sex in grapevines placed on a molecular-based genetic map. *Genome* **2000**, *43*, 333–340. [CrossRef]
77. Riaz, S.; Krivanek, A.F.; Xu, K.; Walker, M.A. Refined mapping of the Pierce's disease resistance locus, PdR1, and Sex on an extended genetic map of *Vitis rupestris* × *V. arizonica*. *Theor. Appl. Genet.* **2006**, *113*, 1317–1329. [CrossRef]
78. Marguerit, E.; Boury, C.; Manicki, A.; Donnart, M.; Butterlin, G.; Ne'morin, A.; Wiedemann-Merdinoglu, S.; Merdinoglu, D.; Ollat, N.; Decroocq, S. Genetic dissection of sex determinism, inflorescence morphology and downy mildew resistance in grapevine. *Theor. Appl. Genet.* **2009**, *118*, 1261–1278. [CrossRef] [PubMed]
79. Coito, J.L.; Silva, H.G.; Ramos, M.J.; Cunha, J.; Eiras-Dias, J.; Amâncio, S.; Costa, M.M.; Rocheta, M. *Vitis* flower types: From the wild to crop plants. *PeerJ* **2019**, *7*, e7879. [CrossRef] [PubMed]
80. Fechter, I.; Hausmann, L.; Daum, M.; Sörensen, T.R.; Viehöver, P.; Weisshaar, B.; Töpfer, R. Candidate genes within a 143 kb region of the flower sex locus in *Vitis*. *Mol. Genet. Genom.* **2012**, *287*, 247–259. [CrossRef]
81. Cunha, J.; Ibáñez, J.; Teixeira-Santos, M.; Brazão, J.; Fevereiro, P.; Martínez-Zapater, J.M.; Eiras-Dias, J.E. Genetic relationships among Portuguese cultivated and wild *Vitis vinifera* L. germplasm. *Front. Plant Sci.* **2020**, *11*, 127. [CrossRef]
82. Albizuri, A.A.; Cano, A.G.; López, M.; Pérez, M.; Ocete, R. Distribución, ecología y caracterización in situ de la vid silvestre en la Reserva de la Biosfera de Urdaibai (Bizkaia). *Munibe* **2009**, *57*, 15–34.
83. Laguna, E. Sobre las formas naturalizadas de *Vitis* en la Comunidad Valenciana. I. Las especies. *Flora Montiberica* **2003**, *23*, 46–82. Available online: https://dialnet.unirioja.es/descarga/articulo/1394799.pdf (accessed on 7 January 2025).
84. Laguna, E. Datos foliares de las especies e híbridos alóctonos de vides (género *Vitis*) en el territorio valenciano. *Toll Negre* **2004**, *3*, 11–25.

85. Laguna, E. American and hybrid grapevines (*Vitis* spp.): A new concept of invasive plants to Europe. In Proceedings of the 4th European Conference on the Conservation of the Wild Plants—A Workshop on the Implementation of the Global Strategy for Plant Conservation in Europe, Valencia, Spain, 17–20 August 2004; Aguilella, A., Ibars, A., Laguna, E., Pérez-Rocher, B., Eds.; CD Rom. Generalitat Valenciana: Valencia, Spain, 2006.
86. De Andrés, M.T.; Benito, A.; Pérez-Rivera, G.; Ocete, R.; Lopez, M.A.; Gaforio, L.; Muñoz, G.; Cabello, F.; Martínez, J.; Arroyo-García, R. Genetic diversity of wild grapevine populations in Spain and their genetic relationships with cultivated grapevines. *Mol. Ecol.* **2012**, *21*, 800–816. [CrossRef]
87. Ocete, R.; López, M.Á.; Gallardo, A.; Arnold, C. Comparative analysis of wild and cultivated grapevine (*Vitis vinifera*) in the Basque Region of Spain and France. *Agric. Ecosyst. Environ.* **2008**, *123*, 95–98. [CrossRef]
88. Zecca, G.A.; De Mattia, F.; Lovicu, G.; Labra, M.; Sala, F.; Grassi, F. Wild grapevine: Silvestris, hybrids or cultivars that escaped from vineyards? Molecular evidence in Sardinia. *Plant Biol.* **2010**, *12*, 558–562. [CrossRef]
89. Grassi, F.; Labra, M.; Imazio, S.; Spada, A.; Sgorbati, S.; Scienza, A.; Sala, F. Evidence of a secondary grapevine domestication centre detected by SSR analysis. *Theor. Appl. Genet.* **2003**, *107*, 1315–1320. [CrossRef] [PubMed]
90. Kikvadze, M.; Valera, J.; Obón, C.; Ocete, R.; Ocete, C.A.; Kikilashvili, S.; Maghradze, D.; Rivera-Obón, D.J.; Rivera, D. Wild grapevines of Georgia and their relationship with local cultivars: A Bayesian analysis of seed morphology. *Vitis* **2024**, *63*, 9. [CrossRef]
91. Terpó, A. A *Vitis sylvestris* C.C.Gmel. Magyar középhegy ségi termőhelyi viszonyainak vizsgálata [Investiga tions on ecological conditions of habitats of *Vitis sylvestris* C.C.Gmel. in Central Hungarian Mountains]. *Bot. Kőzlem.* **1969**, *56*, 27–35.
92. Terpó, A. Origin and distribution of *Vitis sylvestris* C.C.Gmel. In Proceedings of the II International Symposium on the problems of Balkan Flora and Vegetation, Istanbul, Turkey, 3–10 July 1978; pp. 1–5.
93. Ocete, R.; Arroyo-Garcia, R.; Morales, M.L.; Cantos, M.; Gallardo, A.; Pérez, M.A.; Gómez, I.; López, M.A. Characterization of *Vitis vinifera* L. subspecies *sylvestris* (Gmelin) Hegi in the Ebro River Basin (Spain). *Vitis* **2011**, *50*, 11–16.
94. Popescu, C.; Dejeu, L.; Ocete, R. Preliminary Characterization of Wild Grapevine Populations (*Vitis vinifera* ssp. *sylvestris*) Grown Along the Danube River. *Not. Bot. Horti Agrobot. Cluj-Napoca* **2013**, *41*, 472–477. [CrossRef]
95. Ardenghi, N.M.; Galasso, G.; Banfi, E.; Cauzzi, P. *Vitis* × *nova-angliae* (*Vitaceae*): Systematics, distribution and history of an "illegal" alien grape in Europe. *Wiildenowia* **2015**, *45*, 197–207. [CrossRef]
96. Ardenghi, N.M.; Banfi, E.; Galasso, G. A taxonomic survey of the genus *Vitis* L. (Vitaceae) in Italy, part II: The 'Euro-American' hybrids. *Phytotaxa* **2015**, *24*, 232–246. [CrossRef]
97. Stinca, A. The Genus *Vitis* L. (Vitaceae) in Campania (Southern Italy), with emphasis on alien units. *Ann. Bot.* **2019**, *9*, 107–112. [CrossRef]
98. Karasik, A.; Rahimi, O.; David, M.; Weiss, E.; Drori, E. Development of a 3D seed morphological tool for grapevine variety identification, and its comparison with SSR analysis. *Sci. Rep.* **2018**, *8*, 6545. [CrossRef]
99. Bouby, L.; Bonhomme, V.; Ivorra, S.; Bacilieri, R.; Ben Makhad, S.; Bonnaire, E.; Cabanis, M.; Derreumaux, M.; Dietsch-Sellami, M.; Durand, F.; et al. Seed morphometrics unravels the evolutionary history of grapevine in France. *Sci. Rep.* **2024**, *14*, 22207. [CrossRef]

horticulturae

MDPI

Article

Origin and Possible Members of the 'Malvasia' Family: The New Fuencaliente de La Palma Hypothesis on the True 'Malvasia'

Francesca Fort [1,*,†], Luis Ricardo Suárez-Abreu [1,†], Qiying Lin-Yang [1], Juancho Asenjo [2], Leonor Deis [1], Joan Miquel Canals [1] and Fernando Zamora [1]

1 Oenological Technology Group (TECNENOL), Department of Biochemistry and Biotechnology, Faculty of Oenology, Rovira i Virgili University, Sescelades Campus, C/Marcel·lí Domingo, 1. E-43007 Tarragona, Spain; luisricardo.suarez@estudiants.urv.cat (L.R.S.-A.); qiying.lin@estudiants.urv.cat (Q.L.-Y.); leonor.deis@urv.cat (L.D.); jmcanals@urv.cat (J.M.C.); fernando.zamora@urv.cat (F.Z.)

2 Suero de Quinones, 13 (1°A) E-28002 Madrid, Spain; asenjojuancho62@gmail.com

* Correspondence: mariafrancesca.fort@urv.cat

† These authors contributed equally to this work.

Abstract: The name 'Malvasia' and its various spellings has historically been associated with a type of sweet and/or aromatic wine. However, a definitive association with a specific grape variety remains unconfirmed. In fact, up to 413 different grape variety names (cultivar name (synonym name) and/or first name) are related to the term "Malvasia". The question arises: are all of these truly Malvasia? To answer this question, our research group presents a hypothesis. We worked with 43 genetic profiles that various scientific groups have published over decades and that are stored in the world's largest grape database, the Vitis International Variety Catalogue (VIVC). The known molecular profiles were obtained using the SSR (Simple Sequence Repeats) or microsatellite technique. Various population structure programs were applied, information on the possible origin or area where each of the varieties was mostly grown was used, and historical information was used to explain the results obtained. Therefore, it can be concluded that the current varieties best positioned to define the concept of grape and/or wine variety "Malvasia" would be (1) Malvasia Dubrovacka, Malvasia bianca lunga, and Malvasia del Cilento, by genetic proximity; (2) Malvasia volcanica, Malvasia babosa, Malvasia nera di Basilicata, Malvasia nera di Brindisi, Vitovska, Pelena, Prunesta (false), and Lagorthi, by crosses; and (3) Malvasia di Sardegna Rosada, by mutation. The rest of the candidate varieties to be part of the 'Malvasia' family are dismissed because they result from crosses with members of the Muscat family or crosses with other varieties (known or unknown) that, in any case, are not related historically, genetically, or geographically (with the exception of Malvasia istriana and Malvasia Župska) to the hypothetical members of the 'Malvasia' family.

Keywords: *Vitis vinifera* L.; SSR; microsatellite; true Malvasia; muscat; origin; synonym

1. Introduction

At present, the global wine industry is confronted with a number of unresolved questions regarding grape varieties that bear the term 'Malvasia' in their names. These aspects range from their unknown geographical origin to whether all the varieties identified with this term are truly 'Malvasia' [1]. Currently, there are hundreds of varieties that use the term 'Malvasia' (or one of its variants in different languages), in either their main or synonym names [2,3]. Conversely, in today's market, the perception of quality is often linked to the prestige of a specific variety, taking into account both its intrinsic attributes and its historical and cultural relevance. For this reason, it would be highly interesting to restrict the term 'Malvasia' to a smaller number of varieties that meet specific conditions [4].

It appears that in 1214, a written document first mentioned a wine called Malvasia (Monovasia or Monemvasias). The term Malvasia derives from 'Monemvasia', an old commercial port in Greece, which at that time belonged to the Serenissima Republic of Venice (10th–18th centuries). Most of the commercial activity of the Mediterranean Basin during this period was controlled by Venetian merchants, who promoted a sweet and aromatic wine called 'Vinum de Malvasias' (in 1278). The success of this wine and its high demand led Venetian merchants to expand the cultivation of its variety/varieties to different areas of the eastern Mediterranean, and later to the rest of the Mediterranean Basin, eventually reaching the Atlantic [1,5,6]. An example of the expansion of one of the best-known Malvasias is presented in Figure 1, specifically the case of Malvasia Dubrovacka [7]. Moreover, the first written record referencing this variety dates was found to be from 1385 (Archive of the Republic of Dubrovnik, Croatia) [1].

Figure 1. Expansion of the Malvasia Dubrovacka variety during the 14th–18th centuries, from the eastern Mediterranean Basin to the Atlantic Ocean (map created by Crespan M., 2010) [7].

Another indication of the renown of Malvasia is the fact that the term has been translated (adapted) into different languages, according Vitis International Variety Catalogue (VIVC). The grape variety is known by a number of names, including Malvazija in Croatian, Malvasia or Mal-vagia in Italian, Malvasía in Spanish, Malvoisie or Malvasie in French, Malmsey in English-speaking countries, and Malvasier in German-speaking countries, along with other spellings like Malvagia, Malvasika, and Malvasiya [4].

It is widely accepted among experts in the field that the name 'Malvasia' derives from a port city in present-day Greece, and that the origin of the Malvasia variety or varieties lies in the eastern Mediterranean Basin. Nevertheless, the precise geographical origin of this variety, and indeed of any varieties, remains to be ascertained. The location of the site is uncertain, but it may be along the Balkan coasts of the Adriatic Sea, in today's Greece, or in both. It is also known that grapevines of Balkan origin (including present-day Greece) emerged about 8070 years ago from a wild vine that colonized the eastern Mediterranean Basin (with an introgression from the western wild grapevine), and that these were small-berried vines intended for winemaking. Before them (around 10,500 years ago), also originating from the same wild grapevine, the Muscat family emerged, with large berries more suitable for fresh consumption. The Muscat family's origin is in Anatolia (modern-day Turkey) and nearby areas of Asia Minor (northern Mesopotamia (Iraq) and northern Persia (Iran)) [8]. These regions were inhabited by more advanced civilizations that had already initiated viticulture.

Other major mysteries include knowing (1) how many 'Malvasia' varieties exist and (2) whether all the varieties bearing the name 'Malvasia' are truly 'Malvasia'. A major problem with the term Malvasia is its frequent use alongside the names of other cultivars, such as Torrontés, Tinta, Tintilla, Ugni blac etc., which constitutes a clear case of homonymy. A homonym is a name used, knowingly or not, to refer to different grapevine varieties. This obviously leads to errors and confusion. In the case of Spanish 'Malvasia', different research groups were already raising this problem in 2002. Thus, J. Borrego et al. published a study titled 'Genetic Study of Malvasia and Torrontes Groups through Molecular Markers' [9] and in 2009, a study titled 'Synonyms and homonyms of 'Malvasía' cultivars (*Vitis vinifera* L.) existing in Spain' by I. Rodríguez-Torres et al. was published [10]. In any case, these authors highlighted that the term 'Malvasia' was a homonym, i.e., there were varieties called 'Malvasia' that were actually different varieties, such as the Spanish Alarije or Cayetana blanca or the French Chasselas blanc, among others. Another conclusion from this study is that Spain grows two true 'Malvasia' varieties—Malvasia aromatica (Prime Name (PN): Malvasia Dubrovacka), Malvasía de Lanzarote (PN: Malvasia volcanica)—as well as a pink mutation of Malvasia Dubrovacka, known as Malvasia rosada. Therefore, according to this study, the remaining Spanish 'Malvasia' varieties would not be considered true 'Malvasia' by these authors.

At present, several Mediterranean countries and their outermost regions produce 'Malvasia' wines (especially Croatia, Greece, Italy, Spain, and Portugal. In addition there are two Russian 'Malvasia' varieties, as well as one Argentine and one French. In reality, due to the fame and popularity associated with the name 'Malvasia', the term has been used throughout history and in various parts of the world to name very different varieties [1,4].

The present theoretical–bibliographic study on the subject of 'Malvasia' worldwide is intended to identify any potential evidence that may assist in resolving the aforementioned uncertainties concerning the global 'Malvasia' group. For the purpose of this study, data were drawn from the Vitis International Variety Catalogue (VIVC), which is recognized as the world's most extensive database [2,3]. In order to achieve the objectives of the study, both the ampelographic section and the molecular profiles section were consulted, with the relevant microsatellites or SSR (Simple Sequence Repeats) being used.

2. Materials and Methods

2.1. The Term 'Malvasia' in the Cultivar and Prime Names

The study of the term 'Malvasia' and its derived spellings began by consulting these terms in the ampelographic section of the VIVC (Table S1). The different names by which this family is known in various languages were entered. The "Cultivar Name" tab was used first, and then the operation was repeated in the "Prime Name" tab.

2.2. SSR or Microsatellites Used

The molecular study on the genetic profiles of the varieties that make up the 'Malvasia' was based on the SSR or microsatellite technique. The 9 internationally recognized SSR markers were used, i.e., those accepted as reference SSR markers by the global scientific community: VVS2 [11] VVMD5, VVMD7, VVMD28 [12], VVMD25, VVMD27, VVMD32 [13], VrZAG62, VrZAG79 [14] (Table S2).

The genetic profiles of these 9 SSRs were obtained from the VIVC database, as it holds the largest number of genetic profiles for 'Malvasia' varieties [2,3]. The study started with all the Molecular Profiles (SSR-MPs) of 'Malvasia' stored in this database, although not all accessions found in the ampelographic section had SSR-MPs.

2.3. Data Analysis

The population structure of 'Malvasia' was studied using the program Structure 2.3 [15,16], a clustering method based on Bayesian models. In this model, it is assumed that several ancestral populations (K) are present, each characterized by a set of allele frequencies at each locus. The individuals in the sample are assigned to specific populations (clusters), or to multiple populations if their genotypes suggest admixture. All loci are assumed to be independent, and each population K is in Hardy–Weinberg equilibrium. Posterior probabilities were estimated using the Markov Chain Monte Carlo (MCMC) method. The MCMC chains were run with a burn-in period of 100,000 iterations, followed by 1,000,000 iterations, using a model that allows for admixture and correlated allele frequencies. Structure was run at least ten times, configuring K from 1 to 7, and a mean likelihood value, L(K), was calculated across all runs for each K. The mean log-likelihood of the data was calculated for each K to determine the most appropriate number of clusters and the K value for which this probability was highest. Then, ΔK was calculated using the method described by Evanno et al. [17]. ΔK is a quantity based on the rate of change and in the log probability of the data between successive K values. The parameter q defines what proportion of an individual's genome belongs to each predefined cluster (K). Membership in a cluster was accepted for mean q values ≥ 0.74.

The program GenAlEx 6.5 was used for two purposes [18,19]. First, it allowed the performance of assignment tests to evaluate the goodness of fit of a given sample distribution into different populations. GenAlEx 6.5 bases this strategy on the allele frequency of each accession. It calculates a log-likelihood value for this accession for each subpopulation using the allele frequencies of the respective subpopulations. An individual is assigned to the subpopulation with the highest log-likelihood value [20]. Second, based on the standardized covariance of the genetic distances calculated for codominant markers, GenAlEx 6.5 allowed us to generate two-dimensional Principal Coordinate (PCoA) graphical representations for a set of populations and for a set of individuals from different populations.

The programs Python Data [21], applying the Matplotlib strategy, and MEGA version 7 [22], using the Neighbor-Joining strategy [23], were used to create three-dimensional PCoA plots and phylogenetic trees, respectively.

3. Results

3.1. The Names of the Term 'Malvasia'

To begin this study, it was first necessary to determine how many varieties used the term 'Malvasia' and/or its derived spellings in their primary names or synonyms. Twelve different terms appeared (Table 1). For this, the ampelographic section of the VIVC was consulted. Specifically, each of the spellings was entered into the "Cultivar Name" field (where all accession names of a given variety that include the entered spelling are stored): (1) MALVASIA (Spanish), with 290 items; (2) MALVOISIE (French), with 52 items; (3) MALVASIER (English), with 15 items; (4) MALVAZIYA (Russian), with 14 items; (5) MALVAZIJA (Croatian), with 13 items; (6) MALVAZIA (English), with 11 items; (7) MALMSEY (English), with 6 items; (8) MALVAGIA (Italian), with 4 items; (9) MALVASIJA (Latvian), with 4 items; (10) MALVASIE (French), with 2 items; (11) MALVASIKA (Croatian), with 1 item; and (12) MALVASIYA (?), with 1 item. In total, 413 accession names were found that included the term 'Malvasia' or its derived spellings, in all possible combinations (a single name or combination of several names).

Table 1. No. of cultivars names and PNs presenting the 12 derived spellings of the term 'Malvasia'.

NAME	CULTIVARS NAME (No. of Accesions)	PRINCIPAL NAME (No. of Accessions)
MALVASIA	290	203
MALVOISIE	52	13
MALVASIER	15	3
MALVAZIYA	14	8
MALVAZIJA	13	9
MALVAZIA	11	2
MALMSEY	6	3
MALVAGIA	4	4
MALVASIJA	4	4
MALVASIE	2	0
MALVASIKA	1	1
MALVASIYA	1	0
TOTAL No. OF NAMES	413	250

Then, the next step was to enter the previously used spellings into the "Prime Name" field in order to select only those terms where the word "Malvasia" or its derived spellings appeared. Thus, the following results were obtained: (1) MALVASIA was reduced to 203 prime names; (2) MALVOISIE showed only 13 items; (3) MALVASIER had 3 items; (4) MALVAZIYA had 7 items; (5) MALVAZIJA had 9 items; (6) MALVAZIA had 2 items; (7) MALMSEY had 3 items; (8) MALVAGIA had 4 items; (9) MALVASIJA had 4 items; (10) MALVASIE had 0 items; (11) MALVASIKA had 1 item; and (12) MALVASIYA had 0 items. The final count of PNs including the term "Malvasia" was reduced to 250.

Finally, the number of unique PNs including the term "Malvasia" or one of its derived spellings was narrowed down, definitively, to 68 unique and cataloged varieties in the ampelographic section of the VIVC (Supplementary Material 1 (Table S1)). These, classified by their origin (Table S3), correspond to (a) 14 without an assigned country of origin (although the institution that holds the sample described in the ampelographic section of the VIVC is reported); (b) 1 Argentinian; (c) 1 French; (d) 2 Croatian; (e) 2 Russian; (f) 9 Spanish; (g) 11 Portuguese; and (h) 28 Italian.

3.2. *Genetic Structure of the Population of the "Malvasia" Group Varieties (Genetic Strategy)*

Once the 68 unique varieties that included the term 'Malvasia' in their prime name were identified, their molecular profiles (SSR-MPs) were obtained from the SSR or Microsatellite section of the VIVC. Only 43 SSR-MPs (Table S4) out of the 68 described varieties were available, leaving 25 varieties excluded. The main characteristics and genetic profiles for the 9 published SSRs of these 43 SSR-MPs are shown in Table S1.

To carry out the study of the population structure of the group formed by the unique SSR-MP 'Malvasia' varieties efficiently and effectively, data normalization was performed (Table S1). First, 'sports' were eliminated. Thus, four prime names and their corresponding

SSR-MPs were excluded from this study: Malvasia di Sardegna rosada, Malvasia fina roxa, Malvasia preta roxa, and Malvasia rosa. Then, varieties resulting from directed crosses by a breeder were excluded; this was the case of the varieties Malvasia branca de Sao Jorge and Malvasia moscatel Fonte grande. Finally, four varieties were excluded whose natural crosses were known and whose parents (and their crosses) corresponded to known varieties with no relevant aroma and unrelated to either the 'Malvasia' group or the Muscat group. These were (1) Malvasia (Malvasia de Colares); (2) Malvasia fina; (3) Malvasia preta; and (4) Malvoisie de madeira. Therefore, the population structure study of the 'Malvasia' group used 33 SSR-MPs. Varieties for which the cross or one of the parents was unknown were not excluded.

To achieve the best distribution of this population of 33 individuals into different ancestral populations (K), the program Structure 2.3 was used. Population distributions from one to seven were tested, with the best result being the distribution into three populations (K = 3) (Figure S1). Each individual in the 'Malvasia' group population was assigned to one of these three proposed populations based on a statistical parameter q (which indicates the percentage of its inferred genome that belongs to one of these populations). In our study, an arbitrary threshold of 74% was chosen [24], such that values of $q \geq 74\%$ corresponded to pure individuals belonging to a given population, while those with $q < 74\%$ were considered admixed individuals for that population and were excluded from further analysis. Figure 2 shows the best distribution for the 'Malvasia' group population into three populations (POP1, POP2, POP3), with the varieties ordered according to the q parameter (Table S5).

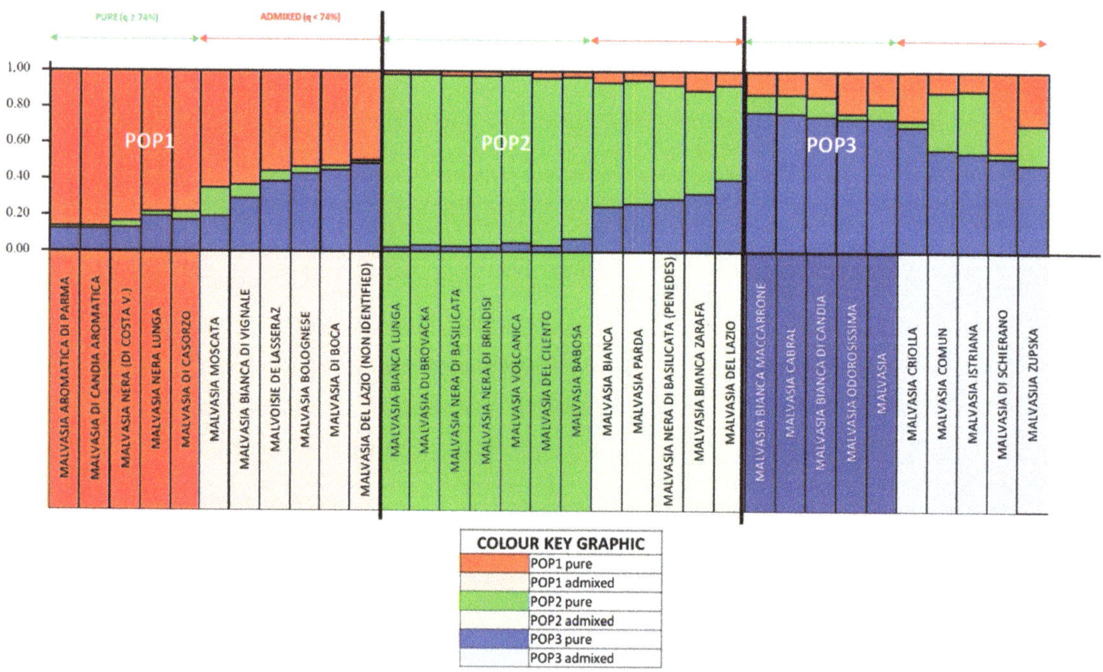

Figure 2. 'Malvasia' grapevine varieties population (unique molecular profiles). Structure 2.3. diagram: K = 3 distribution for pure and admixed individuals.

Table 2 shows the most relevant information for each of the 33 members of this group, following the same classification criteria defined by the Structure 2.3 program.

Table 2. Relevant information (VIVC) on the main characteristics of each of the 33 members of the "Malvasia" group analyzed in this study, distributed into three populations and ordered according to the q parameter (Table S5).

GENERAL INFORMATION								PEDIGREE	
PRIME NAME	COUNTRY ORIGIN	CHLOROTYP	SEED FOR-MATION	SEX	TASTE	COLOR	USE	PARENT 1	PARENT 2
MALVASIA AROMATICA DI PARMA	FRA139 /MEX062/USA 06/USA028	D	COMPLETE	FEMALE	MUSCAT	W	W	MUSCAT A PETITS GRAINS BLANCS	NEBBIOLO x BOTTAGERA (FALSE)
MALVASIA DI CANDIA AROMATICA	ITALY	D	COMPLETE	HERMAFRODITE	MUSCAT	W	T/W	MALVASIA AROMATICA DI PARMA (MUSCAT A PETITS GRAINS BLANCS x NEBBIOLO X BOTTAGERA (FALSE))	?
MALVASIA NERA (DI COSTA V.)	ITALY	D	COMPLETE	HERMAFRODITE	MUSCAT	B	?	MALVASIA AROMATICA DI PARMA (MUSCAT A PETITS GRAINS BLANCS x NEBBIOLO X BOTTAGERA (FALSE))	LAMBRUSCA DI ALESSANDRIA (CROVIN x NERETTO DI MARENGO (LAMBRUSCA DI ALESSANDRIA x ?))
MALVASIA NERA LUNGA	ITALY	D	COMPLETE	HERMAFRODITE	ARMATIC	B	W	MALVASIA AROMATICA DI PARMA (MUSCAT A PETITS GRAINS BLANCS x NEBBIOLO X BOTTAGERA (FALSE))	FREISA (NEBBIOLO x ?)
MALVASIA DI CASORZO	ITALY	D	COMPLETE	HERMAFRODITE	AROMATIC	B	W	MALVASIA AROMATICA DI PARMA (MUSCAT A PETITS GRAINS BLANCS x NEBBIOLO X BOTTAGERA (FALSE))	LAMBRUSCA DI ALESSANDRIA (CROVIN x NERETTO DI MARENGO (LAMBRUSCA DI ALESSANDRIA x ?))
MALVASIA MOSCATA	ITALY	D	COMPLETE	HERMAFRODIT	MUSCAT	W	W	MALVASIA AROMATICA DI PARMA (MUSCAT A PETITS GRAINS BLANCS x NEBBIOLO X BOTTAGERA (FALSE))	?

Table 2. *Cont.*

GENERAL INFORMATION								PEDIGREE	
PRIME NAME	COUNTRY ORIGIN	CHLOROTYP	SEED FORMATION	SEX	TASTE	COLOR	USE	PARENT 1	PARENT 2
MALVASIA BIANCA DI VIGNALE	ITALY	D	COMPLETE	HERMAFRODITI	AROMATIC	W	W	MALVASIA AROMATICA DI PARMA (MUSCAT A PETITS GRAINS BLANCS x NEBBIOLO X BOTTAGERA (FALSE))	COCCALONA BIANCA
MALVOISIE DE LASSERAZ	FRANCE	?	COMPLETE	FEMALE	NONE	W	W	?	?
MALVASIA BOLOGNESE	ITALY	?	COMPLETE	?	?	W	W	?	?
MALVASIA DI BOCA	ITALY	?	COMPLETE	HERMAFRODITI	AROMATIC	W	W	?	?
MALVASIA DEL LAZIO (NON-IDENTIFIED)	ITALY	?	COMPLETE	HERMAFRODITI	MUSCAT	W	W	MUSCAT OF ALEXANDRIA (MUSCAT A PETITS GRAINS BLANCS x HEPTALIKO)	SCHIAVA GROSSA
MALVASIA BIANCA LUNGA	ITALY	D	COMPLETE	HERMAFRODITE	NONE	W	W	?	?
MALVASIA DUBROVACKA	SPAIN	A	COMPLETE	HERMAFRODITE	AROMATIC	W	W	?	?
MALVASIA NERA DI BASILICATA	ITALY	?	COMPLETE	HERMAFRODITE	ARMATIC	B	W	MALVASIA BIANCA LUNGA	SOMARELLO NERO (UVA SACRA (ACHLADI x ?) x GARGANEGA)
MALVASIA NERA DI BRINDISI	ITALY	?	COMPLETE	HERMAFRODITE	NONE	B	T/W	MALVASIA BIANCA LUNGA	NEGRO AMARO (? x MAIOLICA (? x VISPAROLA))
MALVASIA VOLCANICA	SPAIN	?	COMPLETE	HERMAFRODITE	OTHER FLAVOR THAN MUSCAT, FOXY OR HERBACEOUS	W	W	MALVASIA DUBROVACKA	BERMEJUELA

Table 2. *Cont.*

	GENERAL INFORMATION							PEDIGREE	
PRIME NAME	COUNTRY ORIGIN	CHLOROTYP	SEED FORMATION	SEX	TASTE	COLOR	USE	PARENT 1	PARENT 2
MALVASIA DEL CILENTO	ITALY	?	?	?	?	B	W	?	?
MALVASIA BABOSA	PORTUGAL	A	COMPLETE	?	?	W	W	HEBEN	MALVASIA DUBROVACKA
MALVASIA BIANCA	ITALY	?	COMPLETE	HERMAFRODITI	NONE	W	W	SCIACCARELLO	?
MALVASIA PARDA	PORTUGAL	?	COMPLETE	HERMAFRODITI	NONE	W	W	?	?
MALVASIA NERA DI BASILICATA (PENEDES)	SPAIN	?	COMPLETE	?	NONE	B	W	?	?
MALVASIA BIANCA ZARAFA	ITALY	?	COMPLETE	?	MUSCAT	W	W	?	?
MALVASIA DEL LAZIO	ITALY	?	COMPLETE	HERMAFRODITI	NONE	W	W	?	?
MALVASIA BIANCA MACCARRONE	ITALY	?	COMPLETE	?	MUSCAT	W	W	?	?
MALVASIA CABRAL	PRT051	?	COMPLETE	HERMAFRODITE	NONE	Rs	W	?	?
MALVASIA BIANCA DI CANDIA	ITALY	D	COMPLETE	HERMAFRODITE	NONE	W	W	GARGANEGA	?
MALVASIA ODOROSISSIMA	ITALY	?	COMPLETE	FEMALE	MUSCAT	W	W	?	?
MALVASIA	ITALY	?	COMPLETE	?	?	B	W	?	?

Table 2. *Cont.*

GENERAL INFORMATION								PEDIGREE	
PRIME NAME	COUNTRY ORIGIN	CHLOROTYP	SEED FORMATION	SEX	TASTE	COLOR	USE	PARENT 1	PARENT 2
MALVASIA CRIOLLA	ARGENTINA	?	COMPLETE	HERMAFRODITI	MUSCAT	W	W	LISTAN PRIETO	MUSCAT OF ALEXANDRIA (MUSCAT A PETITS GRAINS BLANCS x HEPTALIKO)
MALVASIA COMUN	SPAIN	A	COMPLETE	HERMAFRODITI	NONE	W	W	HEBEN	?
MALVASIA ISTRIANA	CROATIA	D	COMPLETE	HERMAFRODITI	NONE	W	W	?	?
MALVASIA DI SCHIERANO	ITALY	D	COMPLETE	HERMAFRODITI	MUSCAT	B	W	MUSCAT A PETITS GRAINS BLANCS	?
MALVASIJA ZUPSKA	CROATIA	D	COMPLETE	?	?	W	W	HEUNISCH WEISS	?

1: Red; POP1 pure; light red: POP1 admixed; green: POP2 pure; light green: POP2 admixed; blue: POP3 pure; light blue: POP3 admixed. 2: x is the crossing (pedigree).

POP1 is composed of 11 varieties (33.33%), of which 5 are pure and the remaining 6 are admixed. Virtually all members of this cluster population are Italian varieties (except for Malvasia aromatica di Parma (of unknown origin) and Malvasia Lasseraz (French)), white (except for Malvasia nera lunga, which is pure), and intended for winemaking (except for Malvasia di Candia aromatica, which is assigned a dual purpose). Most have hermaphroditic flowers (except for Malvasia aromatica di Parma (Italian) and Malvasia Lasseraz (French) which are female). All known chlorotypes are either D or unknown, and all known crosses involve a member of the Muscat family. Most of its components exhibit a muscat or aromatic aroma.

POP2 consists of 12 components (36.36%), of which 7 are considered pure varieties and 5 are admixed, originating from the Italian and Iberian Peninsulas (7 Italian, 3 Spanish, and 2 Portuguese). This group is characterized by having most individuals with an unknown chlorotype, except for 2 chlorotype A (Malvasia Dubrovacka and Malvasia Babosa) and 1 chlorotype D (Malvasia bianca lunga). Among its 12 components, 4 are red varieties, and only Malvasia nera di Brindisi (also widely known as Malvasia di Lecce) is intended for dual use (table/winemaking); the rest are for winemaking. Most are hermaphroditic, and for the remaining ones, the floral sex has not been described. There are representatives with aromatic, non-aromatic, and unknown aroma profiles, and one with a muscat aroma (Malvasia Zarafa). Four complete crosses are known in which either Malvasia Dubrovacka or Malvasia bianca lunga is involved, and one incomplete cross: Malvasia bianca (with only one known parent, Sciaccarello). The parentage of the remaining varieties is unknown.

POP3 includes 10 components (30.30%), 50% of which are pure and the remaining 50% admixed. Most are Italian, although there are also two varieties from Croatia, one Spanish, one from Argentina, and one of unknown origin. Chlorotype D predominates, although there are also members with unknown chlorotypes and one specimen with chlorotype A (Malvasia común). Most are hermaphroditic; three have flowers of unknown sex, and one variety, Malvasia odorosissima, has female flowers. Four members have been described as having a muscat aroma, four more are non-aromatic, and for the remaining two, no information on this trait is available. All are intended for winemaking and are white, except for Malvasia and Malvasia di Schierano, which are red, and Malvasia Cabral, which is pink. Regarding their crosses, it is worth noting that only one complete cross is known, that of Malvasia criolla; for four crosses, only one parent is known; and for the remaining five, both parents are unknown. Two of the crosses involve Muscat varieties, and three incomplete crosses involve two Italian varieties and one Spanish.

As previously mentioned, the admixed members were removed, and the 'Malvasia' population was reduced to 17 components, distributed across three populations, with a goodness of fit of 82% for each population (Figure S2). The members of each population were then reassigned until a goodness of fit of 100% was achieved (Figure S3). In this process, the distribution of the members of the 'Malvasia' group was reduced to two populations. Figure 3 shows the distribution of the 'Malvasia' population through Principal Coordinates Analysis (PCoA) in two and three dimensions, in addition to the phylogenetic tree derived from this distribution. In Figure 3a, corresponding to the two-dimensional PCoA plot (with a goodness of fit of 35.66%), it can be seen that Coordinate 1, with a goodness of fit of 21.47%, separates the two populations, placing POP1 in the right quadrants and POP2 in the left quadrants. Looking at both populations as a whole, it can be observed that they are mostly arranged around the axis of Coordinate 2 (with some exceptions). This coordinate, which has a goodness of fit of 14.19%, slightly divides both POP1 and POP2 into two subpopulations each. In the upper quadrant of POP1 (right side), the varieties Malvasia nera lunga and Malvasia di Candia aromatica are located, while in the lower quadrant, the remaining four varieties are found (Malvasia aromatica di Parma,

Malvasia odorosissima (originating from POP3), Malvasia nera (di Costa v.), and Malvasia di Casorzo). In the upper left quadrant of POP2, the most distant variety is Malvasia Cabral; closer to the center and near the axis of Coordinate 1, we find Malvasia, Malvasia bianca di Candia, and further down, Malvasia bianca Maccarrone, all of them coming from POP3. In the more central and lower part of this quadrant, Malvasia Dubrovacka and its progeny (Malvasia volcánica and Malvasia Babosa) can be observed. In contrast, Malvasia bianca lunga and its progeny, along with Malvasia del Cilento, are located in the lower left quadrant. The three-dimensional PCoA plot shows an explanatory goodness of fit of 45.71% (Figure 3b). In this plot, with slight variations, the same pattern described above is repeated. Finally, the phylogenetic tree (Figure 3c) clearly shows three branches emerging from the root. In red, the branch corresponding to POP1 includes the five previously described components, highlighting the split of Malvasia odorosissima (originating from POP3). POP2 (in lime green) and its initial components remain grouped on another branch (on the right side of the tree), and with the same color, but now located on the left branch of the phylogenetic tree, the four 'Malvasia' varieties from the former POP3 are grouped together.

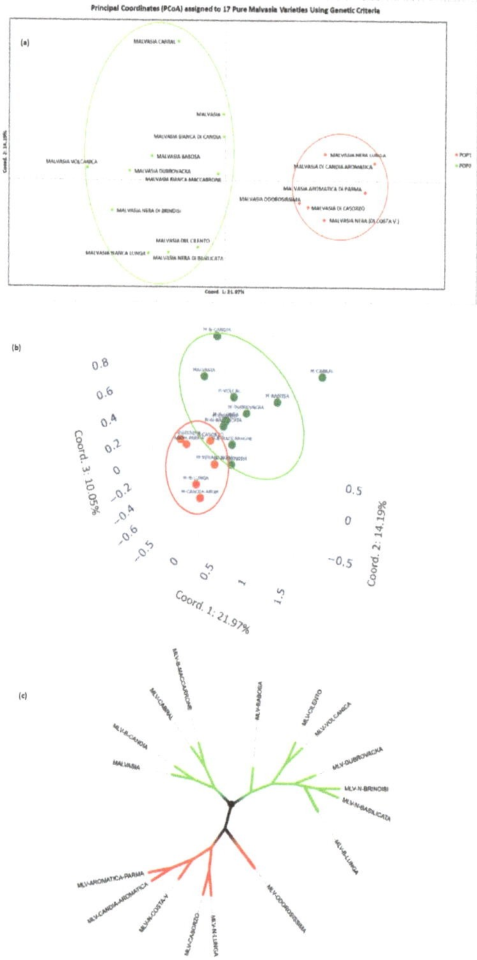

Figure 3. Graphical representation of the distribution of the 17 pure varieties ($q \geq 74\%$), using PCoA of their individuals in two (**a**) and three (**b**) dimensions, and also through the representation of a phylogenetic tree (**c**).

In Figure 4, a map is shown depicting the approximate geographical locations assigned to the 17 pure varieties analyzed in this genetic study. It can be observed that POP1 is located in the northern regions of the Italian Peninsula. In contrast, POP2 encompasses the Balkan areas bordering the Adriatic Sea, as well as the southern half of Italy, parts of Portugal, and the outermost regions of Spain and Portugal (Canary Islands and Madeira Archipelago) [5,25,26].

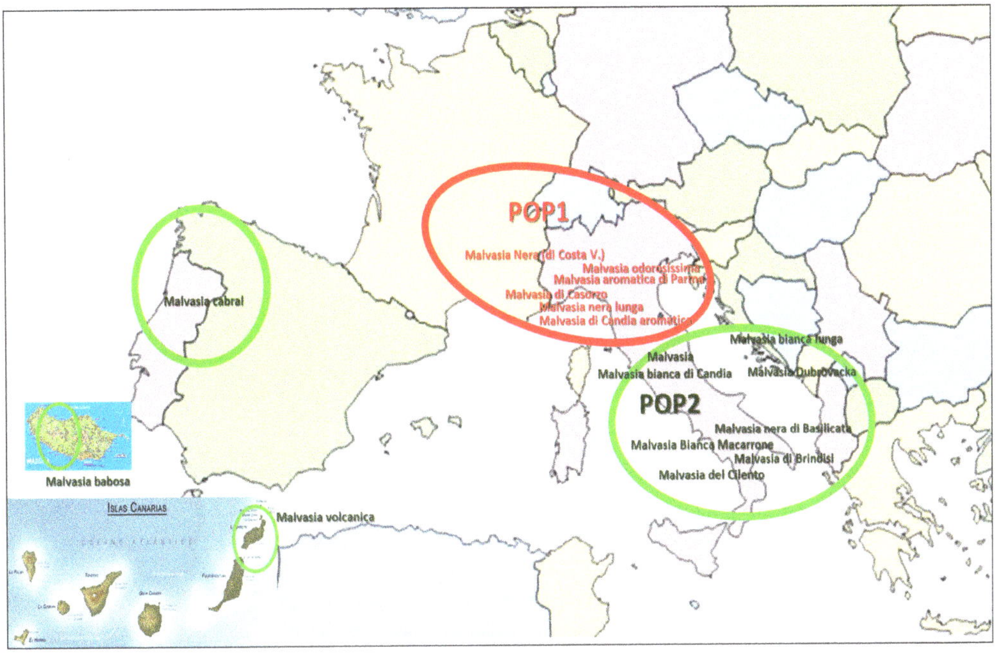

Figure 4. Map of the approximate geographical location ([5,25–27]) of the 17 pure 'Malvasia' varieties. Genetic strategy. Goodness of fit: 100%.

3.3. Distribution of the 'Malvasia' Variety Group Population According to Geographical Criteria

In this section, a geographical criterion was applied in order to observe the composition of the populations and the membership of the groups in question. Evidently, a genetic element could not be disregarded, as it would contribute to the final organization of the populations resulting from this new arrangement.

The process was initiated with the 33 unique and non-redundant profiles of varieties that could potentially be candidates to form part of the 'Malvasia' group. Initially, two populations were created based on whether the variety originated from the Italian Peninsula (POP1) or its surroundings, or from the Iberian Peninsula (POP2) and its areas of influence. Table S6 presents this initial distribution according to geographical criteria (origin as described in the VIVC), and some important trends were already observable. For instance, when the chlorotype of a variety was known, it was type D in POP1, while in POP2, it was type A. Another noteworthy detail was the presence of Muscat family members as progenitors in most of the crosses of varieties from the POP1 area (for this reason, the Argentine variety Malvasia criolla was placed in this group). This distribution showed a goodness of fit of 76%. No misassigned individuals were excluded, as the goal was to observe their behavior; instead, they were reassigned until a 100% goodness of fit was achieved. Table 3 shows the final distribution obtained. In the 'Prime Name' column, the final distribution of the 'Malvasia' group is presented, with varieties assigned to POP1 in red and those assigned to POP2 in green. In the 'Country Origin' column, the initial distribution is shown (with a goodness of fit of 76%). POP1, which previously had 26 components, now has 17, and POP2 increases by 9 components, reaching 16 members. Regarding chlorotypes, in POP1, the known chlorotypes remain D, while in POP2, which has seen a significant incorporation of varieties from POP1 (Italy and nearby areas), there are three A-type chlorotypes (those initially present) and two D-type chlorotypes. The crosses follow the same distribution pattern; varieties whose progenitors belong to the Muscat family are in POP1, and in POP2 there is no evidence of Muscat parentage. However, two distinct lineages can be observed in POP2: one related to Malvasia Dubrovacka and the other to Malvasia bianca lunga and their respective progenies. Regarding aromas, POP1 includes both muscat-flavored varieties (which are the most abundant) and varieties defined as aromatic. In POP2, aromatic varieties are also observed, along with two muscat-flavored ones, while the rest are either neutral or of unknown aroma.

The graphical representation of the PCoA in two and three dimensions, along with the corresponding phylogenetic tree, is presented in Figure 5. In Figure 5a, the distribution of the "Malvasia" population by individuals in two dimensions is shown using PCoA. This representation has a goodness of fit of 24.47%. Once again, Coordinate 1, with 13.63% of goodness of fit, separates most individuals whose crosses involve a member of the Muscat family (group of 'Malvasia' from northern Italy) from the group led by the 'Malvasia' varieties from the Iberian Peninsula, the southern half of the Italian Peninsula, and the Balkan coasts of the Adriatic Sea (Figure 6). Coordinate 2, with 10.84% of goodness of fit, does not play a very defined role in the distribution of the two populations. In this representation, it is worth highlighting how Malvasia, Malvasia bianca di Candia, and Malvasia Zupska, which belong to POP1 (red), are positioned on the right quadrant (near the axis), together with members of POP2. The three-dimensional representation, with a goodness of fit of 32.63% (Figure 5b), does not bring major new insights either. However, it does show that POP1 is positioned above POP2, which appears in the lower part of the graph. It also highlights the ' Malvasia' varieties that show a more distant SSR profile compared to the rest. These include Malvasia cabral, Malvasia babosa, Malvasia istriana, Malvasia nera di Basilicata (from Penedès), Malvasia di Boca, Malvasia di Schierano, Malvasia de Zarafa, and Malvasia moscata. In Figure 5c, we present the distribution adopted by the 'Malvasia' population for the two initial populations with a 100% goodness of fit. This tree is divided into three branches from the point of origin, as POP2 splits into two sub-branches. Another noteworthy aspect is the transfer of Malvasia istriana (originally from Croatia) into POP1.

Table 3. Final distribution of 'Malvasia' varieties based on their origin as described in the VIVC. Goodness of fit of 100% (Prime Name column). No misassigned varieties were eliminated; they were reassigned.

GENERAL INFORMATION								PEDIGREE	
PRIME NAME	COUNTRY ORIGIN	CHLOROTY	SEED FOR-MATION	SEX	TASTE	COLOR	USE	PARENT 1	PARENT 2
MALVASIA AROMATICA DI PARMA	ITA	D	COMPLETE	FEMALE	MUSCAT	W	W	MUSCAT A PETITS GRAINS BLANCS	NEBBIOLO x BOTTAGERA (FALSE)
MALVASIA DI CANDIA AROMATICA	ITA	D	COMPLETE	HERMAFRODIT	MUSCAT	W	T/W	MALVASIA AROMATICA DI PARMA (MUSCAT A PETITS GRAINS BLANCS x NEBBIOLO X BOTTAGERA (FALSE))	?
MALVASIA NERA LUNGA	ITA	D	COMPLETE	HERMAFRODIT	ARMATIC	B	W	MALVASIA AROMATICA DI PARMA (MUSCAT A PETITS GRAINS BLANCS x NEBBIOLO X BOTTAGERA (FALSE))	FREISA (NEBBIOLO x ?)
MALVASIA BIANCA DI VIGNALE	ITA	D	COMPLETE	HERMAFRODIT	AROMATIC	W	W	MALVASIA AROMATICA DI PARMA (MUSCAT A PETITS GRAINS BLANCS x NEBBIOLO X BOTTAGERA (FALSE))	COCCALONA BIANCA
MALVASIA DI CASORZO	ITA	D	COMPLETE	HERMAFRODIT	AROMATIC	B	W	MALVASIA AROMATICA DI PARMA (MUSCAT A PETITS GRAINS BLANCS x NEBBIOLO X BOTTAGERA (FALSE))	LAMBRUSCA DI ALESSANDRIA (CROVIN x NERETTO DI MARENGO (LAMBRUSCA DI ALESSANDRIA x ?))
MALVASIA NERA (DI COSTA V.)	ITA	D	COMPLETE	HERMAFRODIT	MUSCAT	W	?	MALVASIA AROMATICA DI PARMA (MUSCAT A PETITS GRAINS BLANCS x NEBBIOLO X BOTTAGERA (FALSE))	LAMBRUSCA DI ALESSANDRIA (CROVIN x NERETTO DI MARENGO (LAMBRUSCA DI ALESSANDRIA x ?))

Table 3. *Cont.*

GENERAL INFORMATION								PEDIGREE	
PRIME NAME	COUNTRY ORIGIN	CHLOROTY	SEED FOR-MATION	SEX	TASTE	COLOR	USE	PARENT 1	PARENT 2
MALVASIA MOSCATA	ITA	D	COMPLETE	HERMAFRODITI	MUSCAT	W	W	MALVASIA AROMATICA DI PARMA (MUSCAT A PETITS GRAINS BLANCS x NEBBIOLO X BOTTAGERA (FALSE))	?
MALVASIA DI SCHIERANO	ITA	D	COMPLETE	HERMAFRODITI	MUSCAT	B	W	MUSCAT A PETITS GRAINS BLANCS	?
MALVASIA DEL LAZIO (NON-IDENTIFIED)	ITA	?	COMPLETE	HERMAFRODITI	MUSCAT	W	W	MUSCAT OF ALEXANDRIA (MUSCAT A PETITS GRAINS BLANCS x HEPTALIKO)	SCHIAVA GROSSA
MALVASIA CRIOLLA	ARG	?	COMPLETE	HERMAFRODITI	MUSCAT	W	W	LISTAN PRIETO	MUSCAT OF ALEXANDRIA (MUSCAT A PETITS GRAINS BLANCS x HEPTALIKO)
MALVASIA BOLOGNESE	ITA	?	COMPLETE	?	?	W	W	?	?
MALVASIA DI BOCA	ITA	?	COMPLETE	HERMAFRODITI	AROMATIC	W	W	?	?
MALVOISIE DE LASSERAZ	FRA	?	COMPLETE	FEMALE	NONE	W	W	?	?
MALVASIA ODOROSISSIMA	ITA	?	COMPLETE	FEMALE	MUSCAT	W	W	?	?
MALVASIJA ZUPSKA	HRV	D	COMPLETE	?	?	W	W	HEUNISCH WEISS	?
MALVASIA	ITA	?	COMPLETE	?	?	B	W	?	?
MALVASIA BIANCA DI CANDIA	ITA	D	COMPLETE	HERMAFRODITI	NONE	W	W	GARGANEGA	?

Table 3. *Cont.*

PRIME NAME	COUNTRY ORIGIN	CHLOROTY	SEED FOR-MATION	SEX	TASTE	COLOR	USE	PARENT 1	PARENT 2
GENERAL INFORMATION								PEDIGREE	
MALVASIA DUBROVACKA	ESP	A	COMPLETE	HERMAFRODITI	AROMATIC	W	W	?	?
MALVASIA BABOSA	PRT	A	COMPLETE	?	?	W	W	HEBEN	MALVASIA DUBROVACKA
MALVASIA VOLCANICA	ESP	?	COMPLETE	HERMAFRODITI	OTHER FLAVOR THAN MUSCAT, FOXY OR HERBA-CEOUS	W	W	MALVASIA DUBROVACKA	BERMEJUELA
MALVASIA BIANCA LUNGA	ITA	D	COMPLETE	HERMAFRODITI	NONE	W	W	?	?
MALVASIA NERA DI BRINDISI	ITA	?	COMPLETE	HERMAFRODITI	NONE	B	T/W	MALVASIA BIANCA LUNGA	NEGRO AMARO (? x MAIOLICA (? x VISPAROLA))
MALVASIA NERA DI BASILICATA	ITA	?	COMPLETE	HERMAFRODITI	ARMATIC	B	W	MALVASIA BIANCA LUNGA	SOMARELLO NERO (UVA SACRA (ACHLADI x ?) x GARGANEGA)
MALVASIA DEL CILENTO	ITA	?	?	?	?	B	W	?	?
MALVASIA ISTRIANA	HRV	D	COMPLETE	HERMAFRODITI	NONE	W	W	?	?
MALVASIA COMUN	ESP	A	COMPLETE	HERMAFRODITI	NONE	W	W	HEBEN	?
MALVASIA BIANCA	ITA	?	COMPLETE		NONE	W	W	SCIACCARELLO	?
MALVASIA BIANCA ZARAFA	ITA	?	COMPLETE	?	MUSCAT	W	W	?	?
MALVASIA PARDA	PRT	?	COMPLETE	HERMAFRODITI	NONE	W	W	?	?
MALVASIA BIANCA MACCARRONE	ITA	?	COMPLETE	?	MUSCAT	W	W	?	?
MALVASIA CABRAL	PRT051	?	COMPLETE	HERMAFRODITI	NONE	Rs	W	?	?
MALVASIA NERA DI BASILICATA (PENEDES)	ESP	?	COMPLETE	?	NONE	B	W	?	?
MALVASIA DEL LAZIO	ITA	?	COMPLETE	HERMAFRODITI	NONE	W	W	?	?

1: Red; POP1 pure; light red: POP1 admixed; green: POP2 pure; light green: POP2 admixed; blue: POP3 pure; light blue: POP3 admixed. 2 x is the crossing (pedigree).

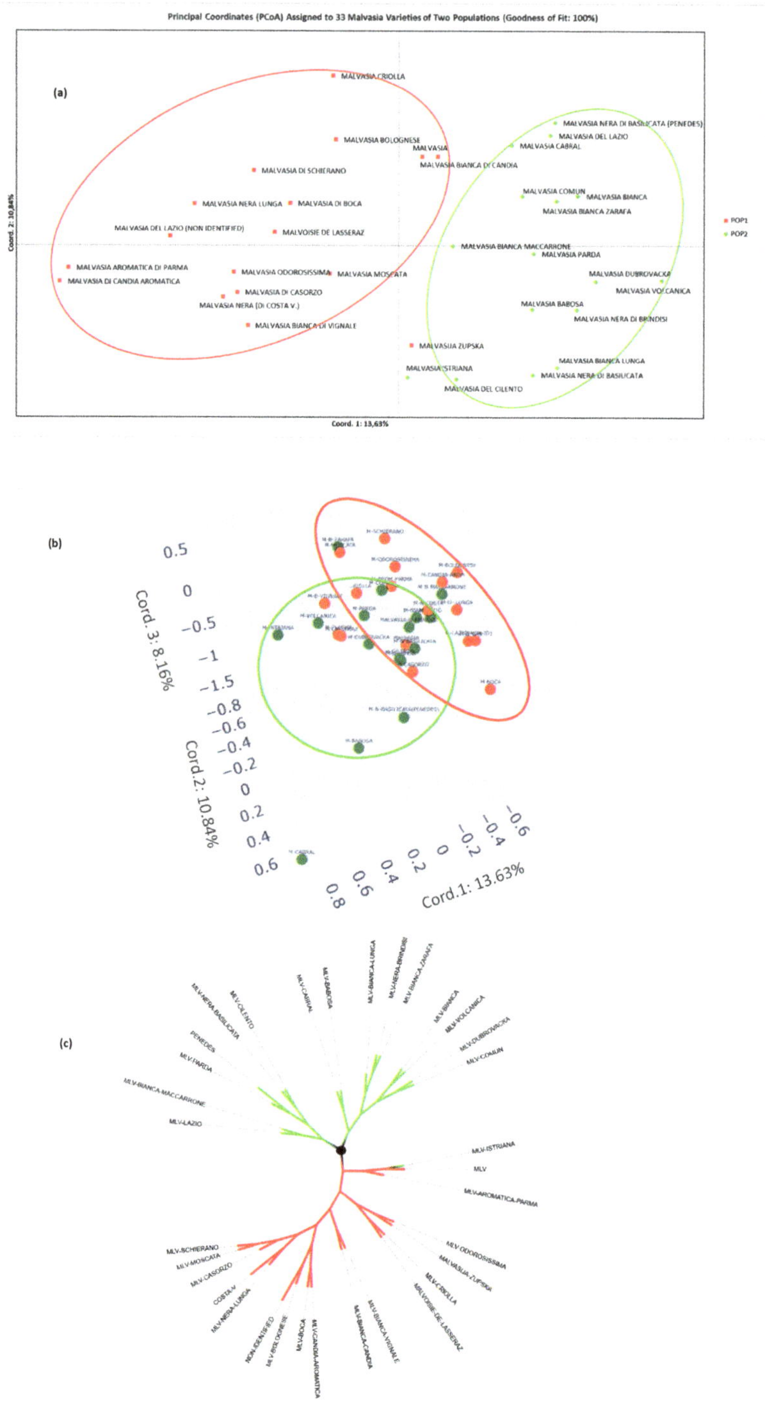

Figure 5. Graphical representation of the distribution of the 33 varieties under the geographical strategy, using PCoA of their individuals in 2D (**a**) and 3D (**b**), as well as a phylogenetic tree (**c**).

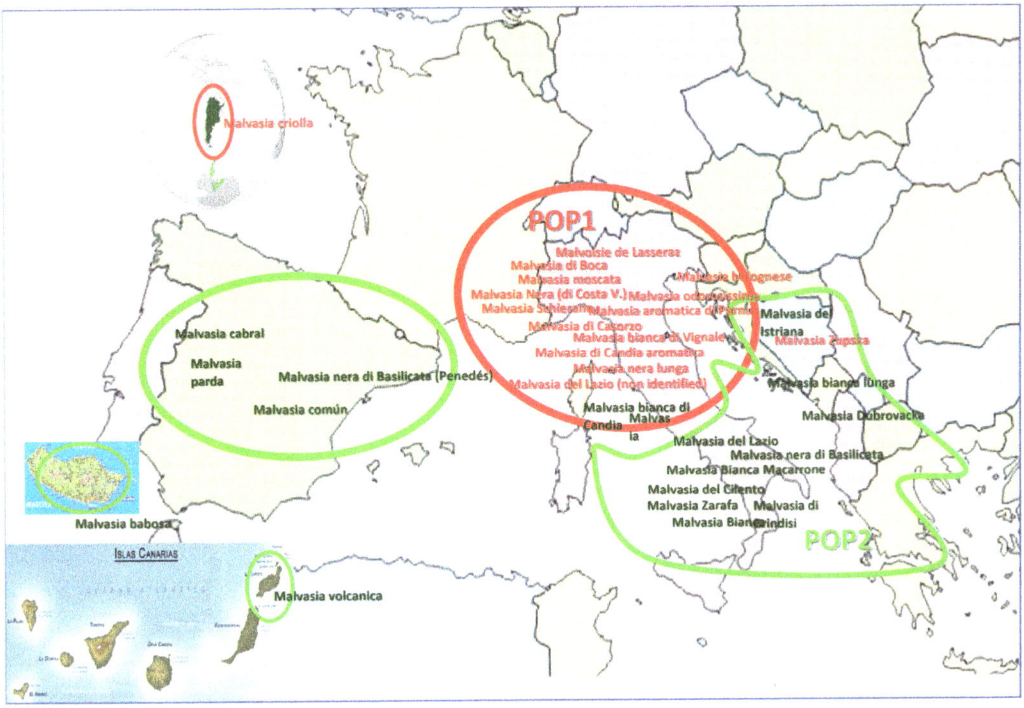

Figure 6. Map of the geographical location (approximate [5,25–27]) of the 33 'Malvasia' varieties under the geographical strategy with two populations. Goodness of fit: 100%.

The 33 members of the 'Malvasia' group were then distributed into three populations (Figures 3c and 5c) to observe the results obtained. In this way, POP2 was split into (1) the group of 'Malvasia' varieties led by Malvasia Dubrovacka and Malvasia bianca lunga and their descendants, along with Malvasia del Cilento, which often appeared close to the latter, as well as the Croatian 'Malvasia' varieties Malvasia istriana and Malvasia Zupska, and (2) the remaining 'Malvasia' group. Table 4 shows the main characteristics of this new distribution following the geographical origin strategy. This distribution achieved a goodness of fit of 97.5%.

The only misassigned variety is the Argentinian Malvasia criolla, which oscillates between POP1 and POP3, meaning this distribution never reaches a 100% goodness of fit. Thus, this misassigned (admixed) variety was left in POP1, where the progenies of the Muscat family are located. POP1 mostly includes 'Malvasia' varieties resulting from crosses with the Muscat family, and in POP2 and POP3 this pattern is not observed. POP1 contains 14 members, mostly Italian (with the exception of one French and one Argentinian variety), and with chlorotype D. POP2 includes individuals (nine varieties) from the Iberian, Italian, and Balkan Peninsulas, with chlorotypes A and D. Finally, POP3, with 10 members, is also characterized by having chlorotypes A and D, and includes members from the Iberian and Italian Peninsulas. It should be noted that this last population results from the division of POP2.

Table 4. Final distribution of 'Malvasia' varieties based on their origin as described in the VIVC, divided into three populations. Goodness of fit: 97.5%. Misassigned varieties were not eliminated, but reassigned.

	GENERAL INFORMATION							PEDIGREE	
PRIME NAME	COUNTRY ORIGIN	CHLOROTYP	SEED FORMATION	SEX	TASTE	COLOR	USE	PARENT 1	PARENT 2
MALVASIA DUBROVACKA	ESP	A	COMPLETE	HERMAFRODIT	AROMATIC	W	W	?	?
MALVASIA BABOSA	PRT	A	COMPLETE	?	?	W	W	HEBEN	MALVASIA DUBROVACKA
MALVASIA VOLCANICA	ESP	?	COMPLETE	HERMAFRODIT	OTHER FLAVOR THAN MUSCAT, FOXY OR HERBACEOUS	W	W	MALVASIA DUBROVACKA	BERMEJUELA
MALVASIA BIANCA LUNGA	ITA	D	COMPLETE	HERMAFRODIT	NONE	W	W	?	?
MALVASIA NERA DI BRINDISI	ITA	?	COMPLETE	HERMAFRODIT	NONE	B	T/W	MALVASIA BIANCA LUNGA	NEGRO AMARO (? x MAIOLICA (? x VISPAROLA))
MALVASIA NERA DI BASILICATA	ITA	?	COMPLETE	HERMAFRODIT	ARMATIC	B	W	MALVASIA BIANCA LUNGA	SOMARELLO NERO (UVA SACRA (ACHLADI x ?) x GARGANEGA)
MALVASIA DEL CILENTO	ITA	?	?	?	?	B	W	?	?
MALVASIA ISTRIANA	HRV	D	COMPLETE	HERMAFRODIT	NONE	W	W	?	?
MALVASIJA ZUPSKA	HRV	D	COMPLETE	?	?	W	W	HEUNISCH WEISS	?
MALVASIA AROMATICA DI PARMA	ITA	D	COMPLETE	FEMALE	MUSCAT	W	W	MUSCAT A PETITS GRAINS BLANCS	NEBBIOLO x BOTTAGERA (FALSE)

Table 4. *Cont.*

	GENERAL INFORMATION							PEDIGREE	
PRIME NAME	COUNTRY ORIGIN	CHLOROTYP	SEED FORMATION	SEX	TASTE	COLOR	USE	PARENT 1	PARENT 2
MALVASIA DI CANDIA AROMATICA	ITA	D	COMPLETE	HERMAFRODITI	MUSCAT	W	T/W	MALVASIA AROMATICA DI PARMA (MUSCAT A PETITS GRAINS BLANCS x NEBBIOLO X BOTTAGERA (FALSE))	?
MALVASIA NERA LUNGA	ITA	D	COMPLETE	HERMAFRODITI	ARMATIC	B	W	MALVASIA AROMATICA DI PARMA (MUSCAT A PETITS GRAINS BLANCS x NEBBIOLO X BOTTAGERA (FALSE))	FREISA (NEBBIOLO x ?)
MALVASIA BIANCA DI VIGNALE	ITA	D	COMPLETE	HERMAFRODITI	AROMATIC	W	W	MALVASIA AROMATICA DI PARMA (MUSCAT A PETITS GRAINS BLANCS x NEBBIOLO X BOTTAGERA (FALSE))	COCCALONA BIANCA
MALVASIA DI CASORZO	ITA	D	COMPLETE	HERMAFRODITI	AROMATIC	N	W	MALVASIA AROMATICA DI PARMA (MUSCAT A PETITS GRAINS BLANCS x NEBBIOLO X BOTTAGERA (FALSE))	LAMBRUSCA DI ALESSANDRIA (CROVIN x NERETTO DI MARENGO (LAMBRUSCA DI ALESSANDRIA x ?))

Table 4. *Cont.*

GENERAL INFORMATION								PEDIGREE	
PRIME NAME	COUNTRY ORIGIN	CHLOROTYP	SEED FORMATION	SEX	TASTE	COLOR	USE	PARENT 1	PARENT 2
MALVASIA NERA (DI COSTA V.)	ITA	D	COMPLETE	HERMAFRODIT	MUSCAT	W	?	MALVASIA AROMATICA DI PARMA (MUSCAT A PETITS GRAINS BLANCS x NEBBIOLO X BOTTAGERA (FALSE))	LAMBRUSCA DI ALESSANDRIA (CROVIN x NERETTO DI MARENGO (LAMBRUSCA DI ALESSANDRIA x ?))
MALVASIA MOSCATA	ITA	D	COMPLETE	HERMAFRODIT	MUSCAT	W	W	MALVASIA AROMATICA DI PARMA (MUSCAT A PETITS GRAINS BLANCS x NEBBIOLO X BOTTAGERA (FALSE))	?
MALVASIA DI SCHIERANO	ITA	D	COMPLETE	HERMAFRODIT	MUSCAT	B	W	MUSCAT A PETITS GRAINS BLANCS	?
MALVASIA DEL LAZIO (NON-IDENTIFIED)	ITA	?	COMPLETE	HERMAFRODIT	MUSCAT	W	W	MUSCAT OF ALEXANDRIA (MUSCAT A PETITS GRAINS BLANCS x HEPTALIKO)	SCHIAVA GROSSA
MALVASIA CRIOLLA	ARG	?	COMPLETE	HERMAFRODIT	MUSCAT	W	W	LISTAN PRIETO	MUSCAT OF ALEXANDRIA (MUSCAT A PETITS GRAINS BLANCS x HEPTALIKO)
MALVASIA BOLOGNESE	ITA	?	COMPLETE	?	?	W	W	?	?
MALVASIA DI BOCA	ITA	?	COMPLETE	HERMAFRODIT	AROMATIC	W	W	?	?
MALVOISIE DE LASSERAZ	FRA	?	COMPLETE	FEMALE	NONE	W	W	?	?
MALVASIA ODOROSISSIMA	ITA	?	COMPLETE	FEMALE	MUSCAT	W	W	?	?

Table 4. *Cont.*

GENERAL INFORMATION								PEDIGREE	
PRIME NAME	COUNTRY ORIGIN	CHLOROTYP	SEED FORMATION	SEX	TASTE	COLOR	USE	PARENT 1	PARENT 2
MALVASIA COMUN	ESP	A	COMPLETE	HERMAFRODIT	NONE	W	W	HEBEN	?
MALVASIA BIANCA	ITA	?	COMPLETE	HERMAFRODIT	NONE	W	W	SCIACCARELLO	?
MALVASIA BIANCA ZARAFA	ITA	?	COMPLETE	?	MUSCAT	W	W	?	?
MALVASIA	ITA	?	COMPLETE	?	?	B	W	?	?
MALVASIA BIANCA DI CANDIA	ITA	D	COMPLETE	HERMAFRODIT	NONE	W	W	GARGANEGA	?
MALVASIA PARDA	PRT	?	COMPLETE	HERMAFRODIT	NONE	W	W	?	?
MALVASIA BIANCA MACCARRONE	ITA	?	COMPLETE	?	MUSCAT	W	W	?	?
MALVASIA CABRAL	PRT051	?	COMPLETE	HERMAFRODIT	NONE	Rs	W	?	?
MALVASIA NERA DI BASILICATA (PENEDES)	ESP	?	COMPLETE	?	NONE	B	W	?	?
MALVASIA DEL LAZIO	ITA	?	COMPLETE	HERMAFRODIT	NONE	W	W	?	?

1: Red; POP1 pure; light red: POP1 admixed; green: POP2 pure; light green: POP2 admixed; blue: POP3 pure; light blue: POP3 admixed. 2 x is the crossing (pedigree).

Figure 7, following the same approach as in the previous cases, presents the results of the graphical representations using PCoA in two and three dimensions, along with the corresponding phylogenetic tree. In Figure 7a, corresponding to the 2D PCoA representation (with a goodness of fit of 24.47%), the three populations are clearly distinguished: POP1 is located in the left quadrants, POP2 in the upper right quadrant, and POP3 in the lower right quadrant. The addition of one more dimension (Figure 7b) and the corresponding increase in the PCoA representation's goodness of fit (32.63%) show the same pattern, with POP1 and POP3 appearing in the front upper part of the graph and POP2 in the inner part, behind POP1 and POP3. The corresponding phylogenetic tree (Figure 7c), with three branches, one for each population, again defines the distribution into three populations.

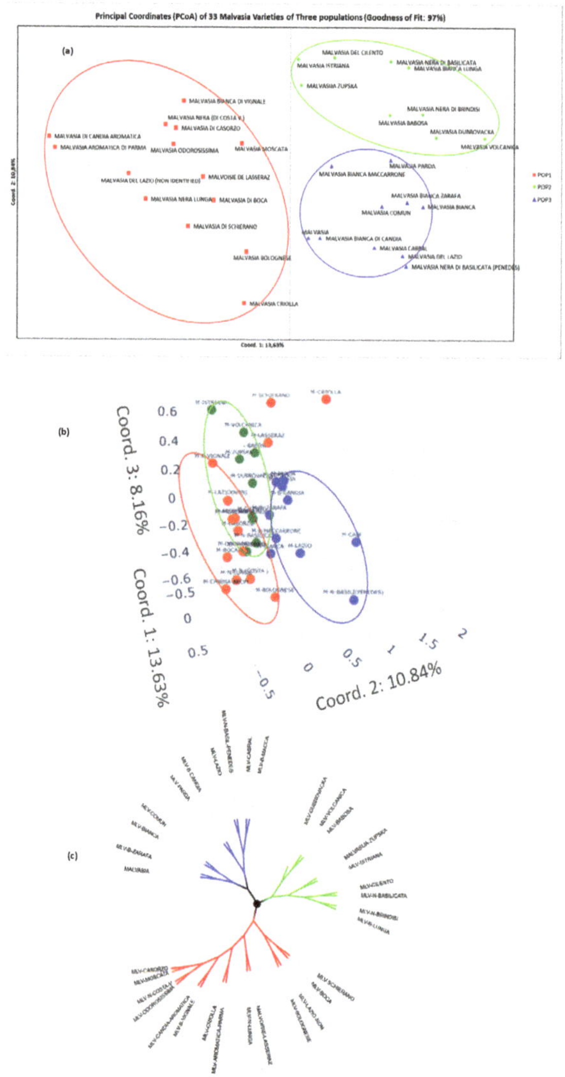

Figure 7. Graphical representation of the distribution of the 33 varieties in three populations from the geographical strategy, using PCoA of their individuals in two (**a**) and three (**b**) dimensions, as well as the representation of a phylogenetic tree (**c**).

The map showing the geographical location corresponding to this distribution is presented in Figure 8. In this image, POP1 covers the entire northern half of the Italian Peninsula, while POP2 is located in the Balkan Peninsula, ultraperipheral regions of the Iberian Peninsula, and the southern half of the Italian Peninsula bordering the Adriatic Sea. Finally, POP3 would be located in the southern half bordering the Mediterranean Sea and in the Iberian Peninsula.

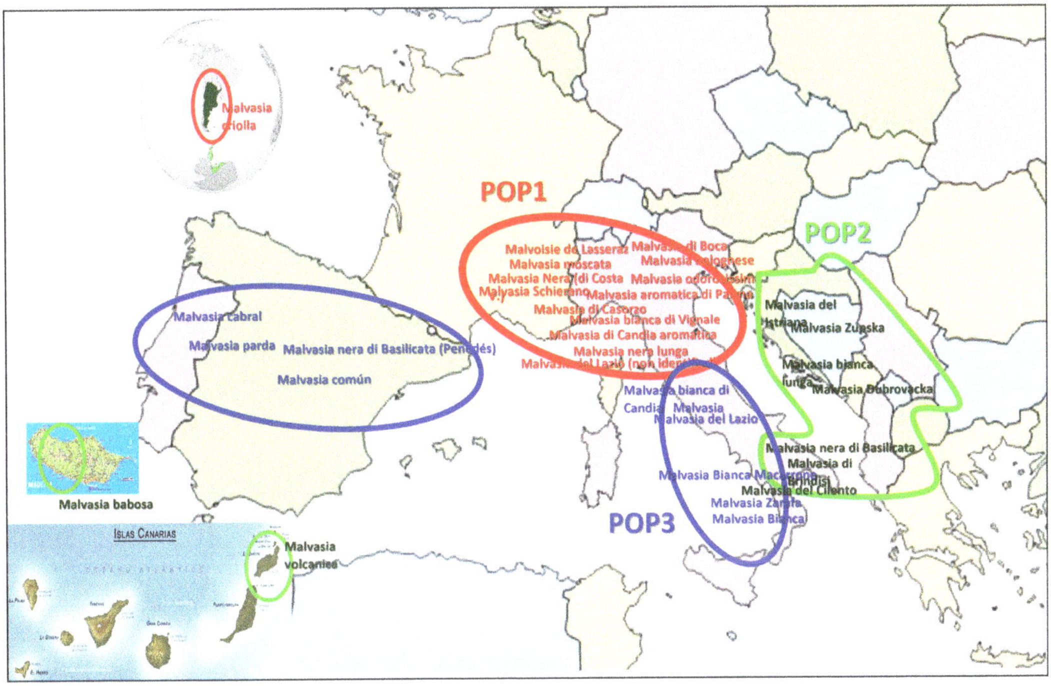

Figure 8. Map of the geographical location (approximate [5,25–27]) of the 33 'Malvasia' varieties from the geographical strategy in three populations. Goodness of fit of 97.5%.

4. Discussion

For most authors, the 'Malvasia' family is a heterogeneous group of varieties [4,28]. There is no agreement regarding their origin, nor ampelographic (Figure S4) [27,29], nor genetic traits [1,28].

Today, the quality and prestige of a wine are usually associated with a specific variety [4]. That is why we find it somewhat unusual when the opposite occurs. 'Malvasia' is a wine that gives its name to many varieties and is characterized by being sweet, aromatic, and alcoholic, although other versions exist in today's market [1].

The merchants of the Most Serene Republic of Venice successfully marketed this type of wine, which reached its peak splendor during the Middle Ages and lasted until the 18th century. They positioned it as one of the great treasures of the trade at the time due to its exquisite quality and the value it acquired among the most discerning consumers (Asenjo, J., personal communication). For that reason, its demand increased substantially [1,4].

Thus, it can be hypothesized that different varieties were possibly used to meet market requirements, always characterized by being more or less aromatic. This has surely been the main reason why so many varieties have incorporated the term "Malvasia" in their names, and why this diversity of organoleptic and ampelographic profiles has been generated for the "Malvasia" family [1].

Our research group has a database with over 1300 accessions, but it contains only 10 samples of possible 'Malvasia' candidates [30–36]. Therefore, to carry out a study of these characteristics, the most extensive and complete database to date was used. In the VIVC, 413 accessions were found that include the term 'Malvasia' or its corresponding spellings in different languages in their names (either as the main name or as a synonym) (Table 1). These accessions correspond to 250 prime names, many of them repeated, since if a prime name has various synonymous names, it will appear as many times as synonyms it has. After correcting this redundancy, the number of unique prime names was reduced to 68. In this way, the ampelographic section of the VIVC contains the description of 68 'Malvasia' varieties, of which 14 have an unknown origin. For this reason, in Table S1, where the main characteristics of these 68 'Malvasia' candidates are presented, an alternative indication is also provided of the grapevine collection (Institution) that hosts the given sample. This helps to form a hypothesis regarding its possible area of origin, although it is not always indicative (Tables S1 and S3). Table S3 shows that the country currently contributing the highest number of 'Malvasia' candidates is Italy, with 28 candidates, followed by Portugal with 11 accessions, Spain with 9 entries, and then Russia (2), Croatia (2), France (1), and Argentina (1). Out of all these, only 43 MP-SSRs are stored in the SSR section of the same database, which will therefore be the starting point of this study.

To carry out the study of population structure, the initial data were normalized to avoid redundant profiles and other artifacts that could bias and therefore interfere with the final statistical results. Thus, four varieties that turned out to be color 'sports' were excluded (variations or mutations in color and/or hairiness that have an identical genetic profile to the main variety). In this population, they were pink mutations (Malvasia di Sardegna rosada, Malvasia fina roxa, Malvasia preta roxa, and Malvasia rosa). The SSR-MPs of varieties resulting from non-natural crosses were also excluded, as they would bias the final result due to being directed crosses. This was the case for Malvasia branca de Sao Jorge and Malvasia moscatel Fonte grande, both Portuguese and designed by the breeders J. L. Ferreira de Almeida and A.P. Soares Franco, respectively in Vitis International Variety Catalogue [3] (Table S1). The varieties Malvasia (Malvasia de Colares), Malvasia fina, Malvasia preta, and Malvoisie de madera were also excluded. These four varieties were the result of natural crosses with varieties from the Iberian Peninsula, France, or Turkey, and thus presumably had no relationship with aromatic varieties (presumed 'Malvasia') or with Muscats (Table S1). This led to the exclusion of 10 varieties, leaving a final study population of 33 SSR-MPs corresponding to their respective varieties.

It is also important to note an aspect concerning the graphical representations of PCoA. Considering that we are working with 9 SSRs and that each SSR has 2 alleles, we will use 18 numerical values (measuring the length of each microsatellite in base pairs (bp)) to define the profile of an individual. Its accurate graphical representation would require the use of 18 dimensions, and that, at present and with our resources, is impossible. Normally, we work in two dimensions and, exceptionally, in three dimensions. This dimensional reduction involves an error that the scientific community must accept, and does accept, which is why we often refer to trends.

In this work, two strategies have been addressed. The first, known as the genetic strategy, consists of distributing the varieties into several populations based on the parameter q, which indicates the percentage of their inferred genome that belongs to one of these populations (Structure 2.3. software). The second is the geographic strategy, in which the distribution of varieties into different populations according to their allelic frequency is conditioned by their geographical origin (GenAlEx 6.5 software (assignment)). Observing the results of these two strategies will allow us to refine the hypothesis that is the subject of this study.

The result obtained in the genetic strategy is presented in Figure 2. The Structure 2.3 software proposes, as the best distribution of the 'Malvasia' population, dividing it into three populations (Figure S1), where varieties are grouped based on their genetic proximity as defined, in this case, by the q parameter (Table S5). A color has been assigned to each population, and at a glance, one can distinguish the so-called 'pure' varieties ($q \geq 74\%$) in which the corresponding color predominates, and the varieties referred to as 'admixed' ($q < 74\%$), where the primary color of the population is shared with the other two colors corresponding to the other populations. Thus, the admixed varieties were eliminated, and the analysis continued with the 17 pure varieties. In this distribution, with a goodness of fit of 82%, POP2 was the most consistent and closely related group (Table S5), with a range of q values for pure varieties between 95.9% and 89.2% and represented by the color green. This was because this cluster was composed of two main varieties (Malvasia Dubrovacka and Malvasia bianca lunga) and their respective progeny, in addition to Malvasia del Cilento, which was always very close to these families (Figure 3). This group was characterized by not having any members related to the Muscat family (only Malvasia bianca Zarafa, which was removed for being admixed, had a Muscat aroma). Interestingly, almost all the 'Malvasia' candidates with one or two Muscat family progenitors were located in POP1 (Table 2). These were not as closely related to one another as those in POP2, and their group color was red (the range of group membership for pure varieties was between 86.2% and 78% (Table S5)). In POP3, the most genetically distant varieties were grouped, with a range for pure varieties between 77.7% and 74.1%, and this population was represented by the color blue (Table S5). When reassignment was performed to reach 100% goodness of fit, POP3, the most distant group, was split between the other two. Malvasia odorosissima ended up in POP1, and Malvasia Maccarrone, Malvasia cabral, Malvasia bianca di Candia, and Malvasia were assigned to POP2. The PCoA graphical representations in two (Figure 3a) and three dimensions (Figure 3b) supported this new redistribution into two populations clearly, with a noticeable distance observed for Malvasia cabral, Malvasia bianca di Candia, and Malvasia from the rest of POP2. However, Malvasia Maccarrone (despite having a Muscat aroma (Table 2)) appeared well integrated into group POP2, and Malvasia odorosissima into POP1. Thus, in the phylogenetic tree, the former POP3 members appeared distant from their respective new groupings to the extent that the four varieties integrated into POP2 formed a main branch on their own (Figure 3c), demonstrating the distance of POP3 members both from each other and from their adoptive populations. Looking at Figure 4, the approximate geographical distribution of these two populations (after reassignment) shows POP1 located in the northern regions of the Italian Peninsula, while POP2 is spread across the southern zone of the Italian Peninsula and the Adriatic area of the Balkan Peninsula (also in its southern part), in addition to overseas areas and the Iberian Peninsula. This distribution of POP2 could have a historical explanation, due to human movements across the Mediterranean Basin and the conquest of the Atlantic Ocean [7,37].

In the geographic strategy, we chose not to eliminate the misassigned varieties (corresponding to the admixed ones), knowing that the behavior of the graphical representations might not be as clear and strong as in the previous case. However, this approach was expected to allow the monitoring of all candidates. Curiously, it turned out that when the 33 varieties were grouped according to the geographic area listed in the VIVC and their related and/or nearby zones, two populations were also formed (Table S6). POP1 included 26 varieties corresponding to the Italian Peninsula and nearby areas, plus the exception of Malvasia criolla (from Argentina). This Argentinean variety was included in POP1 since all the varieties with natural crosses involving members of the Muscat family were found here. Thus, most of the varieties in this cluster had Muscat-like aromas or were classified as aromatic. Likewise, it could be observed that the known chlorotypes of their components were D, a chlorotype widely spread in the Italian and Balkan Peninsulas [24,38].

In addition, when examining the flower sex column, it was found that three 'Malvasia' varieties exhibited female sex, while the vast majority of cultivated grapevine varieties were hermaphroditic. These were the northern Italian varieties, Malvasia aromatica di Parma, Malvasia odorosissima, and the French Alpine variety Malvoisie de Lasseraz. Wild grapevine (*Vitis vinifera* L. ssp. *sylvestris* (C.C. Gmel.) Beger & Hegi) is dioecious, meaning it has either male or female individuals, a strategy that was selected for being favorable to colonizing new territories once this subspecies began expanding from Mediterranean refugia [39].

In contrast, during the domestication process of wild grapevine for cultivation, the selected flower type was hermaphroditic, significantly reducing dioecy. Currently, there are very few grapevine varieties that exhibit female flowers [40]. POP2, with seven members corresponding to the Iberian Peninsula (and outermost territories), presented crosses without members of the Muscat family, and therefore without Muscat aromas, with any variety having female flowers and with known chlorotypes of type A, which is very common in varieties from the Iberian Peninsula and much less present in the Italian and Balkan Peninsulas [24,38]. This distribution had a goodness of fit of 76%. The misassigned varieties were reassigned until reaching a goodness of fit of 100%. Interestingly, as expected, the varieties that in the genetic strategy had constituted POP3 (Figure 2) were divided between POP1 and POP2 in the geographic strategy. The pure varieties Malvasia odorosissima, Malvasia, and Malvasia bianca di Candia were placed in POP1, along with the admixed varieties Malvasia di Schierano, Malvasia criolla, and Malvasija Zupska. In POP2, the pure varieties Malvasia Maccarrone and Malvasia cabral were assigned, as well as the admixed varieties Malvasia istriana and Malvasia comun. It can be observed that the presence of the admixed candidates changed the previous placement of some of the pure varieties from POP3. In the PCoA representations in two and three dimensions (Figure 5a,b), in the phylogenetic tree representation (Figure 5c), and in the approximate geographical distribution presented in Figure 6, it can be stated that, globally and generally, the behavior observed in the genetic strategy for two populations is repeated. Once again, the phylogenetic tree suggests a third population formed by the individuals from POP3 of the genetic strategy (Figure 2), which are mostly now located in POP2 of this new geographic distribution.

Based on all the observations up to this point, we decided to configure a new population within the geographic strategy, composed of all the varieties located in the upper right quadrant of Figure 5a, along with two varieties located along the axis of Coordinate 2 within this quadrant (most of which corresponded to POP3 from the previous distribution under the genetic strategy). The objective was to divide POP2, as suggested by the two phylogenetic trees from the two strategies studied so far (Figures 3c and 5c). Now, POP3 would be composed of the following 10 members: Malvasia comun, Malvasia bianca, Malvasia, Malvasia Zarafa, Malvasia bianca di Candia, Malvasia parda, Malvasia Maccarrone, Malvasia cabral, Malvasia nera de Basilicata (Penedès), and Malvasia del Lazio. Once this new geographic distribution into three populations was carried out (Table 4), the corresponding assignment test was conducted, which resulted in a goodness of fit of 97.5%. A full 100% was never reached because the Argentine variety Malvasia criolla, which shifted back and forth between POP1 and POP3, was ultimately placed in POP1, since one of its parents belonged to the Muscat family. Figure 7a,b show the graphical representations using PCoA in two and three dimensions, respectively. In them, the three populations are clearly distinguishable. As has consistently occurred, POP1 includes the varieties related to the Muscat family, POP2 includes varieties with a possible origin in the Balkan Peninsula, and POP3 contains a group without an apparent relationship, possibly composed of varieties whose crosses involve Italian and Iberian representatives with some ancestral link resulting from human migrations that we cannot determine [7,37]. These results are consistent with those found by Lacombe et al. [4]. In their study, a clear dichotomy is also established between the varieties whose name includes the term 'Malvasia' and descend from crosses with members of the Muscat family, in contrast to another group composed of varieties whose origin is possibly the Balkan Peninsula, along with some groups of varieties without apparent connection. Meneghetti et al. [28] also studied a group of varieties belonging to the 'Malvasia' family, with only 10 representatives. Their results would also be in line with this. From their work, it is worth highlighting that the Malvasia Istriana alone formed a significant sub-branch of the dendrogram they presented, underscoring a trend that is also observed in the present study. Malvasia Istriana, although of Balkan origin [41], is often found either distant from the grouping of Balkan varieties or associated with individuals from POP1. The genetic study (Figure 2) identified it as an admixed individual from POP3, just like the other Balkan variety, Malvasia Zubska, which also appeared as a member of POP1 on some occasions (Figure 6).

Before moving on to the conclusions from the previous paragraphs, the authors of this work would like to make one final observation regarding the Muscat family. If one consults the ampelographic section of the VIVC and enters the term 'Moscatel' in the 'Prime Name' field, a list of Muscats appears. In this list, it can be seen that the variety Muscat d'Istanbul is a natural cross between the Spanish variety Beba and Muscat of Alexandria (Muscat à petits grains blancs x Heptaliko) and that the variety Muscat fleur d'oranger is a natural cross between Muscat à petits grains blancs and the French variety Chasselas. It is logical to think that the offspring of crosses between members of the Muscat family would also carry the term 'Muscat' in their compound names, since they may retain the characteristic aromas of this family. From this reflection arises the question: why, when a cross is made between a member of the Muscat family and another variety, be it Spanish, Italian, Portuguese, etc., is the term 'Malvasia' included in the name and not 'Muscat'? (see, for example, the crosses in Table 2).

For the authors of this work, such a practice is illogical, and the results of these crosses cannot be considered members of the 'Malvasia' family. Therefore, all varieties that are descendants of the Muscat family would be excluded from being part of the 'Malvasia' family. On the other hand, candidates that could be considered part of this family would be the progeny of Malvasia Dubrovacka and Malvasia bianca lunga. In the case of Malvasia Dubrovacka, the crosses in which it is involved, as described in the VIVC, correspond to Malvasia volcanica and Malvasia babosa. Both are daughter varieties that already include the term 'Malvasia' in their names and can thus be considered members of this family. In the case of Malvasia bianca lunga, its progeny includes a mix. There are varieties that have retained the term 'Malvasia' in their names, such as Malvasia nera di Basilicata and Malvasia nera di Brindisi. However, the Slovenian variety Vitovska and the Italian variety Pelena (Glera (Vulpea (Visparola x ?) x ?) x Malvasia bianca lunga), the Italian variety Prunesta (false) (Malvasia bianca lunga x ?), and the Greek variety Lagorthi (known in Italy as Verdeca) (Malvasia bianca lunga x ?), all descendants of Malvasia bianca lunga, have not retained the term 'Malvasia' in their names, yet in our opinion they would still be members of the 'Malvasia' family [2,3,41].

Based on everything discussed so far, it is clear that POP2 (the green population) contains the strongest candidates to belong to the 'Malvasia' family, both in the genetic and the geographical strategies, and in all their combinations. That said, it can be hypothesized that the leading candidates to represent the 'Malvasia' family would be the following seven varieties (Figure 2, Table S5): (a) Malvasia Dubrovacka (with q = 94.2%) and its progeny, Malvasia volcanica (q = 93.2%) and Malvasia Babosa (q = 89.2%); (b) Malvasia bianca lunga

(q = 95.9%) and its progeny, Malvasia nera di Basilicata (q = 94.2%) and Malvasia nera di Brindisi (q = 93.6%); (c) Malvasia del Cilento (q = 92%). Let us not forget either that the 'sport' Malvasia di Sardegna rosada (pink Malvasia), with an identical MP-SSR to Malvasia Dubrovacka, would also be a notable member of the 'Malvasia' family.

As for hypothesizing about the origin of this or these varieties, it can be said that they most likely originated in the Balkan Peninsula (including present-day Greece), even though it has been stated that, for example, the genotype of Malvasia Dubrovacka (one of the strongest candidates to be a true 'Malvasia') showed no connection with individuals in Greek databases, nor could its allele frequencies be related to Croatian, Greek, Italian, Spanish, or Portuguese varieties [1]. The fact remains that a record of Malvasia Dubrovacka exists in the Archives of the Republic of Dubrovnik (Croatia) as early as the year 1385 [1]. In the case of Malvasia bianca lunga, another candidate to be part of this highly selective group of varieties, D'Onofrio et al. [41] proposed its origin in the Balkan Peninsula. These clues, along with the origin of the term that defines 'Malvasia', as discussed previously in the Introduction to this work, reinforce the possibility of a Balkan origin for the varieties used to produce 'Malvasia' wine.

5. Conclusions

It is a fact that the term "Malvasia" and its versions in different languages have been associated throughout history with grape varieties and high-quality wines. It is probably for this very reason that it has even been used to name varieties that have nothing to do with those that are very possibly the true members of this family. Of the 413 variety names containing the term "Malvasia", only 68 remained as candidates to form part of this family. In the MP-SSR section of the VIVC, the genetic profiles of only 43 of them were found. After data normalization, the study was reduced to 33 MP-SSRs. Finally, using different genetic and geographical population structure strategies, it can be concluded that a small group of seven varieties and a color mutation of one of them were chosen as candidates to constitute the "Malvasia" family. In other published works, a small group of four varieties is also proposed as true "Malvasia", although their names do not include this term.

Essentially, this would be a starting group composed of two "reference" varieties genetically linked to each other and their derived strains, either through crosses or mutations. These would be Malvasia Dubrovacka and Malvasia bianca lunga. Another variety that is also very genetically similar to the previous varieties, Malvasia del Cilento, would also be included in this family. Through crossing Malvasia Dubrovacka with other varieties, the varieties Malvasia volcanica (ESP) and Malvasia babosa (PRT) are considered "Malvasia", and through mutation, Malvasia di Sardegna rosé (ESP). If the cross in question involved Malvasia bianca lunga, the varieties to be incorporated into this family would be Malvasia nera di Basilicata (ITA) and Malvasia nera di Brindisi (ITA). Furthermore, the following varieties would also be part of the offspring of Malvasia bianca lunga and therefore also considered "Malvasia" the varieties Vitovska (SVN), Pelena (ITA), Prunesta (false) (ITA), and Lagorthi (GRC). The remaining varieties studied as candidates for "Malvasia" status were rejected, either because they originated from crosses with Muscats or because they derived from crosses with other local varieties from each geographical area in question.

Substantially reducing the number of candidates for this family ensures, on the one hand, the family prestige, and on the other, ensures that consumers associate unique organoleptic characteristics with the varieties proposed as candidates for the "Malvasia" family.

Supplementary Materials: The following supporting information can be downloaded at: https://www.mdpi.com/article/10.3390/horticulturae11060561/s1, Table S1. General information on the 68 unique varieties which include the term 'Malvasia' in their PNs and are registered in the ampelographic section of the VIVC; Table S2. List of "international" primers used by different authors for the amplification of the selected satellite regions. Main characteristics; Table S3. Information on the origin of the 68 unique varieties that include the term 'Malvasia' in their prime name and are registered in the ampelographic section of the VIVC; Table S4. Summary of the general information on the 43 unique varieties with published SSR-MP (Table S1) that include the term 'Malvasia' in their prime name; Figure S1. The four steps of the graphical method of Evanno et al. (2005) [17], allowing the estimation of the true number of ancestral K groups for a population with 33 individuals from 'Malvasia' group; Table S5. Details of the distribution into three populations of the 33 unique varieties and the proportion of pure and admixed individuals based on the q value (pure: $q \geq 74\%$; admixed: $q < 74\%$), as proposed by the Structure 2.3 program; Figure S2. Result of the assignment test for the genetic strategy with K = 3, that is, for three populations; Figure S3. Result of the assignment test for the geographical strategy with two populations; Table S6. Initial distribution of 'Malvasia' varieties based on their origin as described in the VIVC. Goodness of fit: 76%. POP1 (in red) corresponds to the population from the Italian Peninsula and surrounding areas, and POP2 (in green) corresponds to the population from the Iberian Peninsula and areas of influence (with the exception of the Argentine variety Malvasia criolla); Figure S4. Examples of leaves and bunches of different varieties of the "Malvasia" family, corresponding to "true" components. Leaf and bunch shape polymorphism exhibited by genetically related varieties.

Author Contributions: Conceptualization, F.F. and L.R.S.-A.; methodology, L.R.S.-A., Q.L.-Y. and F.F.; software, L.R.S.-A., Q.L.-Y. and F.F.; validation, F.F. and L.R.S.-A.; formal analysis, F.F., Q.L.-Y. and L.R.S.-A.; investigation, F.F., J.A., Q.L.-Y. and L.R.S.-A.; resources, F.F., J.A. and L.R.S.-A.; data curation, F.F. and L.R.S.-A.; writing—original draft preparation, F.F., L.R.S.-A. and L.D.; writing—review and editing, L.D. and L.R.S.-A.; visualization, J.A., J.M.C. and F.Z.; supervision, F.F., Q.L.-Y., L.R.S.-A., J.A., L.D., J.M.C. and F.Z. All authors have read and agreed to the published version of the manuscript.

Funding: This research received no external funding.

Data Availability Statement: Data are not available because they are confidential.

Acknowledgments: The authors would like to thank Javier Ibáñez for his valuable advice. We also wish to express our gratitude to Rosa Pastor, Braulio Esteve-Zarzoso, Laia Fañanás, Cristina Domènech, and Iris Ginés for all their support in achieving these results. Finally, we extend our thanks to Carlos Lozano, Pedro Ballesteros, Amaya Cervera, and Ana María Martín for their support in developing the idea for this publication.

Conflicts of Interest: The authors declare no conflicts of interest.

References

1. Crespan, M.; Cabello, F.; Giannetto, S.; Ibáñez, J.; Kontic, J.K.; Maletic, E.; Peji'c, I.; Rodriguez-Torres, I.; Antonacci, D. Malvasia delle Lipari, Malvasia di Sardegna, Greco di Gerace, Malvasia de Sitges and Malvasia dubrovacka-synonyms of an old and famous grape cultivar. *Vitis* **2006**, *45*, 69–73. [CrossRef]
2. Maul, E.; Röckel, R. *Vitis* International Variety Catalogue (VIVC): A cultivar database referenced by genetic profiles and morphology. In *BIO Web of Conferences, Proceedings of the 38th World Congress of Vine and Wine, Mainz, Germany, 5–10 July 2015*; EDP Sciences: Hulis, France, 2015; Volume 5, p. 01009. [CrossRef]
3. *Vitis* International Variety Catalogue. Available online: http://www.vivc.de (accessed on 1 March 2025).
4. Lacômbe, T.; Boursiquot, J.M.; Laucou, V.; Dechesne, F.; Varès, D.; This, P. Relationships and genetic diversity within the accessions related to Malvasia held in the Domaine de Vassal grape germplasm repository. *Am. J. Enol. Vitic.* **2007**, *58*, 124–131. [CrossRef]
5. Galet, P. *Dictionnaire Encylcopédique Des Cépages*, 1st ed.; Hachette: Paris, France, 2000.
6. Calò, A.; Costacurta, A.; Scienza, A. *Vitigni d'Italia*; Calderini Edagricole: Milano, Italy, 2001.
7. Crespan, M. Exploration and Evaluation of Grapevine Biodiversity Using Molecular Markers. *Mittellungen Klosterneuberg* **2010**, *60*, 310–315. Available online: https://www.researchgate.net/publication/298851588_Exploration_and_evaluation_of_grapevine_biodiversity_using_molecular_markers (accessed on 3 March 2025).
8. Dong, Y.; Duan, S.; Xia, Q.; Liang, Z.; Dong, X.; Margaryan, K.; Musayev, M.; Goryslavets, S.; Zdunić, G.; Bert, P.-F.; et al. Dual domestications and origin of traits in grapevine evolution. *Science* **2023**, *379*, 892–901. [CrossRef]

9. Borrego, J.; de Andres, M.T.; Gomez, J.L.; Ibáñez, J. Genetic study of Malvasia and Torrontes groups through molecular markers. *Am. J. Enol. Vitic.* **2002**, *53*, 125–130. [CrossRef]
10. Rodriguez-Torres, I.; Ibáñez, J.; De Andrés, M.T.; Rubio, C.; Borrego, J.; Cabello, F.; Zerolo, J.; Muñoz-Organero, G. Synonyms and homonyms of 'Malvasía' cultivars (*Vitis vinifera* L.) existing in Spain. *Span. J. Agric. Res.* **2009**, *7*, 563–571. [CrossRef]
11. Thomas, M.R.; Scott, N.S. Microsatellite repeats in grapevine reveal DNA polymorphisms when analyzed as sequence-tagged sites (STSs). *Theor. Appl. Genet.* **1993**, *86*, 985–990. [CrossRef]
12. Bowers, J.E.; Dangl, G.S.; Vignani, R.; Meredith, C.P. Isolation and characterization of new polymorphic simple sequence repeat loci in grape (*Vitis vinifera* L.). *Genome* **1996**, *39*, 628–633. [CrossRef]
13. Bowers, J.E.; Dangl, G.S.; Meredith, C.P. Development and characterization of additional microsatellite DNA markers for grape. *Am. J. Enol. Vitic.* **1999**, *50*, 243–246. [CrossRef]
14. Sefc, K.M.; Regner, F.; Turetschek, E.; Glössl, J.; Steinkellner, H. Identification of microsatellite sequences in Vitis riparia and their applicability for genotyping of different Vitis species. *Genome* **1999**, *42*, 367–373. [CrossRef]
15. Pritchard, J.K.; Stephens, M.; Donnelly, P. Inference of population structure using multilocus genotype data. *Genetics* **2000**, *155*, 945–959. [CrossRef] [PubMed]
16. Falush, D.; Stephens, M.; Pritchard, J.K. Inference of population structure using multilocus genotype data: Linked loci and correlated allele frequencies. *Genetics* **2003**, *164*, 1567–1587. [CrossRef] [PubMed]
17. Evanno, G.; Regnaut, S.; Goudet, J. Detecting the number of clusters of individuals using the software structure: A simulation study. *Mol. Ecol.* **2005**, *14*, 2611–2620. [CrossRef] [PubMed]
18. Peakall, R.; Smouse, P.E. GenAlEx 6.5: Genetic analysis in Excel. Population genetic software for teaching and research-an update. *Bioinformatics* **2012**, *28*, 2537–2539. [CrossRef] [PubMed]
19. Paetkau, D.; Calvert, W.; Stirling, I.; Strobeck, C. Microsatellite analysis of population structure in Canadian polar bears. *Mol. Ecol.* **1995**, *4*, 347–354. [CrossRef]
20. Paetkau, D.; Slade, R.; Burden, M.; Estoup, A. Genetic assignment methods for the direct, real-time estimation of migration rate: A simulation-based exploration of accuracy and power. *Mol. Ecol.* **2004**, *13*, 55–65. [CrossRef]
21. VanderPlas, J. Three-Dimensional Plotting in Matplotlib. In *Python Data Science Handbook*; O'Reilly Media, Inc.: Sevastopol, CA, USA, 2016. Available online: https://jakevdp.github.io/PythonDataScienceHandbook/04.12-three-dimensional-plotting.html (accessed on 21 August 2023).
22. Kumar, S.; Stecher, G.; Tamura, K. MEGA7: Molecular Evolutionary Genetics Analysis version 7.0 for bigger datasets. *Mol. Biol. Evol.* **2016**, *33*, 1870–1874. [CrossRef]
23. Saitou, N.; Nei, M. The neighbor-joining method: A new method for reconstructing phylogenetic threes. *Mol. Biol. Evol.* **1987**, *4*, 406–425. [CrossRef]
24. Žulj Mihaljević, M.; Maletić, E.; Preiner, D.; Zdunić, G.; Bubola, M.; Zyprian, E.; Pejić, I. Genetic Diversity, Population Structure, and Parentage Analysis of Croatian Grapevine Germplasm. *Genes* **2020**, *11*, 737. [CrossRef]
25. Italian Vitis Database. Available online: https://vitisdb.it/varieties/list?page=8 (accessed on 8 March 2025).
26. Vine to Wine Cercle (Portugal and Spain). Available online: http://www.vinetowinecircle.com/en/ (accessed on 8 March 2025).
27. Rodriguez-Torres, I. Albillo Forastero. In *Variedades de vid Cultivadas en Canarias. Descriptores Morfológicos. Caracterización Morfológica, Molecular, Agronómica y Enológica*, 2nd ed.; Instituto Canario de Investigaciones Agrarias, Ed.; Gobierno de Canarias: Santa Cruz de Tenerife, Spain, 2018. Available online: https://www.icia.es/icia/download/Publicaciones/Variedades_Vid_Canarias.pdf (accessed on 4 March 2025).
28. Meneghetti, S.; Bavaresco, L.; Calò, A.; Costacurta, A. Inter- and Intra-Varietal Genetic Variability in *Vitis vinifera* L. In *The Mediterrean Genetic Code-Grapevine and Olive*; Sladonja, B., Ed.; IntechOpen: London, UK, 2013; ISBN 978-953-51-4251-5. [CrossRef]
29. Mendez, J.J. Acerca del Canary Wine. In *Compendio de la Vitivinicultura del Archipiélago Canario*, 2nd ed.; Asociación de Viticultores y Bodegueros de Canarias AVIBO, Ed.; Centro de la Cultura Popular Canaria (CCPC): Canary Island, Spain, 2024.
30. Marsal, G.; Mateo, J.M.; Canals, J.M.; Zamora, F.; Fort, F. SSR analysis of 338 accessions planted in Penedes (Spain) reveals 28 unreported molecular profiles of *Vitis vinifera* L. *Am. J. Enol. Vitic.* **2016**, *67*, 466–470. [CrossRef]
31. Marsal, G.; Bota, J.; Martorell, A.; Canals, J.M.; Zamora, F.; Fort, F. Local cultivars of *Vitis vinifera* L. in Spanish Islands: Balearic Archipelago. *Sci. Hortic.* **2017**, *226*, 122–132. [CrossRef]
32. Marsal, G.; Mendez, J.J.; Mateo-Sanz, J.M.; Ferrer, S.; Canals, J.M.; Zamora, F.; Fort, F. Molecular characterization of *Vitis vinifera* L. local cultivars from volcanic areas (the Canary Islands and Madeira) using SSR markers. *OENO One* **2019**, *53*, 667–680. [CrossRef]
33. Fort, F.; Marsal, G.; Mateo-Sanz, J.M.; Pena, V.; Canals, J.M.; Zamora, F. Molecular characterisation of the current cultivars of *Vitis vinifera* L. in Lanzarote (Canary Islands, Spain) reveals nine individuals which correspond to eight new varieties and two new sports. *OENO One* **2022**, *56*, 281–295. [CrossRef]
34. Fort, F.; Lin-Yang, Q.; Suárez-Abreu, L.R.; Sancho-Galán, P.; Canals, J.M.; Zamora, F. Study of Molecular Biodiversity and Population Structure of *Vitis vinifera* L. ssp. *vinifera* on the Volcanic Island of El Hierro (Canary Islands, Spain) by Using Microsatellite Markers. *Horticulturae* **2023**, *9*, 1297. [CrossRef]

35. Fort, F.; Lin-Yang, Q.; Valls, C.; Sancho-Galán, P.; Canals, J.M.; Zamora, F. Characterisation and Identification of Vines from Fuerteventura (Canary Volcanic Archipelago (Spain)) Using Simple Sequence Repeat Markers. *Horticulturae* **2023**, *9*, 1301. [CrossRef]
36. Fort, F.; Lin-Yang, Q.; Valls, C.; Sancho-Galán, P.; Canals, J.M.; Zamora, F. Analysis of the Diversity Presented by *Vitis vinifera* L. in the Volcanic Island of La Gomera (Canary Archipelago, Spain) Using Simple Sequence Repeats (SSRs) as Molecular Markers. *Horticulturae* **2024**, *10*, 14. [CrossRef]
37. García-Muñoz, S.; Lacômbe, T.; De Andrés, M.T.; Gaforio, L.; Muñoz-Organero, G.; Laucou, V.; This, P.; Cabello, F. Grape varieties (*Vitis vinifera* L.) from the Balearic Islands: Genetic characterization and relationship with Iberian Peninsula and Mediterranean Basin. *Genet. Resour. Crop Evol.* **2012**, *59*, 589–605. [CrossRef]
38. Arroyo-Garcia, R.; Ruiz-García, L.; Bolling, L.; Ocete, R.; López, M.A.; Arnold, C.; Ergul, A.; Söylemezoğlu, G.; Uzun, H.I.; Cabello, F.; et al. Multiple origins of cultivated grape-vine (*Vitis vinifera* L. ssp. sativa) based on chloroplast DNA polymorphisms. *Mol. Ecol.* **2006**, *15*, 3707–3714. [CrossRef]
39. Méndez, J.J.; Marsal, G.; Canals, N.; Canals, J.M.; Zamora, F.; Fort, F. Del Origen de la familia Vitaceae hasta la aparición de la especie *Vitis vinifera* L. In *Jornadas Históricas "Canary Wine"*; Cátedra de Agroturismo y Enoturismo de Canarias, Ed.; Instituto Canario de Calidad Agroalimentaria y Universidad de La Laguna: Santa Cruz de Tenerife, Spain, 2021; pp. 88–89.
40. Massonnet, M.; Cochetel, N.; Minio, A.; Vondras, A.M.; Lin, J.; Muyle, A.; Garcia, J.F.; Zhou, Y.; Delledonne, M.; Riaz, S.; et al. The genetic basis of sex determination in grapes. *Nat. Commun.* **2020**, *11*, 2902. [CrossRef]
41. D'Onofrio, C.; Tumino, G.; Gardiman, M.; Crespan, M.; Bignami, C.; de Palma, L.; Barbagallo, M.G.; Muganu, M.; Morcia, C.; Novello, V.; et al. Parentage atlas of Italian grapevine varieties as inferred from SNP genotyping. *Front. Plant Sci.* **2021**, *11*, 605934. [CrossRef]

horticulturae

MDPI

Article

Preliminary Clonal Characterization of Malvasia Volcanica and Listan Prieto by Simple Sequence Repeat (SSR) Markers in Free-Phylloxera Volcanic Vineyards (Lanzarote and Fuerteventura (Canary Island, Spain))

Francesca Fort [1,*,†], Luis Ricardo Suárez-Abreu [1,†], Qiying Lin-Yang [1], Leonor Deis [1,2], Joan Miquel Canals [1] and Fernando Zamora [1]

[1] Grupo de Tecnología Enológica (TECNENOL), Department of Biochemistry and Biotechnology, Faculty of Oenology, Rovira i Virgili University, Sescelades Campus, C/Marcel·lí Domingo, 1, E-43007 Tarragona, Spain; luisricardo.suarez@estudiants.urv.cat (L.R.S.-A.); qiying.lin@estudiants.urv.cat (Q.L.-Y.); leonor.deis@urv.cat (L.D.); jmcanals@urv.cat (J.M.C.); fernando.zamora@urv.cat (F.Z.)

[2] Plant Physiology, Faculty of Agricultural Sciences, National University of Cuyo, Mendoza CPA M5528AHB, Argentina

* Correspondence: mariafrancesca.fort@urv.cat

† These authors contributed equally to this work.

Abstract

Climate change is usually recognized as the most significant challenge facing viticulture in the 21st century. As a result, experts are increasingly emphasizing the need to explore the biodiversity within the species *Vitis vinifera* L. In this context, the present study investigated the intra-varietal biodiversity of two widely cultivated grapevine varieties on the Canary Islands of Lanzarote and Fuerteventura (Spain). These islands, characterized by desert-like climates, strong winds, volcanic soils, and phylloxera-free conditions, have presented uninterrupted grapevine cultivation for the past three to five centuries. Intra-varietal variability was detected in 93.46% of the 107 accessions analyzed. The most divergent samples were a Malvasia Dubrovacka (LNZ-87) and a Listan prieto (FTV-8), each exhibiting five distinct variations. Another Listan prieto accession (FTV-13) showed four variations. A group of seven individuals displayed three variations including two Malvasia volcanica accessions (LNZ-12, LNZ-72) and five Listan prieto accessions (FTV-1, FTV-2, FTV-7, FTV-9, FTV-12). A set of 100 SSR markers was used to analyze this grapevine collection, of which 17 revealed variability. The most informative markers were VChr15b, VVIp34, VVMD32, VChr9b, VVMD5, VVMD28, and VMC4F3, while the least informative was VVNTM1, which detected no variation. The parentage of Malvasia volcanica (Malvasia Dubrovacka × Bermejuela) was supported by all SSR markers, assuming that three of them may involve a mutated parent.

Keywords: *Vitis vinifera* L.; microsatellite; clones; intra-varietal variability; Malvasia Dubrovacka (Malvasia aromatica); Bermejuela (Marmajuelo); pedigree

1. Introduction

The grapevine (*Vitis vinifera* L.) is one of the world's most economically important and historical crops [1,2]. In 2022, the global vineyard surface was approximately 7.3 million hectares, and wine production exceeded 258 million hectoliters. That same year, the international wine trade generated an estimated value of approximately EUR 37.6 billion [3].

There are about 6000 to 10,000 different varieties (unique genotypes) of *Vitis vinifera* L., of which around 1100 are commercial cultivars used exclusively for winemaking. Remarkably, only 12 of these account for nearly 80% of the total cultivated vineyard area (the 12 varieties that represent the 1% of total genetic variability are Cabernet Sauvignon, Chardonnay, Merlot, Pinot noir, Syrah, Sauvignon blanc, Riesling, Muscat à petits grains blanc, Gewürztraminer, Viognier, Pinot blanc, and Pinot gris). This genetic variability is unlikely to increase in the near future; in fact, it may continue to decline due to several factors. This reduction in varietal diversity (genetic erosion) is driven by several factors: market globalization, the standardization of commercial wines, legal frameworks and the *Appellation d'Origine Contrôlée/Protégée* (AOC/AOP), and climate change [4].

Nevertheless, contemporary viticulture places increasing emphasis on clonal selection, that is, the propagation of specific genotypes within a variety that exhibit desirable characters such as higher yields, greater disease resistance, unique aromatic profiles, or improved tolerance to water stress [5]. This intra-varietal diversity arises primarily from somatic mutations that have accumulated over time and been vegetatively propagated from the original cultivar. Some of these mutations are clearly expressed, for example in pigment biosynthesis pathways, while others are more discreet and phenotypically indistinguishable [6].

To detect, differentiate, and characterize clones within a single grapevine variety, a wide array of molecular techniques has been developed, among these, microsatellite or simple sequence repeat (SSR) markers. Despite being the primary tool for inter-varietal identification in grapevines due to their high polymorphic information content, multiallelic, and availability of standardized protocols, this technique presents a relatively low mutation rate and genetic uniformity is observed amongst most clones, limiting their ability to detect intra-varietal differences, as demonstrated in diverse studies [7–10]. Subsequently, more sensitive approaches have been introduced to detect polymorphisms potentially undetectable by SSRs such as amplified fragment length polymorphism (AFLP) and its variants [10–12]. The development of retrotransposon-based marker systems, including sequence specific amplified polymorphism (S-SAP), retrotransposon-microsatellite amplified polymorphism (REMAP), and inter-retrotransposon amplified polymorphism (IRAP), has further enhanced the resolution of clonal differentiation by enabling the detection of polymorphisms associated with transposable DNA elements [11,13]. Nevertheless, these techniques (AFLP, S-SAP, REMAP, IRAP) rely on dominant markers, which exclude the distinction between homozygous and heterozygous individuals. Additionally, they require meticulous standardization to ensure reproducibility across laboratories. Consequently, in comparison to SSR, these methods are generally considered less suitable for genetic analysis in grapevines [11,14,15].

It is also worth noting that other techniques, such as methylation-sensitive amplified polymorphism (MSAP), have revealed that some clones exhibit distinct DNA methylation patterns despite displaying identical SSR or AFLP profiles [7].

In the past decade, next-generation sequencing (NGS)-based techniques have emerged as the most effective approaches for clonal discrimination in grapevines. (1) Whole-genome resequencing (WGR) and whole-genome sequencing (WGS) techniques enable the detection of point mutations, insertions, deletions, and structural variants that are specific to each clone, although their high cost currently limits their widespread application [14–16]. (2) Genome reduction techniques, such as double digest restriction-site associated DNA sequencing (ddRADseq) [7] and genotyping-by-sequencing (GBS), have offered a more cost-effective and scalable solution, enabling the identification of thousands of informative SNPs [17–19]. (3) High-throughput amplicon sequencing (AmpSeq) has also proven to be a reliable tool for validating clonal SNPs across multiple accessions, increasing both the genetic resolution and clonal traceability [20]. (4) However, commercial SNP chips, which are designed for varietal-level differentiation, often fail to detect rare mutations among closely related clones [13]. To address this limitation, SNP panels adapted to a specific variety are being developed through the resequencing of representative clones [14].

Although, as previously mentioned, standard SSR markers typically fail to detect intra-varietal variation in most clones due to their relatively low mutation rate [7,9], the study by Migliaro et al. demonstrated that when the locus selection is refined *"using an upgraded core set"* of SSR markers, it is possible to detect genetic differences even in subtle intra-varietal mutations such as those affecting berry color [9]. Similarly, other studies have highlighted the utility of SSRs for identifying clonal diversity under specific conditions: Jahnke et al. successfully distinguished closely related Pinot noir clones using a carefully selected SSR set [21], and Meneghetti et al. confirmed that combining SSRs with other molecular markers could effectively capture relevant intra-varietal variation [12].

Viticulture across large regions of the globe is facing significant challenges as a result of climate change including rising temperatures, prolonged droughts, increased ultraviolet radiation, and a growing frequency of extreme weather events. One of the primary concerns is whether these changing climatic conditions will negatively impact the grapevine yield and fruit quality. Among the key strategies to address these challenges, although still relatively underexplored, is varietal selection, which harnesses the existing genetic diversity within *Vitis vinifera* L. Across the broad range of grapevine cultivars, we find substantial variation in agronomic traits including tolerance to cold, heat, and water stress as well as differences in phenological development. To achieve high-quality wine production, the plant's phenology must align with the climatic conditions of the growing region, in addition to demonstrating resilience to both abiotic and biotic stresses. As climate patterns shift, the phenological requirements of grapevines must also adapt, which can be addressed through the selection of optimal varietal and clonal combinations suited to the local environmental conditions. Indeed, the concept of *terroir* fundamentally encompasses this alignment between grapevine genotype and regional climate [4,22].

The Canary Islands, with their unique geographical features and long-standing viticultural tradition, represent a remarkable case of biodiversity preservation and generation. Due to its isolation and phylloxera-free status, the archipelago has accumulated a high number of somatic mutations over more than five centuries, primarily due to the continuous vegetative propagation of grapevine material. The region can thus be considered as a true hotspot for the generation of *Vitis vinifera* L. biodiversity, offering a wide repertoire of both varietal and clonal diversity that contributes to avoiding the homogenization of wine profiles. This is particularly relevant in light of the fact that, as previously mentioned, just 12 grapevine varieties account for approximately 80% of the global vineyard surface area, a trend that, when combined with the pressures of climate change, exacerbates the genetic erosion of the species [4,23,24].

Accordingly, the islands of Lanzarote and Fuerteventura (Canary Islands; Figure 1) represent a unique natural laboratory for evaluating clonal diversity developed under extreme environmental conditions.

Lanzarote, commonly known as the "Island of Volcanoes", features soils covered by volcanic *lapilli* (locally known as *rofe or picón*), originating from the Timanfaya eruptions (1730–1736) [27]. The area blanketed by these *lapilli* constitutes what is now known as La Geria (20 km^2) (Figure 2). This unique soil exhibits specific properties, including enhanced water retention and thermal insulation, which facilitate grapevine cultivation on the island [25,28]. Viticulture in Lanzarote is primarily characterized by singular agricultural practices, such as the planting of vines in "planting pits", the use of both natural and artificial sand mulching, and cultivation in trench-like structures called *chabocos*, all of which have been adapted to optimize the limited water availability resulting from the island's low annual precipitation of just 90–120 mm [28].

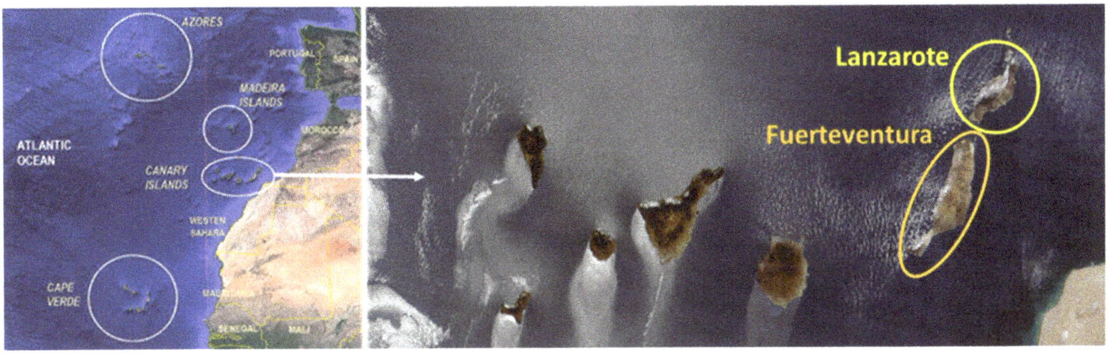

Figure 1. Geographical location of the Macaronesian region (**left**). The Canary Archipelago and detailed view of the islands of Lanzarote and Fuerteventura (**right**) [23,25,26].

Figure 2. Vineyard in La Geria (Lanzarote Island). Soil composed of *picón* or *rofe* (small black volcanic lapilli). Detail of the traditional "hole" planting strategy [28].

On the other hand, Fuerteventura, often referred to as the "Island of Wind", offers an even more arid environment, with annual rainfall dropping below 100 mm in some areas, high solar radiation, and intense wind erosion. In this situation, grapevines only persist in limited areas through a traditional agricultural system known as gavias (Figure 3), which efficiently captures and retains scarce rainwater from occasional precipitation events, enabling cultivation under such extreme climatic conditions [26]. Consequently, the inter- and intra-varietal biodiversity of the Canary Islands stands out as a valuable resource for the identification and selection of varieties and clones with traits related to resilience under global warming scenarios, such as drought tolerance, thus representing a promising long-term strategy for the sustainability of viticulture in Mediterranean and low-rainfall regions [23].

Figure 3. *Gavias* landscape on Fuerteventura Island [26].

Based on the above, the primary objective of this preliminary study was to detect samples exhibiting variation in their genetic profiles within the varieties Malvasia volcanica and Listan prieto. The second objective was to confirm the suitability of SSR markers for revealing differences resulting from the vegetative propagation of grapevine clones in geographically isolated and phylloxera-free areas (Lanzarote and Fuerteventura). Finally, the third objective was to confirm the pedigree of the Malvasia volcanica variety.

2. Materials and Methods

2.1. Plant Material

A total of 107 individuals from the two varieties used in this study (Malvasia volcanica and Listan prieto) were collected from the Lanzarote and Fuerteventura islands.

2.1.1. Lanzarote Island

The population of Malvasia volcanica analyzed in this intra-varietal variability study comprised 86 accessions collected from across the island of Lanzarote. In addition, two individuals of Malvasia (Main Name [MN]: Malvasia Dubrovacka), identified by codes MAR1454 and MARMA467, and one accession of Bermejuela (MARMA467), were included for lineage analysis in the pruning stage (Table S1). As shown, the accessions originated from various points across the island (29°02′06″ N, 13°37′59″ W), representing a diversity of cultivation systems, abiotic conditions, and planting densities between vines, ranging from 1.5 to 2 m in the northern zone, 1.5 m in Tinajo, and up to 4 m in La Geria. From each vine selected by technicians from the Island Council of Lanzarote and the Protected Designation of Origin (PDO) "Vinos de Lanzarote", 4 to 6 vine cuttings of approximately 15 cm were collected, separately bagged, and kept individually throughout the entire study. The plant material was sent to Rovira i Virgili University, where it was examined and labeled with laboratory identification codes, then stored at −20 °C until further use.

2.1.2. Fuerteventura Island

The Listan prieto population analyzed in this intra-varietal variability study consisted of 18 accessions collected from various locations across Fuerteventura Island in the pruning stage (Table S1). The majority of samples were obtained from the Betancuria area, along with two accessions from La Oliva (two samples), one from Puerto del Rosario, and one from Antigua. As in the previous case, 4 to 6 vine cuttings of approximately 15 cm were collected from each vine selected by the technicians and winegrowers of the Majuelo Association. The cuttings were separately bagged and maintained individually throughout the study. These were also sent to Rovira i Virgili University, where they were reviewed, assigned laboratory identification codes, and preserved at −20 °C until further analysis.

2.2. DNA Extraction and Purification

A proprietary method developed by this research group was used for DNA extraction, applicable to both the leaf tissue [29] and recalcitrant tissues [30] as well as RNA extraction from grape berries [31]. All procedures involved three main steps: preparation of the plant material, execution of the extraction protocol, and assessment of the purity index (quality control). The wood DNA extraction protocol [30] employed in this study was optimized through the inclusion of two chloroform washing steps, aiming to more effectively eliminate proteins. Once the DNA was extracted, its purity and concentration were assessed spectrophotometrically using a Thermo Fisher® Scientific NanoDrop™ 1000 Spectrophotometer (Waltham, MA, USA).

2.3. Simple Sequence Repeat (SSR) Markers

After verifying the quality of the DNA, the microsatellite regions of each sample were amplified using 100 SSRs (simple sequence repeat) markers. These markers were selected based on their discriminatory capacity and polymorphism, according to previous studies (Table S2): VVS2, VVS3, VVS29 [32]; VVMD5, VVMD6, VVMD7, VVMD21, VVMD24, VVMD25, VVMD32 [33]; VVMD14, VVMD17, VVMD26, VVMD27, VVMD28, VVMD31, VVMD36 [34]; VrZAG7, VrZAG21, VrZAG25, VrZAG47, VrZAG62, VrZAG64, VrZAG67, VrZAG79, VrZAG83, VrZAG112 [35]; SCU06vv [36]; VMC1b11, VMC4f3 [37]; VvUCH11, VvUCH12, VvUCH19 [38]; VMC3D8, ISV2 (VMC6e1) [39]; VMC6e10 [40]; VVIb01, VVIq52, VVIh54, VVIp60, VVIn16, VVIn61, VVIb66, VVIv37, VVIv67, VVIn73, VVIp31, VVIv33, VVIb09, VVIb32, VVIp22, VVIp34, VVIp37, VVIn57, VVIp77, VVIt60, VVIv04, VVIv17, VVIv51, VVIv70 [41]; VMC4D9.2, VMC4G6 [42]; VRG1, VRG2, VRG3, VRG4, VRG7, VRG9, VRG10, VRG11, VRG13, VRG15, VRG16 [43]; VChr1b, VChr3a, VChr4a, VChr5b, VChr5c, VChr7b, VChr8a, VChr8b, VChr9a, VChr9b, VChr10a, VChr10b, VChr11b, VChr12a, VChr13a, VChr13b, VChr13c, VChr14b, VChr15a, VChr15b, VChr16a, VChr18a, VChr18b, VChr19a, VChr19b [44]; VVNTM1, VVNTM5 [45] (Figure S1). Nine of these markers are widely used as international reference loci by the scientific community [46] and have been selected by the OIV as standard descriptors for grapevine varieties and *Vitis* species (OIV 801–806).

2.4. DNA Amplification

SSR amplification was carried out using polymerase chain reaction (PCR) on various Applied Biosystems 2720 Thermal Cyclers (Foster City, CA, USA).

Each PCR reaction consisted of 4 ng of template DNA and 1 µM of each primer, with the forward primer labeled with a fluorescent dye (6-FAM: VVS3, VVMD7, VVMD24, VVMD25, VVMD28, VVMD32, VVMD36, VrZAG7, VrZAG47, VrZAG62, VrZAG83, VvUCH11, VvUCH19, VMC6e10, VVIb32, VVIp34, VVIp77,VVIv37, VVIq52, VVIn16, VVIp31, VVIb66, VVIv33, VMC4D9.2, VRG2, VRG3, VRG13, VChr1b, VChr5c, VChr8b,

VChr9a, VChr9b, VChr10a, VChr10b, VChr12a, VChr13c, VChr15b and VVNTM1; HEX: VVS2, VVS29, VVMD6, VVMD21, VVMD26, VVMD27, VrZAG21, VrZAG25, VrZAG67, VrZAG79, VrZAG112, VMC4G6, VVIb09, VVIp22, VVIn57, VVIt60, VVIv04, VVIb01, VVIp60, VVIn61, VRG4, VRG9, VRG10, VRG15 VChr3a, VChr4a, VChr5b, VChr7b, VChr14b, VChr16a, VChr18a, VChr19a and VChr19b; NED: VVMD5, VVMD14, VVMD17, VVMD31, VrZAG64, SCU06vv, VvUCH12, VMC1b11, VMC3D8, ISV2 (VMC6e1), VMC4f3, VVNTM5, VVIp37, VVIv17, VVIv51, VVIv70, VVIn73, VVIh54, VVIv67, VRG1, VRG7, VRG11, VRG16 VChr8a, VChr11b, VChr13a, VChr13b, VChr15a, and VChr18b) using the AmpliTaq DNA Polymerase Kit (Applied Biosystems, Foster City, CA). SSR amplification was carried out using the annealing temperatures and thermocycling regimes specified in Table S2. These temperatures and thermocycling regimes were selected based on the previously cited literature and further optimized in the laboratory.

2.5. Amplified Fragments Analysis

Amplified products were mixed with 20 μL of deionized formamide and 0.5 μL of internal size standard (GeneScan 500 ROX, Applied Biosystems, Foster City, CA, USA), and denatured at 95 °C for 3 min. Fragment separation was performed by capillary electrophoresis using an ABI PRISM 3730® Genetic Analyzer (Applied Biosystems, Foster City, CA, USA). The resulting electropherograms were analyzed using *Peak Scanner Software* (Applied Biosystems, Sparta, NJ, USA) to size the amplified fragments. Each accession was analyzed at least twice using DNA from independent extractions to avoid possible errors (Figure S2).

2.6. Sample Identification

The molecular profiles (MP-SSRs) obtained for each SSR marker and accession were compared with published references and existing databases. The international SSR markers VVS2, VVMD5, VVMD7, VVMD27, VVMD28, ZAG62, and ZAG79 were cross-checked with data from the Vine Biology Database of the TECNENOL Research Group [8,23–26,47,48]. In addition, markers VVMD25 and VVMD32 were compared with entries in the global "Vitis International Variety Catalogue" (VIVC) [49]. The TECNENOL database [8,23–26,47,48] also includes MP-SSR profiles for all varieties studied using SSRs such as VVS3, VVS29, VVMD6, VVMD36, ZAG21, ZAG47, ZAG64, ZAG83, UCH11, UCH12, UCH19, SCU06, and VChr19a. The microsatellites VrZAG67 and VrZAG112 are documented in the 2018 book on cultivated grapevine varieties in the Canary Islands [50]. For Muscat of Alexandria, a comparison was also possible with the following markers: VMC1b11, VMC4F3, VVIb01, VVIh54, VVIn16, VVIn73, VVIp31, VVIp60, VVIq52, VVIv37, VVIv67, VVMD21, and VVMD24 [51]. For the remaining SSR markers, MP-SSR profiles were not compared due to limited reference data in the literature, which typically focuses on the aforementioned markers—even when evaluating non-local cultivars such as Malvasia Fina, Malvasia Dubrovacka, or Listan prieto.

3. Results

The results obtained are presented in three sections. The first provides an overview of the accessions that exhibited variations. Next, we describe the main characteristics of the SSR markers used in this study. Finally, we discuss the lineage analysis of Malvasia volcanica.

3.1. Inter-Varietal and Intra-Varietal Variability (Clones Identified: Variations Relative to the Most Common Molecular Profile (Mutations))

The study population consisted of 107 accessions collected from various locations across the islands of Lanzarote and Fuerteventura (Table S1). Specifically, 86 samples of *Malvasia volcanica* from Lanzarote were analyzed to assess the intra-varietal variability, along with 2 accessions of *Malvasia Dubrovacka* [52] and 1 of the local Canarian variety known by its primary name (PN) as *Bermejuela* [23], in order to confirm the pedigree of *Malvasia volcanica* [53]. From Fuerteventura, 18 individuals of the *Listan prieto* variety were examined [26]. While no inter-varietal variability was detected among the *Malvasia Dubrovacka* and *Listan prieto* samples, four accessions within the *Malvasia volcanica* group were identified as belonging to other varieties: *Muscat of Alexandria* (LNZ-18 and LNZ-52), a Greek white variety; *Listan negro* (LNZ-59), a local Canarian red variety; and the Portuguese white variety *Malvasia fina* (LNZ-69).

Intra-varietal variability was detected in 93.46% of the analyzed accessions (100 out of 107 individuals). Tables 2 and 3 present the samples showing differences relative to the most widespread or reference molecular profile (ARP) within the analyzed population, corresponding to 100 accessions. Thus, five individuals of the Malvasia volcanica variety (LNZ-1, LNZ-28, LNZ-46, LNZ-57, LNZ-65), one sample of Listan prieto (FTV-15), and one accession of Muscat of Alexandria (LNZ-52) were identified as non-variable or matching the ARP (Assumed Reference Profile). The accessions exhibiting the greatest number of variations corresponded to a Malvasia Dubrovacka (LNZ-87) (Figure 4) and a Listan prieto (FTV-8) (Figure 5), each differing from the ARP at five SSR loci. Another Listan prieto accession (FTV-13) (Figure S3) showed four variations. The group presenting three variations comprised seven individuals: two belonging to the Malvasia volcanica variety (LNZ-12, LNZ-72), and five to the Listan prieto variety (FTV-1, FTV-2, FTV-7, FTV-9, FTV-12). A total of 36 individuals exhibited two variations: 30 of these corresponded to Malvasia volcanica, five to Listan prieto, and one to Malvasia Dubrovacka (LNZ-88). In 54 samples, only a single variation was detected. Of these, 45 were members of the Malvasia volcanica variety, five belonged to the Listan prieto cluster, and the remaining four were classified as Muscat of Alexandria (LNZ-18), Listan Negro (LNZ-59), Malvasia Fina (LNZ-69), and the local Canarian variety Bermejuela (LNZ-89).

Table 1. Complete list of accessions showing variation in their molecular profile (MP-SSR) compared with the most widespread or reference profile (ARP). Variations typically involve the presence or absence of an allele (homozygosity or heterozygosity), triallelism, tetraallelism, or multiallelism. Some cases also presented numerical variations affecting one or more peak families. List of SSR or microsatellite markers with unique profiles detected in some or all of the analyzed samples.

SSR No.	2	3	4	8	22	23	34	40	47	55	60	69	75	77	80	84	86	87	88	92	97	98	100
Peak size range/ Fluorochrome colour	214–222	168–186	215–270	210–279	278–318	156–230	239–263	217–289	106–158	114–135	350–400	250–300	102–160	90–151	150–200	106–292	250–300	222–250	250–300	176–243	185–292	50–100	145–170
SAMPLE CODE/SSR	VVS3	VVS29	VVMD5	VVMD28	VVIb01	VMC4F3	VRG16	VVMD32	ZAG7	VChr13c	VVIv33	VVIb09	VChr9b	VChr15b	VVIv51	VRG2	VVIn57	VVMD14	VVIt60	VChr14b	VRG3	VVIp34	VChr13b
LNZ-1	LNZ-1	LNZ-1	LNZ-1	LNZ-1	LNZ-1	LNZ-1	LNZ-1	LNZ-1	LNZ-1	LNZ-1	LNZ-1	LNZ-1	LNZ-1	LNZ-1	LNZ-1	LNZ-1	LNZ-1	LNZ-1	LNZ-1	LNZ-1	LNZ-1	LNZ-1	LNZ-1
LNZ-2	LNZ-2*	LNZ-2	LNZ-2	LNZ-2	LNZ-2	LNZ-2	LNZ-2	LNZ-2	LNZ-2	LNZ-2	LNZ-2	LNZ-2	LNZ-2	LNZ-2	LNZ-2	LNZ-2	LNZ-2	LNZ-2	LNZ-2	LNZ-2	LNZ-2	LNZ-2	LNZ-2
LNZ-3	LNZ-3*	LNZ-3	LNZ-3	LNZ-3	LNZ-3	LNZ-3	LNZ-3	LNZ-3	LNZ-3	LNZ-3	LNZ-3	LNZ-3	LNZ-3*	LNZ-3	LNZ-3	LNZ-3	LNZ-3	LNZ-3	LNZ-3	LNZ-3	LNZ-3	LNZ-3	LNZ-3
LNZ-4	LNZ-4*	LNZ-4	LNZ-4	LNZ-4	LNZ-4	LNZ-4	LNZ-4	LNZ-4	LNZ-4	LNZ-4	LNZ-4	LNZ-4	LNZ-4	LNZ-4	LNZ-4	LNZ-4	LNZ-4	LNZ-4	LNZ-4	LNZ-4	LNZ-4	LNZ-4	LNZ-4
LNZ-5	LNZ-5*	LNZ-5*	LNZ-5	LNZ-5	LNZ-5	LNZ-5	LNZ-5	LNZ-5	LNZ-5	LNZ-5	LNZ-5	LNZ-5	LNZ-5	LNZ-5	LNZ-5	LNZ-5	LNZ-5	LNZ-5	LNZ-5	LNZ-5	LNZ-5	LNZ-5	LNZ-5
LNZ-6	LNZ-6*	LNZ-6	LNZ-6	LNZ-6	LNZ-6	LNZ-6	LNZ-6	LNZ-6*	LNZ-6	LNZ-6	LNZ-6	LNZ-6	LNZ-6	LNZ-6	LNZ-6	LNZ-6	LNZ-6	LNZ-6	LNZ-6	LNZ-6	LNZ-6	LNZ-6	LNZ-6
LNZ-7	LNZ-7*	LNZ-7	LNZ-7	LNZ-7	LNZ-7	LNZ-7	LNZ-7	LNZ-7	LNZ-7	LNZ-7	LNZ-7	LNZ-7	LNZ-7	LNZ-7	LNZ-7	LNZ-7	LNZ-7	LNZ-7	LNZ-7	LNZ-7	LNZ-7	LNZ-7	LNZ-7
LNZ-8	LNZ-8*	LNZ-8	LNZ-8	LNZ-8	LNZ-8	LNZ-8	LNZ-8	LNZ-8*	LNZ-8	LNZ-8	LNZ-8	LNZ-8	LNZ-8	LNZ-8	LNZ-8	LNZ-8	LNZ-8	LNZ-8	LNZ-8	LNZ-8	LNZ-8	LNZ-8	LNZ-8
LNZ-9	LNZ-9*	LNZ-9	LNZ-9	LNZ-9	LNZ-9	LNZ-9	LNZ-9	LNZ-9	LNZ-9	LNZ-9	LNZ-9	LNZ-9	LNZ-9	LNZ-9	LNZ-9	LNZ-9	LNZ-9	LNZ-9	LNZ-9	LNZ-9	LNZ-9	LNZ-9	LNZ-9
LNZ-10	LNZ-10*	LNZ-10	LNZ-10	LNZ-10	LNZ-10	LNZ-10	LNZ-10	LNZ-10	LNZ-10	LNZ-10	LNZ-10	LNZ-10	LNZ-10	LNZ-10	LNZ-10	LNZ-10	LNZ-10	LNZ-10	LNZ-10	LNZ-10	LNZ-10	LNZ-10	LNZ-10
LNZ-11	LNZ-11*	LNZ-11	LNZ-11	LNZ-11	LNZ-11	LNZ-11	LNZ-11	LNZ-11	LNZ-11	LNZ-11	LNZ-11	LNZ-11	LNZ-11	LNZ-11	LNZ-11	LNZ-11	LNZ-11	LNZ-11	LNZ-11	LNZ-11	LNZ-11	LNZ-11	LNZ-11
LNZ-12	LNZ-12*	LNZ-12	LNZ-12	LNZ-12	LNZ-12	LNZ-12	LNZ-12	LNZ-12	LNZ-12	LNZ-12	LNZ-12	LNZ-12	LNZ-12	LNZ-12	LNZ-12	LNZ-12	LNZ-12	LNZ-12	LNZ-12	LNZ-12	LNZ-12	LNZ-12	LNZ-12
LNZ-13	LNZ-13*	LNZ-13	LNZ-13	LNZ-13	LNZ-13	LNZ-13	LNZ-13	LNZ-13	LNZ-13	LNZ-13	LNZ-13	LNZ-13	LNZ-13	LNZ-13	LNZ-13	LNZ-13	LNZ-13	LNZ-13	LNZ-13	LNZ-13	LNZ-13	LNZ-13	LNZ-13
LNZ-14	LNZ-14*	LNZ-14	LNZ-14	LNZ-14	LNZ-14	LNZ-14	LNZ-14	LNZ-14	LNZ-14	LNZ-14	LNZ-14	LNZ-14	LNZ-14	LNZ-14	LNZ-14	LNZ-14	LNZ-14	LNZ-14	LNZ-14	LNZ-14	LNZ-14	LNZ-14	LNZ-14
LNZ-15	LNZ-15*	LNZ-15	LNZ-15	LNZ-15	LNZ-15	LNZ-15	LNZ-15	LNZ-15	LNZ-15	LNZ-15	LNZ-15	LNZ-15	LNZ-15	LNZ-15	LNZ-15	LNZ-15	LNZ-15	LNZ-15	LNZ-15	LNZ-15	LNZ-15	LNZ-15	LNZ-15
LNZ-16	LNZ-16	LNZ-16	LNZ-16	LNZ-16	LNZ-16	LNZ-16	LNZ-16	LNZ-16	LNZ-16	LNZ-16	LNZ-16	LNZ-16	LNZ-16	LNZ-16	LNZ-16	LNZ-16	LNZ-16	LNZ-16	LNZ-16	LNZ-16	LNZ-16	LNZ-16*	LNZ-16
LNZ-17	LNZ-17*	LNZ-17	LNZ-17	LNZ-17	LNZ-17	LNZ-17	LNZ-17	LNZ-17*	LNZ-17	LNZ-17	LNZ-17	LNZ-17	LNZ-17*	LNZ-17	LNZ-17	LNZ-17	LNZ-17	LNZ-17	LNZ-17	LNZ-17	LNZ-17	LNZ-17	LNZ-17
LNZ-19	LNZ-19*	LNZ-19	LNZ-19	LNZ-19	LNZ-19	LNZ-19	LNZ-19	LNZ-19	LNZ-19	LNZ-19	LNZ-19	LNZ-19	LNZ-19	LNZ-19*	LNZ-19	LNZ-19	LNZ-19	LNZ-19	LNZ-19	LNZ-19	LNZ-19	LNZ-19	LNZ-19
LNZ-20	LNZ-20*	LNZ-20	LNZ-20	LNZ-20	LNZ-20	LNZ-20	LNZ-20	LNZ-20	LNZ-20	LNZ-20	LNZ-20	LNZ-20	LNZ-20	LNZ-20	LNZ-20	LNZ-20	LNZ-20	LNZ-20	LNZ-20	LNZ-20	LNZ-20	LNZ-20	LNZ-20
LNZ-21	LNZ-21*	LNZ-21	LNZ-21	LNZ-21	LNZ-21	LNZ-21	LNZ-21	LNZ-21	LNZ-21	LNZ-21	LNZ-21	LNZ-21	LNZ-21*	LNZ-21	LNZ-21	LNZ-21	LNZ-21	LNZ-21	LNZ-21	LNZ-21	LNZ-21	LNZ-21	LNZ-21
LNZ-22	LNZ-22*	LNZ-22	LNZ-22	LNZ-22	LNZ-22	LNZ-22	LNZ-22	LNZ-22	LNZ-22	LNZ-22	LNZ-22	LNZ-22	LNZ-22	LNZ-22	LNZ-22	LNZ-22	LNZ-22	LNZ-22	LNZ-22	LNZ-22	LNZ-22	LNZ-22	LNZ-22
LNZ-23	LNZ-23*	LNZ-23	LNZ-23	LNZ-23	LNZ-23	LNZ-23	LNZ-23	LNZ-23	LNZ-23	LNZ-23	LNZ-23	LNZ-23	LNZ-23	LNZ-23	LNZ-23	LNZ-23	LNZ-23	LNZ-23	LNZ-23	LNZ-23	LNZ-23	LNZ-23	LNZ-23
LNZ-24	LNZ-24	LNZ-24*	LNZ-24	LNZ-24	LNZ-24	LNZ-24	LNZ-24	LNZ-24	LNZ-24	LNZ-24	LNZ-24	LNZ-24	LNZ-24*	LNZ-24	LNZ-24	LNZ-24	LNZ-24	LNZ-24	LNZ-24	LNZ-24	LNZ-24	LNZ-24	LNZ-24
LNZ-25	LNZ-25*	LNZ-25	LNZ-25	LNZ-25	LNZ-25	LNZ-25	LNZ-25	LNZ-25	LNZ-25	LNZ-25	LNZ-25	LNZ-25	LNZ-25	LNZ-25	LNZ-25	LNZ-25	LNZ-25	LNZ-25	LNZ-25	LNZ-25	LNZ-25	LNZ-25	LNZ-25
LNZ-26	LNZ-26*	LNZ-26	LNZ-26	LNZ-26	LNZ-26	LNZ-26	LNZ-26	LNZ-26	LNZ-26	LNZ-26	LNZ-26	LNZ-26	LNZ-26	LNZ-26	LNZ-26	LNZ-26	LNZ-26	LNZ-26	LNZ-26	LNZ-26	LNZ-26	LNZ-26	LNZ-26
LNZ-27	LNZ-27*	LNZ-27	LNZ-27	LNZ-27	LNZ-27	LNZ-27	LNZ-27	LNZ-27	LNZ-27	LNZ-27	LNZ-27	LNZ-27	LNZ-27	LNZ-27	LNZ-27	LNZ-27	LNZ-27	LNZ-27	LNZ-27	LNZ-27	LNZ-27	LNZ-27	LNZ-27
LNZ-28	LNZ-28	LNZ-28	LNZ-28	LNZ-28	LNZ-28	LNZ-28	LNZ-28	LNZ-28	LNZ-28	LNZ-28	LNZ-28	LNZ-28	LNZ-28	LNZ-28	LNZ-28	LNZ-28	LNZ-28	LNZ-28	LNZ-28	LNZ-28	LNZ-28	LNZ-28	LNZ-28
LNZ-29	LNZ-29*	LNZ-29	LNZ-29	LNZ-29	LNZ-29	LNZ-29	LNZ-29	LNZ-29	LNZ-29	LNZ-29	LNZ-29	LNZ-29	LNZ-29*	LNZ-29	LNZ-29	LNZ-29	LNZ-29	LNZ-29	LNZ-29	LNZ-29	LNZ-29	LNZ-29	LNZ-29
LNZ-30	LNZ-30*	LNZ-30	LNZ-30	LNZ-30	LNZ-30	LNZ-30	LNZ-30	LNZ-30	LNZ-30	LNZ-30	LNZ-30	LNZ-30	LNZ-30	LNZ-30	LNZ-30	LNZ-30	LNZ-30	LNZ-30	LNZ-30	LNZ-30	LNZ-30	LNZ-30	LNZ-30
LNZ-31	LNZ-31	LNZ-31	LNZ-31	LNZ-31	LNZ-31	LNZ-31	LNZ-31	LNZ-31	LNZ-31	LNZ-31	LNZ-31	LNZ-31	LNZ-31*	LNZ-31	LNZ-31	LNZ-31	LNZ-31	LNZ-31	LNZ-31	LNZ-31	LNZ-31	LNZ-31	LNZ-31
LNZ-32	LNZ-32*	LNZ-32	LNZ-32	LNZ-32	LNZ-32	LNZ-32	LNZ-32	LNZ-32	LNZ-32	LNZ-32	LNZ-32	LNZ-32	LNZ-32*	LNZ-32	LNZ-32	LNZ-32	LNZ-32	LNZ-32	LNZ-32	LNZ-32	LNZ-32	LNZ-32	LNZ-32
LNZ-33	LNZ-33*	LNZ-33	LNZ-33	LNZ-33	LNZ-33	LNZ-33	LNZ-33	LNZ-33	LNZ-33	LNZ-33	LNZ-33	LNZ-33	LNZ-33	LNZ-33	LNZ-33	LNZ-33	LNZ-33	LNZ-33	LNZ-33	LNZ-33	LNZ-33	LNZ-33	LNZ-33
LNZ-34	LNZ-34*	LNZ-34*	LNZ-34	LNZ-34	LNZ-34	LNZ-34	LNZ-34	LNZ-34	LNZ-34	LNZ-34	LNZ-34	LNZ-34	LNZ-34	LNZ-34	LNZ-34	LNZ-34	LNZ-34	LNZ-34	LNZ-34	LNZ-34	LNZ-34	LNZ-34	LNZ-34
LNZ-35	LNZ-35*	LNZ-35	LNZ-35	LNZ-35	LNZ-35	LNZ-35	LNZ-35	LNZ-35	LNZ-35	LNZ-35	LNZ-35	LNZ-35	LNZ-35*	LNZ-35	LNZ-35	LNZ-35	LNZ-35	LNZ-35	LNZ-35	LNZ-35	LNZ-35	LNZ-35	LNZ-35
LNZ-36	LNZ-36*	LNZ-36*	LNZ-36	LNZ-36	LNZ-36	LNZ-36	LNZ-36	LNZ-36	LNZ-36	LNZ-36	LNZ-36	LNZ-36	LNZ-36	LNZ-36	LNZ-36	LNZ-36	LNZ-36	LNZ-36	LNZ-36	LNZ-36	LNZ-36	LNZ-36	LNZ-36
LNZ-37	LNZ-37*	LNZ-37	LNZ-37	LNZ-37	LNZ-37	LNZ-37	LNZ-37	LNZ-37	LNZ-37	LNZ-37	LNZ-37	LNZ-37	LNZ-37	LNZ-37	LNZ-37	LNZ-37	LNZ-37	LNZ-37	LNZ-37	LNZ-37	LNZ-37	LNZ-37	LNZ-37
LNZ-38	LNZ-38*	LNZ-38	LNZ-38	LNZ-38	LNZ-38	LNZ-38	LNZ-38	LNZ-38	LNZ-38	LNZ-38	LNZ-38	LNZ-38	LNZ-38	LNZ-38	LNZ-38	LNZ-38	LNZ-38	LNZ-38	LNZ-38	LNZ-38	LNZ-38	LNZ-38	LNZ-38
LNZ-39	LNZ-39*	LNZ-39	LNZ-39	LNZ-39	LNZ-39	LNZ-39	LNZ-39	LNZ-39	LNZ-39	LNZ-39	LNZ-39	LNZ-39	LNZ-39	LNZ-39	LNZ-39	LNZ-39	LNZ-39	LNZ-39	LNZ-39	LNZ-39	LNZ-39	LNZ-39	LNZ-39
LNZ-40	LNZ-40	LNZ-40	LNZ-40	LNZ-40	LNZ-40	LNZ-40	LNZ-40	LNZ-40	LNZ-40	LNZ-40	LNZ-40	LNZ-40	LNZ-40*	LNZ-40	LNZ-40	LNZ-40	LNZ-40	LNZ-40	LNZ-40	LNZ-40	LNZ-40	LNZ-40	LNZ-40
LNZ-41	LNZ-41	LNZ-41*	LNZ-41	LNZ-41	LNZ-41	LNZ-41	LNZ-41	LNZ-41	LNZ-41	LNZ-41	LNZ-41	LNZ-41	LNZ-41	LNZ-41	LNZ-41	LNZ-41	LNZ-41	LNZ-41	LNZ-41	LNZ-41	LNZ-41	LNZ-41	LNZ-41
LNZ-42	LNZ-42*	LNZ-42*	LNZ-42	LNZ-42	LNZ-42	LNZ-42	LNZ-42	LNZ-42	LNZ-42	LNZ-42	LNZ-42	LNZ-42	LNZ-42	LNZ-42	LNZ-42	LNZ-42	LNZ-42	LNZ-42	LNZ-42	LNZ-42	LNZ-42	LNZ-42	LNZ-42
LNZ-43	LNZ-43*	LNZ-43	LNZ-43	LNZ-43	LNZ-43	LNZ-43	LNZ-43	LNZ-43	LNZ-43	LNZ-43	LNZ-43	LNZ-43	LNZ-43	LNZ-43	LNZ-43	LNZ-43	LNZ-43	LNZ-43	LNZ-43	LNZ-43	LNZ-43	LNZ-43	LNZ-43
LNZ-44	LNZ-44*	LNZ-44	LNZ-44	LNZ-44	LNZ-44	LNZ-44	LNZ-44	LNZ-44	LNZ-44	LNZ-44	LNZ-44	LNZ-44	LNZ-44	LNZ-44	LNZ-44	LNZ-44	LNZ-44	LNZ-44	LNZ-44	LNZ-44	LNZ-44	LNZ-44	LNZ-44
LNZ-45	LNZ-45*	LNZ-45	LNZ-45	LNZ-45	LNZ-45	LNZ-45	LNZ-45	LNZ-45	LNZ-45	LNZ-45	LNZ-45	LNZ-45	LNZ-45*	LNZ-45	LNZ-45	LNZ-45	LNZ-45	LNZ-45	LNZ-45	LNZ-45	LNZ-45	LNZ-45	LNZ-45
LNZ-46	LNZ-46	LNZ-46	LNZ-46	LNZ-46	LNZ-46	LNZ-46	LNZ-46	LNZ-46	LNZ-46	LNZ-46	LNZ-46	LNZ-46	LNZ-46	LNZ-46	LNZ-46	LNZ-46	LNZ-46	LNZ-46	LNZ-46	LNZ-46	LNZ-46	LNZ-46	LNZ-46
LNZ-47	LNZ-47*	LNZ-47	LNZ-47	LNZ-47	LNZ-47	LNZ-47	LNZ-47	LNZ-47	LNZ-47	LNZ-47	LNZ-47	LNZ-47	LNZ-47*	LNZ-47	LNZ-47	LNZ-47	LNZ-47	LNZ-47	LNZ-47	LNZ-47	LNZ-47	LNZ-47	LNZ-47
LNZ-48	LNZ-48*	LNZ-48	LNZ-48	LNZ-48	LNZ-48	LNZ-48	LNZ-48	LNZ-48	LNZ-48	LNZ-48	LNZ-48	LNZ-48	LNZ-48*	LNZ-48	LNZ-48	LNZ-48	LNZ-48	LNZ-48	LNZ-48	LNZ-48	LNZ-48	LNZ-48	LNZ-48
LNZ-49	LNZ-49*	LNZ-49	LNZ-49	LNZ-49	LNZ-49	LNZ-49	LNZ-49	LNZ-49	LNZ-49	LNZ-49	LNZ-49	LNZ-49	LNZ-49	LNZ-49	LNZ-49	LNZ-49	LNZ-49	LNZ-49	LNZ-49	LNZ-49	LNZ-49	LNZ-49	LNZ-49
LNZ-50	LNZ-50*	LNZ-50	LNZ-50	LNZ-50	LNZ-50	LNZ-50	LNZ-50	LNZ-50	LNZ-50	LNZ-50	LNZ-50	LNZ-50	LNZ-50*	LNZ-50	LNZ-50	LNZ-50	LNZ-50	LNZ-50	LNZ-50	LNZ-50	LNZ-50	LNZ-50	LNZ-50
LNZ-51	LNZ-51*	LNZ-51	LNZ-51	LNZ-51	LNZ-51	LNZ-51	LNZ-51	LNZ-51	LNZ-51	LNZ-51	LNZ-51	LNZ-51	LNZ-51	LNZ-51	LNZ-51	LNZ-51	LNZ-51	LNZ-51	LNZ-51	LNZ-51	LNZ-51	LNZ-51	LNZ-51
LNZ-53	LNZ-53*	LNZ-53	LNZ-53	LNZ-53	LNZ-53	LNZ-53	LNZ-53	LNZ-53	LNZ-53	LNZ-53	LNZ-53	LNZ-53	LNZ-53	LNZ-53	LNZ-53	LNZ-53	LNZ-53	LNZ-53	LNZ-53	LNZ-53	LNZ-53	LNZ-53	LNZ-53
LNZ-54	LNZ-54*	LNZ-54	LNZ-54	LNZ-54	LNZ-54	LNZ-54	LNZ-54	LNZ-54	LNZ-54	LNZ-54	LNZ-54	LNZ-54	LNZ-54*	LNZ-54	LNZ-54	LNZ-54	LNZ-54	LNZ-54	LNZ-54	LNZ-54	LNZ-54	LNZ-54	LNZ-54
LNZ-55	LNZ-55*	LNZ-55	LNZ-55	LNZ-55	LNZ-55	LNZ-55	LNZ-55	LNZ-55	LNZ-55	LNZ-55	LNZ-55	LNZ-55	LNZ-55*	LNZ-55	LNZ-55	LNZ-55	LNZ-55	LNZ-55	LNZ-55	LNZ-55	LNZ-55	LNZ-55	LNZ-55
LNZ-56	LNZ-56*	LNZ-56	LNZ-56	LNZ-56	LNZ-56	LNZ-56	LNZ-56	LNZ-56	LNZ-56	LNZ-56	LNZ-56	LNZ-56	LNZ-56*	LNZ-56	LNZ-56	LNZ-56	LNZ-56	LNZ-56	LNZ-56	LNZ-56	LNZ-56	LNZ-56	LNZ-56
LNZ-57	LNZ-58	LNZ-57	LNZ-57	LNZ-57	LNZ-57	LNZ-57	LNZ-57	LNZ-57	LNZ-57	LNZ-57	LNZ-57	LNZ-57	LNZ-57	LNZ-57	LNZ-57	LNZ-57	LNZ-57	LNZ-57	LNZ-57	LNZ-57	LNZ-57	LNZ-57	LNZ-57
LNZ-58	LNZ-58*	LNZ-58	LNZ-58	LNZ-58	LNZ-58	LNZ-58	LNZ-58	LNZ-58	LNZ-58	LNZ-58	LNZ-58	LNZ-58	LNZ-58	LNZ-58	LNZ-58	LNZ-58	LNZ-58	LNZ-58	LNZ-58	LNZ-58	LNZ-58	LNZ-58	LNZ-58
LNZ-60	LNZ-60*	LNZ-60	LNZ-60	LNZ-60	LNZ-60	LNZ-60	LNZ-60	LNZ-60	LNZ-60	LNZ-60	LNZ-60	LNZ-60	LNZ-60	LNZ-60	LNZ-60	LNZ-60	LNZ-60	LNZ-60	LNZ-60	LNZ-60	LNZ-60	LNZ-60*	LNZ-60
LNZ-61	LNZ-61*	LNZ-61	LNZ-61	LNZ-61	LNZ-61	LNZ-61	LNZ-61	LNZ-61	LNZ-61	LNZ-61	LNZ-61	LNZ-61	LNZ-61	LNZ-61	LNZ-61	LNZ-61	LNZ-61	LNZ-61	LNZ-61	LNZ-61	LNZ-61	LNZ-61	LNZ-61
LNZ-62	LNZ-62*	LNZ-62	LNZ-62	LNZ-62	LNZ-62	LNZ-62	LNZ-62	LNZ-62	LNZ-62	LNZ-62	LNZ-62	LNZ-62	LNZ-62	LNZ-62	LNZ-62	LNZ-62	LNZ-62	LNZ-62	LNZ-62	LNZ-62	LNZ-62	LNZ-62	LNZ-62
LNZ-63	LNZ-63*	LNZ-63	LNZ-63	LNZ-63	LNZ-63	LNZ-63	LNZ-63	LNZ-63	LNZ-63	LNZ-63	LNZ-63	LNZ-63	LNZ-63	LNZ-63	LNZ-63	LNZ-63	LNZ-63	LNZ-63	LNZ-63	LNZ-63	LNZ-63	LNZ-63	LNZ-63
LNZ-64	LNZ-64*	LNZ-64	LNZ-64	LNZ-64	LNZ-64	LNZ-64	LNZ-64	LNZ-64	LNZ-64	LNZ-64	LNZ-64	LNZ-64	LNZ-64	LNZ-64	LNZ-64	LNZ-64	LNZ-64	LNZ-64	LNZ-64	LNZ-64	LNZ-64	LNZ-64	LNZ-64

Table 2. *Cont.*

SSR No.	2	3	4	8	22	23	34	40	47	55	60	69	75	77	80	84	86	87	88	92	97	98	100
Peak size range/ Fluorochrome colour	214–222	168–186	215–270	210–279	278–318	156–230	239–263	217–289	106–158	114–135	350–400	250–300	102–160	90–151	150–200	106–292	250–300	222–250	250–300	176–243	185–292	50–100	145–170
SAMPLE CODE/SSR	VVS3	VVS29	VVMD5	VVMD28	VVIb01	VMC4F3	VRG16	VVMD32	ZAG7	VChr13c	VVIv33	VVIb09	VChr9b	VChr15b	VVIv51	VRG2	VVIn57	VVMD14	VVIt60	VChr14b	VRG3	VVIp34	VChr13b
LNZ-65	LNZ-65	LNZ-65	LNZ-65	LNZ-65	LNZ-65	LNZ-65	LNZ-65	LNZ-65	LNZ-65	LNZ-65	LNZ-65	LNZ-65	LNZ-65	LNZ-65	LNZ-65	LNZ-65	LNZ-65	LNZ-65	LNZ-65	LNZ-65	LNZ-65	LNZ-65	LNZ-65
LNZ-66	LNZ-66*	LNZ-66*	LNZ-66	LNZ-66	LNZ-66	LNZ-66	LNZ-66	LNZ-66	LNZ-66	LNZ-66	LNZ-66	LNZ-66	LNZ-66	LNZ-66	LNZ-66	LNZ-66	LNZ-66	LNZ-66	LNZ-66	LNZ-66	LNZ-66	LNZ-66	LNZ-66
LNZ-67	LNZ-67*	LNZ-67	LNZ-67	LNZ-67	LNZ-67	LNZ-67	LNZ-67	LNZ-67	LNZ-67	LNZ-67	LNZ-67	LNZ-67	LNZ-67	LNZ-67	LNZ-67	LNZ-67	LNZ-67	LNZ-67	LNZ-67	LNZ-67	LNZ-67	LNZ-67	LNZ-67
LNZ-68	LNZ-68*	LNZ-68	LNZ-68	LNZ-68	LNZ-68	LNZ-68	LNZ-68	LNZ-68	LNZ-68	LNZ-68	LNZ-68	LNZ-68	LNZ-68*	LNZ-68	LNZ-68	LNZ-68	LNZ-68	LNZ-68	LNZ-68	LNZ-68	LNZ-68	LNZ-68	LNZ-68
LNZ-70	LNZ-70*	LNZ-70	LNZ-70	LNZ-70	LNZ-70	LNZ-70	LNZ-70	LNZ-70	LNZ-70	LNZ-70	LNZ-70	LNZ-70	LNZ-70	LNZ-70	LNZ-70	LNZ-70	LNZ-70	LNZ-70	LNZ-70	LNZ-70	LNZ-70	LNZ-70	LNZ-70
LNZ-71	LNZ-71*	LNZ-71	LNZ-71	LNZ-71	LNZ-71	LNZ-71	LNZ-71	LNZ-71	LNZ-71	LNZ-71	LNZ-71	LNZ-71	LNZ-71	LNZ-71	LNZ-71	LNZ-71	LNZ-71	LNZ-71	LNZ-71	LNZ-71	LNZ-71	LNZ-71	LNZ-71
LNZ-72	LNZ-72*	LNZ-72	LNZ-72	LNZ-72	LNZ-72	LNZ-72	LNZ-72	LNZ-72*	LNZ-72	LNZ-72*	LNZ-72	LNZ-72	LNZ-72	LNZ-72	LNZ-72	LNZ-72	LNZ-72	LNZ-72	LNZ-72	LNZ-72	LNZ-72	LNZ-72	LNZ-72
LNZ-73	LNZ-73*	LNZ-73	LNZ-73	LNZ-73	LNZ-73	LNZ-73	LNZ-73	LNZ-73	LNZ-73	LNZ-73	LNZ-73	LNZ-73*	LNZ-73	LNZ-73	LNZ-73	LNZ-73	LNZ-73	LNZ-73	LNZ-73	LNZ-73	LNZ-73	LNZ-73	LNZ-73
LNZ-74	LNZ-74*	LNZ-74	LNZ-74	LNZ-74	LNZ-74	LNZ-74	LNZ-74	LNZ-74	LNZ-74	LNZ-74	LNZ-74	LNZ-74	LNZ-74	LNZ-74	LNZ-74	LNZ-74	LNZ-74	LNZ-74	LNZ-74	LNZ-74	LNZ-74	LNZ-74	LNZ-74
LNZ-75	LNZ-75*	LNZ-75	LNZ-75	LNZ-75	LNZ-75	LNZ-75	LNZ-75	LNZ-75	LNZ-75	LNZ-75	LNZ-75	LNZ-75	LNZ-75	LNZ-75	LNZ-75	LNZ-75	LNZ-75	LNZ-75	LNZ-75	LNZ-75	LNZ-75	LNZ-75	LNZ-75
LNZ-76	LNZ-76*	LNZ-76	LNZ-76	LNZ-76	LNZ-76	LNZ-76*	LNZ-76	LNZ-76	LNZ-76	LNZ-76	LNZ-76	LNZ-76	LNZ-76	LNZ-76	LNZ-76	LNZ-76	LNZ-76	LNZ-76	LNZ-76	LNZ-76	LNZ-76	LNZ-76	LNZ-76
LNZ-77	LNZ-77*	LNZ-77	LNZ-77	LNZ-77	LNZ-77	LNZ-77	LNZ-77	LNZ-77	LNZ-77	LNZ-77	LNZ-77	LNZ-77	LNZ-77	LNZ-77	LNZ-77	LNZ-77	LNZ-77	LNZ-77	LNZ-77	LNZ-77	LNZ-77	LNZ-77	LNZ-77
LNZ-78	LNZ-78*	LNZ-78	LNZ-78	LNZ-78	LNZ-78	LNZ-78	LNZ-78	LNZ-78	LNZ-78	LNZ-78	LNZ-78	LNZ-78	LNZ-78*	LNZ-78	LNZ-78	LNZ-78	LNZ-78	LNZ-78	LNZ-78	LNZ-78	LNZ-78	LNZ-78	LNZ-78
LNZ-79	LNZ-79*	LNZ-79	LNZ-79	LNZ-79	LNZ-79	LNZ-79	LNZ-79	LNZ-79	LNZ-79	LNZ-79	LNZ-79	LNZ-79	LNZ-79	LNZ-79	LNZ-79	LNZ-79	LNZ-79	LNZ-79	LNZ-79	LNZ-79	LNZ-79	LNZ-79	LNZ-79
LNZ-80	LNZ-80*	LNZ-80	LNZ-80	LNZ-80	LNZ-80	LNZ-80	LNZ-80	LNZ-80	LNZ-80	LNZ-80	LNZ-80	LNZ-80	LNZ-80*	LNZ-80	LNZ-80	LNZ-80	LNZ-80	LNZ-80	LNZ-80	LNZ-80	LNZ-80	LNZ-80	LNZ-80
LNZ-81	LNZ-81	LNZ-81	LNZ-81	LNZ-81	LNZ-81	LNZ-81	LNZ-81	LNZ-81	LNZ-81	LNZ-81	LNZ-81	LNZ-81	LNZ-81*	LNZ-81	LNZ-81	LNZ-81	LNZ-81	LNZ-81	LNZ-81	LNZ-81	LNZ-81	LNZ-81	LNZ-81
LNZ-82	LNZ-82*	LNZ-82	LNZ-82	LNZ-82	LNZ-82	LNZ-82	LNZ-82	LNZ-82	LNZ-82	LNZ-82	LNZ-82	LNZ-82	LNZ-82	LNZ-82	LNZ-82	LNZ-82	LNZ-82	LNZ-82	LNZ-82	LNZ-82	LNZ-82	LNZ-82	LNZ-82
LNZ-83	LNZ-83*	LNZ-83	LNZ-83	LNZ-83	LNZ-83	LNZ-83	LNZ-83	LNZ-83	LNZ-83	LNZ-83	LNZ-83	LNZ-83	LNZ-83*	LNZ-83	LNZ-83	LNZ-83	LNZ-83	LNZ-83	LNZ-83	LNZ-83	LNZ-83	LNZ-83	LNZ-83
LNZ-84	LNZ-84*	LNZ-82	LNZ-84	LNZ-84	LNZ-84	LNZ-84	LNZ-84	LNZ-84	LNZ-84	LNZ-84	LNZ-84	LNZ-84	LNZ-84*	LNZ-84	LNZ-84	LNZ-84	LNZ-84	LNZ-84	LNZ-84	LNZ-84	LNZ-84	LNZ-84	LNZ-84
LNZ-85	LNZ-85*	LNZ-85	LNZ-85	LNZ-85	LNZ-85	LNZ-85	LNZ-85	LNZ-85	LNZ-85	LNZ-85	LNZ-85	LNZ-85	LNZ-85*	LNZ-85	LNZ-85	LNZ-85	LNZ-85	LNZ-85	LNZ-85	LNZ-85	LNZ-85	LNZ-85	LNZ-85
LNZ-86	LNZ-86*	LNZ-86	LNZ-86	LNZ-86	LNZ-86	LNZ-86	LNZ-86	LNZ-86	LNZ-86	LNZ-86	LNZ-86	LNZ-86	LNZ-86	LNZ-86	LNZ-86	LNZ-86	LNZ-86	LNZ-86	LNZ-86	LNZ-86	LNZ-86	LNZ-86	LNZ-86
FTV-1	FTV-1	FTV-1	FTV-1	FTV-1*	FTV-1	FTV-1	FTV-1	FTV-1	FTV-1	FTV-1	FTV-1	FTV-1	FTV-1*	FTV-1*	FTV-1	FTV-1	FTV-1	FTV-1	FTV-1	FTV-1	FTV-1	FTV-1	FTV-1
FTV-2	FTV-2*	FTV-2	FTV-2	FTV-2*	FTV-2	FTV-2	FTV-2	FTV-2	FTV-2	FTV-2	FTV-2	FTV-2	FTV-2	FTV-2*	FTV-2	FTV-2	FTV-2	FTV-2	FTV-2	FTV-2	FTV-2	FTV-2	FTV-2
FTV-3	FTV-3	FTV-3	FTV-3	FTV-3*	FTV-3	FTV-3	FTV-3	FTV-3	FTV-3	FTV-3	FTV-3	FTV-3	FTV-3*	FTV-3	FTV-3	FTV-3	FTV-3	FTV-3	FTV-3	FTV-3	FTV-3	FTV-3	FTV-3
FTV-4	FTV-4	FTV-4	FTV-4	FTV-4*	FTV-4	FTV-4	FTV-4	FTV-4	FTV-4	FTV-4	FTV-4	FTV-4	FTV-4	FTV-4*	FTV-4	FTV-4	FTV-4	FTV-4	FTV-4	FTV-4	FTV-4	FTV-4	FTV-4
FTV-5	FTV-5	FTV-5	FTV-5	FTV-5*	FTV-5	FTV-5	FTV-5	FTV-5	FTV-5	FTV-5	FTV-5	FTV-5	FTV-5	FTV-5	FTV-5	FTV-5	FTV-5	FTV-5	FTV-5	FTV-5	FTV-5	FTV-5	FTV-5
FTV-7	FTV-7	FTV-7	FTV-7	FTV-7*	FTV-7	FTV-7	FTV-7	FTV-7	FTV-7	FTV-7	FTV-7	FTV-7	FTV-7	FTV-7*	FTV-7	FTV-7	FTV-7	FTV-7	FTV-7	FTV-7*	FTV-7	FTV-7	FTV-7
FTV-8	FTV-8	FTV-8	FTV-8	FTV-8	FTV-8	FTV-8	FTV-8	FTV-8	FTV-8*	FTV-8	FTV-8	FTV-8	FTV-8*	FTV-8*	FTV-8	FTV-8	FTV-8	FTV-8	FTV-8	FTV-8*	FTV-8	FTV-8	FTV-8*
FTV-9	FTV-9*	FTV-9	FTV-9	FTV-9*	FTV-9	FTV-9	FTV-9	FTV-9	FTV-9*	FTV-9	FTV-9	FTV-9	FTV-9	FTV-9	FTV-9	FTV-9	FTV-9	FTV-9	FTV-9	FTV-9	FTV-9	FTV-9	FTV-9
FTV-10	FTV-10	FTV-10	FTV-10	FTV-10*	FTV-10	FTV-10	FTV-10	FTV-10	FTV-10	FTV-10	FTV-10	FTV-10	FTV-10*	FTV-10	FTV-10	FTV-10	FTV-10	FTV-10	FTV-10	FTV-10	FTV-10	FTV-10	FTV-10
FTV-11	FTV-11	FTV-11	FTV-11	FTV-11*	FTV-11	FTV-11	FTV-11	FTV-11	FTV-11	FTV-11	FTV-11	FTV-11	FTV-11*	FTV-11	FTV-11	FTV-11	FTV-11	FTV-11	FTV-11	FTV-11	FTV-11	FTV-11	FTV-11
FTV-12	FTV-12	FTV-12	FTV-12	FTV-12*	FTV-12	FTV-12	FTV-12	FTV-12	FTV-12	FTV-12	FTV-12	FTV-12	FTV-12	FTV-12*	FTV-12	FTV-12	FTV-12	FTV-12	FTV-12	FTV-12*	FTV-12	FTV-12	FTV-12
FTV-13	FTV-13*	FTV-13	FTV-13	FTV-13*	FTV-13	FTV-13	FTV-13	FTV-13	FTV-13*	FTV-13	FTV-13	FTV-13	FTV-13	FTV-13*	FTV-13	FTV-13	FTV-13	FTV-13	FTV-13	FTV-13	FTV-13	FTV-13	FTV-13
FTV-14	FTV-14	FTV-14	FTV-14	FTV-14*	FTV-14	FTV-14	FTV-14	FTV-14	FTV-14*	FTV-14	FTV-14	FTV-14	FTV-14	FTV-14	FTV-14	FTV-14	FTV-14	FTV-14	FTV-14	FTV-14	FTV-14	FTV-14	FTV-14
FTV-15	FTV-15	FTV-15	FTV-15	FTV-15	FTV-15	FTV-15	FTV-15	FTV-15	FTV-15	FTV-15	FTV-15	FTV-15	FTV-15	FTV-15	FTV-15	FTV-15	FTV-15	FTV-15	FTV-15	FTV-15	FTV-15	FTV-15	FTV-15
FTV-16	FTV-16	FTV-16	FTV-16	FTV-16	FTV-16	FTV-16	FTV-16	FTV-16	FTV-16*	FTV-16	FTV-16	FTV-16	FTV-16	FTV-16	FTV-16	FTV-16	FTV-16	FTV-16	FTV-16	FTV-16*	FTV-16	FTV-16	FTV-16
FTV-18	FTV-18	FTV-18	FTV-18	FTV-18*	FTV-18	FTV-18	FTV-18	FTV-18	FTV-18	FTV-18	FTV-18	FTV-18	FTV-18	FTV-18	FTV-18	FTV-18	FTV-18	FTV-18	FTV-18	FTV-18	FTV-18	FTV-18	FTV-18
FTV-20	FTV-20	FTV-20	FTV-20	FTV-20*	FTV-20	FTV-20	FTV-20	FTV-20	FTV-20	FTV-20	FTV-20	FTV-20	FTV-20	FTV-20	FTV-20	FTV-20	FTV-20	FTV-20	FTV-20	FTV-20	FTV-20	FTV-20	FTV-20
FTV-32	FTV-32	FTV-32	FTV-32	FTV-32	FTV-32	FTV-32	FTV-32	FTV-32	FTV-32	FTV-32	FTV-32	FTV-32	FTV-32*	FTV-32	FTV-32	FTV-32	FTV-32	FTV-32	FTV-32	FTV-32	FTV-31	FTV-32	FTV-32
LNZ-18	LNZ-18	LNZ-18	LNZ-18*	LNZ-18	LNZ-18	LNZ-18	LNZ-18	LNZ-18	LNZ-18	LNZ-18	LNZ-18	LNZ-18	LNZ-18	LNZ-18	LNZ-18	LNZ-18	LNZ-18	LNZ-18	LNZ-18	LNZ-18	LNZ-18	LNZ-18	LNZ-18
LNZ-52	LNZ-52	LNZ-52	LNZ-52	LNZ-52	LNZ-52	LNZ-52	LNZ-52	LNZ-52	LNZ-52	LNZ-52	LNZ-52	LNZ-52	LNZ-52	LNZ-52	LNZ-52	LNZ-52	LNZ-52	LNZ-52	LNZ-52	LNZ-52	LNZ-52	LNZ-52	LNZ-52
LNZ-87	LNZ-87*	LNZ-87	LNZ-87	LNZ-87	LNZ-87	LNZ-87*	LNZ-87	LNZ-87	LNZ-87	LNZ-87	LNZ-87*	LNZ-87	LNZ-87	LNZ-87	LNZ-87	LNZ-87	LNZ-87*	LNZ-87	LNZ-87	LNZ-87	LNZ-87	LNZ-87	LNZ-87*
LNZ-88	LNZ-88*	LNZ-88	LNZ-88	LNZ-88	LNZ-88	LNZ-88	LNZ-88	LNZ-88	LNZ-88	LNZ-88	LNZ-88	LNZ-88	LNZ-88	LNZ-88	LNZ-88	LNZ-88	LNZ-88	LNZ-88	LNZ-88	LNZ-88	LNZ-88	LNZ-88	LNZ-88*
LNZ-89	LNZ-89	LNZ-89	LNZ-89	LNZ-89	LNZ-89	LNZ-89	LNZ-89*	LNZ-89	LNZ-89	LNZ-89	LNZ-89	LNZ-89	LNZ-89	LNZ-89	LNZ-89	LNZ-89	LNZ-89	LNZ-89	LNZ-89	LNZ-89	LNZ-89	LNZ-89	LNZ-89
LNZ-59	LNZ-59*	LNZ-59	LNZ-59	LNZ-59	LNZ-59	LNZ-59	LNZ-59	LNZ-59	LNZ-59	LNZ-59	LNZ-59	LNZ-59	LNZ-59	LNZ-59	LNZ-59	LNZ-59	LNZ-59	LNZ-59	LNZ-59	LNZ-59	LNZ-59	LNZ-59	LNZ-59
LNZ-69	LNZ-69*	LNZ-69	LNZ-69	LNZ-69	LNZ-69	LNZ-69	LNZ-69	LNZ-69	LNZ-69	LNZ-69	LNZ-69	LNZ-69	LNZ-69	LNZ-69	LNZ-69	LNZ-69	LNZ-69	LNZ-69	LNZ-69	LNZ-69	LNZ-69	LNZ-69	LNZ-69
SAMPLE CODE/SSR	VVS3*	VVS29*	VVMD5*	VVMD28*	VVIb01	VMC4F3*	VRG16*	VVMD32*	ZAG7*	VChr13c*	VVIv33*	VVIb09*	VChr9b*	VChr15b*	VVIv51	VRG2	VVIn57*	VVMD14	VVIt60	VChr14b*	VRG3	VVIp34*	VChr13b*

Sample code	Cultivar	Colour key	Meaning
LNZ-?	MALVASIA VOLCANICA	(dark green)	HETEROZYGOUS INDIVIDUAL (2 PEAK FAMILIES)
FTV-?	LISTAN PRIETO	(light green)	HETEROZYGOUS INDIVIDUAL (2 PEAK FAMILIES). NUMERICAL VARIATION
LNZ-18 LNZ-52	MUSCAT OF ALEXANDRIA	(white)	HOMOZYGOUS INDIVIDUAL (1 PEAK FAMILY)
		(pale green)	HOMOZYGOUS INDIVIDUAL (1 PEAK FAMILY). NUMERICAL VARIATION
LNZ-87 LNZ-88	MALVASIA AROMATICA	(yellow)	TRIALLELIC INDIVIDUAL (3 PEAK FAMILIES)
		(pale yellow)	TRIALLELIC INDIVIDUAL (3 PEAK FAMILIES). NUMERICAL VARIATION
LNZ-89	BERMEJUELA	(orange)	TETRAALLELIC INDIVIDUAL (4 PEAK FAMILIES)
LNZ-59	LISTAN NEGRO	(brown-orange)	MULTIALLELIC INDIVIDUAL (MORE THAN 4 PEAK FAMILIES)
LNZ-69	MALVASIA FINA	*	MUTATED INDIVIDUAL

Table 3. List of accessions showing variations in their molecular profile (MP-SSR), ordered from the highest to the lowest number of variations. Variations typically involve the presence or absence of an allele (homozygosity or heterozygosity), triallelism, or tetraallelism. Some cases also presented numerical variations affecting one or more peak families.

LNZ-87	VVS3	VMC4F3	VVIv33	VVIn57	VChr13b
FTV-8	ZAG7	VChr9b	VChr15b	VChr14b	VChr13b
FTV-13	VVS3	VVMD28	ZAG7	VChr15b	
LNZ-17	VVS3	VVMD32	VChr9b		
LNZ-72	VVS3	VVMD32	VChr13c		
FTV-1	VVMD28	VChr9b	VChr15b		
FTV-2	VVS3	VVMD28	VChr15b		
FTV-7	VVMD28	VChr15b	VChr14b		
FTV-9	VVS3	VVMD28	ZAG7		
FTV-12	VVMD28	VChr15b	VChr14b		
LNZ-3	VVS3	VChr9b			
LNZ-21	VVS3	VChr9b			
LNZ-29	VVS3	VChr9b			
LNZ-32	VVS3	VChr9b			
LNZ-35	VVS3	VChr9b			
LNZ-45	VVS3	VChr9b			
LNZ-47	VVS3	VChr9b			
LNZ-48	VVS3	VChr9b			
LNZ-50	VVS3	VChr9b			
LNZ-54	VVS3	VChr9b			
LNZ-55	VVS3	VChr9b			
LNZ-56	VVS3	VChr9b			
LNZ-68	VVS3	VChr9b			
LNZ-78	VVS3	VChr9b			
LNZ-80	VVS3	VChr9b			
LNZ-83	VVS3	VChr9b			
LNZ-84	VVS3	VChr9b			
LNZ-85	VVS3	VChr9b			
LNZ-5	VVS3	VVS29			
LNZ-34	VVS3	VVS29			
LNZ-36	VVS3	VVS29			
LNZ-42	VVS3	VVS29			
LNZ-66	VVS3	VVS29			
LNZ-6	VVS3	VVMD32			
LNZ-8	VVS3	VVMD32			
LNZ-19	VVS3	VChr15b			
LNZ-87	VVS3	VMC4F3	VVIv33	VVIn57	VChr13b
LNZ-24	VVS29	VChr9b			
LNZ-60	VVS3	VVIp34			
LNZ-73	VVS3	VVIb09			
LNZ-76	VVS3	VMC4F3			

Table 3. *Cont.*

LNZ-87	VVS3	VMC4F3	VVIv33	VVIn57	VChr13b
FTV-4	VVMD28	VChr15b			
FTV-10	VVMD28	VChr9b			
FTV-11	VVMD28	VChr9b			
FTV-14	VVMD28	ZAG7			
FTV-16	ZAG7	VChr14b			
LNZ-88	VVS3	VChr13b			
LNZ-2	VVS3				
LNZ-4	VVS3				
LNZ-7	VVS3				
LNZ-9	VVS3				
LNZ-10	VVS3				
LNZ-11	VVS3				
LNZ-12	VVS3				
LNZ-13	VVS3				
LNZ-14	VVS3				
LNZ-15	VVS3				
LNZ-20	VVS3				
LNZ-22	VVS3				
LNZ-23	VVS3				
LNZ-25	VVS3				
LNZ-26	VVS3				
LNZ-27	VVS3				
LNZ-30	VVS3				
LNZ-33	VVS3				
LNZ-37	VVS3				
LNZ-38	VVS3				
LNZ-39	VVS3				
LNZ-43	VVS3				
LNZ-44	VVS3				
LNZ-49	VVS3				
LNZ-51	VVS3				
LNZ-53	VVS3				
LNZ-58	VVS3				
LNZ-61	VVS3				
LNZ-62	VVS3				
LNZ-63	VVS3				
LNZ-64	VVS3				
LNZ-67	VVS3				
LNZ-70	VVS3				
LNZ-71	VVS3				
LNZ-74	VVS3				
LNZ-75	VVS3				
LNZ-77	VVS3				

Table 3. *Cont.*

LNZ-87	VVS3	VMC4F3	VVIv33	VVIn57	VChr13b
LNZ-79	VVS3				
LNZ-82	VVS3				
LNZ-86	VVS3				
LNZ-16	VVIp34				
LNZ-31	VChr9b				
LNZ-40	VChr9b				
LNZ-81	VChr9b				
LNZ-41	VVS29				
FTV-3	VVMD28				
FTV-5	VVMD28				
FTV-18	VVMD28				
FTV-20	VVMD28				
FTV-32	VChr9b				
LNZ-18	VVMD5				
LNZ-59	VVS3				
LNZ-69	VVS3				
LNZ-89	VRG16				
LNZ-1					
LNZ-28					
LNZ-46					
LNZ-57					
LNZ-65					
FTV-15					
LNZ-52					
LNZ-?	MALVASIA VOLCANICA				
FTV-?	LISTAN PRIETO				
LNZ-18 LNZ-52	MUSCAT OF ALEXANDRIA				
LNZ-87 LNZ-88	MALVASIA AROMATICA				
LNZ-89	BERMEJUELA				
LNZ-59	LISTAN NEGRO				
LNZ-69	MALVASIA FINA				
	HETEROZYGOUS INDIVIDUAL (2 PEAK FAMILIES)				
	HETEROZYGOUS INDIVIDUAL (2 PEAK FAMILIES). NUMERICAL VARIATION				
	HOMOZYGOUS INDIVIDUAL (1 PEAK FAMILY)				
	HOMOZYGOUS INDIVIDUAL (1 PEAK FAMILY). NUMERICAL VARIATION				
	TRIALLELIC INDIVIDUAL (3 PEAK FAMILIES)				
	TRIALLELIC INDIVIDUAL (3 PEAK FAMILIES). NUMERICAL VARIATION				
	TETRAALLELIC INDIVIDUAL (4 PEAK FAMILIES)				

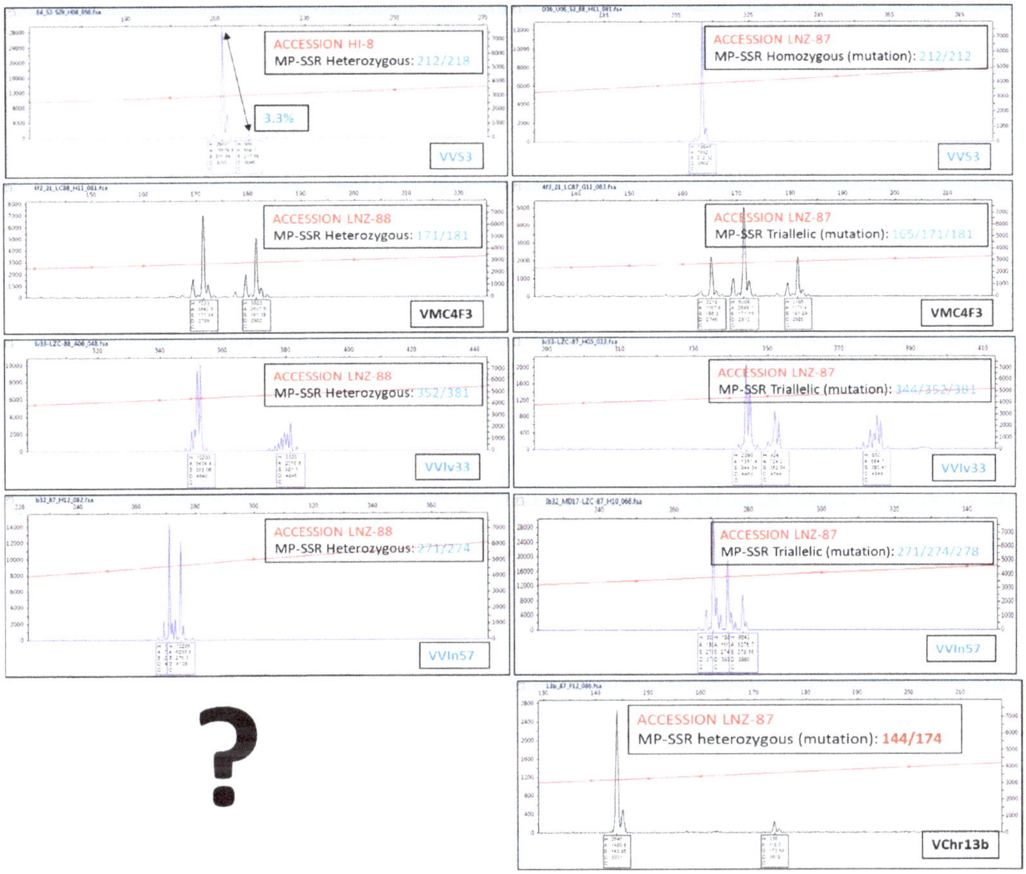

Figure 4. Malvasia Dubrovacka profiles (LNZ-87) showing variations compared with the most widespread or reference molecular profile (ARP). On the left, ARP electropherograms. HI: El Hierro Island; LNZ: Lanzarote Island; FTV: Fuerteventura Island. LNZ-87 is one of the two profiles with the highest number of variations. "?": The ARP profile of this SSR for Malvasia Dubrovacka is unknown (both samples of this variety, LNZ-87 and LNZ-88, showed the same result); based on their pedigree relationship, both are suspected to be mutated. N° %: Examples of peak ratios (percentage indicating the ratio between the smallest and largest peaks), used to disregard extremely small peaks.

As noted in the previous paragraph, the accessions that exhibited the highest number of variations were LNZ-87 and FTV-8 (Tables 2 and 3, Figures 4 and 5). The Malvasia Dubrovacka from Lanzarote (LNZ-87) showed mutations at the SSR loci VVS3, VMC4F3, VVIv33, VVIn57, and VChr13b. The ARPs (Assumed Reference Profiles) for these five SSRs were heterozygous, and the observed variations were as follows: (1) homozygosity at VVS3; (2) triallelism at VMC4F3, VVIv33, and VVIn57; and (3) retention of heterozygosity at VChr13b. The Listan prieto accession from Fuerteventura (FTV-8) exhibited variation at the SSR loci ZAG7, VChr9, VChr15b, VChr14b, and VChr13b. All corresponding ARPs were homozygous, and the observed variations in FTV-8 were as follows: (1) heterozygosity at VChr14b (Figure 5 and Figure S2b) and VChr13b; (2) homozygosity with a change in allele size at VChr9; and (3) multiallelic profiles at ZAG7 and VChr15b, with ZAG7 being triallelic (Figure S2d) and VChr15b tetraallelic.

The accession FTV-13 exhibited four variations (VVS3, VVMD28, ZAG7, and VChr15b) (Tables 2 and 3, Figure S3). This sample of Listan prieto was heterozygous for VVS3 and

VVMD28 in the ARPs, while the corresponding mutated profiles were homozygous. In contrast, ZAG7 and VChr15b had homozygous ARPs, with FTV-13 displaying heterozygosity at ZAG7 and triallelism at VChr15b.

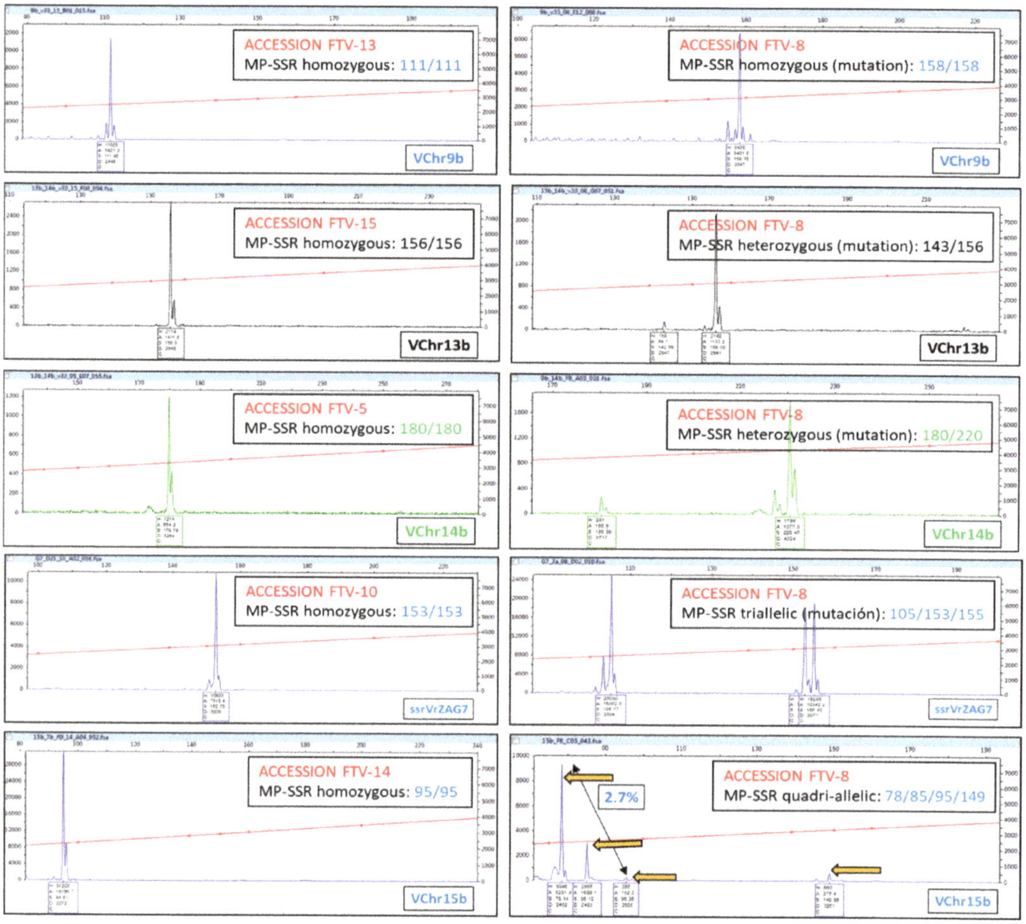

Figure 5. Listan prieto profiles (FTV-8) showing variations compared with the most widespread or reference molecular profile (ARP). On the left, ARP electropherograms. FTV-8 was one of the two profiles with the greatest number of variations.

The group of samples that showed three variations consisted of 7 accessions (Tables 2 and 3): (1) LNZ-17 varied at SSR loci VVS3 and VChr9b (homozygous variation) and VVMD32 (heterozygous mutation with allelic shift); (2) LNZ-72 varied at SSR loci VVS3 and VChr13c (homozygous variation) and VVMD32 (triallelic); (3) FTV-1 showed mutations at SSR loci VVMD28 (homozygous variation), VChr9b (heterozygous variation), and VChr15b (triallelic); (4) FTV-2 varied at SSR loci VVS3 and VVMD28 (homozygous variation) and VChr15b (triallelic); (5) FTV-7 presented mutations at SSR loci VVMD28 (homozygous variation), VChr15b (triallelic), and VChr14b (heterozygous variation); (6) FTV-9 varying at SSR loci VVS3 and VVMD28 (homozygous variation) and ZAG7 (heterozygous variation); and (7) FTV-12 mutated at SSR loci VVMD28 (homozygous variation), VChr15b (triallelic), and VChr14b (heterozygous variation).

The following group of accessions corresponded to those that exhibited two variations (Tables 2 and 3). The samples of *Malvasia volcanica* (LNZ-) 3, 21, 29, 32, 35, 45, 47, 48, 50, 54, 55, 56, 68, 78, 80, 83, 84, and 85 showed variations at SSR loci VVS3 and VChr9b (homozygous variation), while accessions (LNZ-) 5, 34, 36, 42, and 66 presented mutations at SSR loci VVS3 (homozygous variation) and VVS29 (heterozygous variation). LNZ-6 and LNZ-8 varied at SSR loci VVS3 (homozygous variation) and VVMD32 (heterozygous variation). LNZ-19 changed at SSR loci VVS3 and VChr15b (Figure S2a) (homozygous variation). LNZ-24 showed variation at VVS29 (heterozygous variation) and VChr9b (homozygous variation). LNZ-60 exhibited mutation at SSR loci VVS3 (homozygous variation) and VVIp34 (triallelic with allelic shift). LNZ-73 varied at SSR loci VVS3 (homozygous variation) and VVIb09 (triallelic). LNZ-76 mutated at VVS3 (homozygous variation) and VMC4F3 (triallelic). FTV-4 showed variation at SSR loci VVMD28 (homozygous variation) and VChr15b (triallelic). Accessions FTV-10 and FTV-11 mutated at the same SSR loci, VVMD28 (homozygous variation) and VChr9b (homozygous variation with allelic shift). FTV-14 presented mutations at SSR loci VVMD28 (homozygous variation) and ZAG7 (heterozygous variation). FTV-16 varied at SSR loci ZAG7 (heterozygous variation) and VChr14b (heterozygous variation), while the *Malvasia Dubrovacka* variety (LNZ-88) was homozygous at SSR locus VVS3 and heterozygous at SSR locus VChr13

Finally, the following accessions exhibited a single variation: (1) varying at SSR locus VVS3 with homozygosity for the same, the samples of *Malvasia volcanica* (LNZ-) 2, 4, 7, 9, 10, 11, 12, 13, 14, 15, 20, 22, 23, 25, 26, 27, 30, 33, 37, 38, 39, 43, 44, 49, 51, 53, 58, 61, 62, 63, 64, 67, 70, 71, 74, 75, 77, 79, 82, and 86; (2) LNZ-16 exhibited a triallelic profile with allelic shift at SSR VVIp34; (3) accessions LNZ-31, LNZ-40, and LNZ-81 varied at SSR locus VChr9b (homozygous variation), while LNZ-41 showed heterozygous variation at SSR VVS29; (4) FTV-3, FTV-5, FTV-18, and FTV-20 presented homozygous variation at SSR VVMD28; (5) FTV-32 exhibited a heterozygous mutation at SSR VChr9b; (6) *Muscat of Alexandria* (LNZ-18) showed a heterozygous mutation with allelic shift at SSR VVMD5 compared with another sample of the same variety (LNZ-52) (Figure S2c); (7) the local Canary variety *Bermejuela* exhibited homozygous variation at SSR VRG16; (8) finally, a homozygous mutation at SSR VVS3 was observed in representatives of the *Listan negro* (LNZ-59) and *Malvasia fina* (LNZ-69) varieties.

3.2. SSR Markers Used: Effectiveness and Number of Alleles Computed

Out of the 100 SSR markers used in this study, 17 SSRs (VVS3, VVS29, VVMD5, VVMD28, VMC4F3, VRG16, VVMD32, ZAG7, VChr13c, VVIv33, VVIb09, VChr9b, VChr15b, VVIn57, VChr14b, VVIp34, VChr13b) were able to detect intra-varietal variability in at least one of the four varieties (out of the seven analyzed in this study), where such variability is possible due to the presence of more than one clone (Table 2). The remaining six SSRs (VVIb01, VVIv51, VRG2, VVMD14, VVIt60, VRG3) were included in Table 2 because they exhibited noteworthy specificity. For instance, SSRs VVIb01 and VVIt60 showed triallelic profiles in both clones of the Malvasia Dubrovacka variety (LNZ-87 and LNZ-88) while the SSR VVIv51 displayed quadriallelism in the two clones of Muscat of Alexandria (LNZ-18 and LNZ-52) (Figure S2e), and triallelism in the local Canarian variety Bermejuela (LNZ-89). The only representatives of Listan negro and Malvasia fina (LNZ-59 and LNZ-69) exhibited a triallelic MP-SSR for SSR VVMD14. Finally, SSRs VRG2 (Figure S2f) and VRG3 showed multiallelic profiles in Listan prieto; in contrast, VRG2 detected three alleles in Malvasia volcanica and the remaining varieties, while VRG3 detected four alleles in the same samples.

3.2.1. Effectiveness in Detecting Molecular Profiles

Table S3 presented a ranking of SSR markers classified according to their efficiency in detecting both inter-varietal and intra-varietal variability within the sampled population. Two versions of the table are provided: Table S3a highlights the SSRs that differentiated the highest number of MP-SSR profiles, whereas Table S3b focuses on those with the greatest capacity to detect inter-varietal differences. As shown, although the specific markers differed between the two lists, the overall trends remained consistent.

Table S3a shows that the SSR marker distinguishing the highest number of MP-SSR profiles was VChr15b, which detected 11 distinct MP-SSRs, although it did not differentiate between the Listan negro and Malvasia volcanica varieties. Nonetheless, it discriminated six different varietal groups out of the seven present in the population (in this population, the maximum intra-varietal discrimination was 7 MP-SSRs, because in this study, 7 different varieties were described) as well as one mutation in Malvasia volcanica and four variations in Listan prieto. The SSRs VVIp34, VVMD32, and VChr9b followed, each of which distinguished nine MP-SSR profiles. SSR VVIp34 was able to differentiate seven varietal groups and two numerical variations within Malvasia volcanica. In contrast, VVMD32 differentiated six varietal groups, failing to distinguish the MP-SSR of Malvasia Dubrovacka and Bermejuela, although it detected three variations among the Malvasia volcanica samples. A slightly lower discrimination capacity was exhibited by the following 17 SSRs, which did not distinguish between some varieties: VVS2, VVMD6, VVMD7, VVZAG79, VVIp31, ZAG112, VVIn16, VVIb66, VVIp37, VChr9a, VVIv33, VVIb09, VVIv17, VMC4D9.2, VVIp22, VRG9, and VVIn57, each of which detected six MP-SSR profiles. The subsequent group, capable of distinguishing five MP-SSR profiles, comprised 18 SSRs (same as above, duplicated for clarity). Fifteen SSRs detected four MP-SSR profiles (VVZAG83, VRG16, VVIq52, VVIv37, VVMD25, VVMD17, ZAG25, VChr13c, VChr10b, VVMD31, VChr11b, VChr7b, VChr13a, VVIv51, and VChr14b). Seventeen SSRs detected three MP-SSR profiles (VVS3, VVUCH12, VVNTM5, VVIn73, VChr5b, VChr15a, VChr1b, VChr18b, VChr10a, VChr4a, VVMD26, VRG1, ZAG67, VRG7, VRG11, VRG15, and VChr19b). Finally, nine SSRs distinguished only two MP-SSR profiles (VVS29, VChr19a, VRG4, VMC3D8, VChr12a, VChr16a, VRG2, VRG13, and VRG3), and SSR VVNTM1 did not discriminate any MP-SSR profiles.

When the classification criterion was changed to rank the SSRs according to their ability to detect inter-varietal variability (Table S3b), that is, whether they differentiated all seven varieties defined in this study, the ranking shifted. Seventeen SSRs exhibited maximum varietal discrimination, each distinguishing MP-SSR profiles for all seven varieties: VVIp34, VVMD5, VVMD28, VVZAG21, VVIb01, VIn61, VRG10, VVMD24, VVIp60, VVIv67, VVIb32, ISV2, VChr8b, VChr3a, VMC6e10, VVMD14, and VVIt60. Among these, VVIp34, VVMD5, and VVMD28 also exhibited a higher capacity to distinguish MP-SSR profiles, detected between eight and nine profiles including intra-varietal variability. The following group of varieties, which failed to distinguish between two of the seven varieties present in our population, detecting only six varietal groups based on MP-SSR analysis, was composed of 18 SSR markers: VChr15b, VVMD32, VMC4F3, VVS2, VVMD6, VVMD7, VVZAG79, VVIp31, ZAG112, VVIn16, VVIb66, VVIp37, VChr9a, VVIv33, VVIv17, VMC4D9.2, VVIp22, and VRG9. This group included SSR VChr15b, which showed the highest overall discriminatory capacity within the collection, distinguishing 11 MP-SSR profiles as reported in Table S3a as well as VVMD32 and VMC4F3, which distinguished nine and eight profiles, respectively. Ten SSRs distinguished two varietal groups (VChr14b, VVS29, VChr19a, VRG4, VMC3D8, VChr12a, VChr16a, VRG2, VRG13, and VRG3), with two exceptions: VChr14b, which distinguished four MP-SSR profiles, and VVS29, which distinguished three. Finally, SSR VVNTM1 showed no discriminatory power, as also noted in Table S3a.

3.2.2. Variations in Allele Number

The number of alleles detected for a given SSR varied depending on the variety being genotyped, as shown in Table 2, since the MP-SSRs are specific and unique to each variety and even to each individual (as illustrated in Figure 5 with the representative FTV-8 for VChr15b).

The variety Malvasia volcanica showed 9 SSRs that detected 1, 2, 3, and even 4 alleles (Table S4). A total of 41% of the SSRs were homozygous, 56% were heterozygous, 2% were triallelic, and the remaining 1% were tetraallelic. Among the homozygous SSRs, two showed variations in their MP-SSR profiles. These were SSR VVS29, which exhibited a heterozygous variation in accessions LNZ-5/LNZ-24/LNZ-34/LNZ-36/LNZ-41/LNZ-42/LNZ-66, and SSR VVMD32, which showed: (1) heterozygous variation in samples LNZ-6/LNZ-8; (2) heterozygous variation with numerical variation in sample LNZ-17; and (3) a triallelic case in sample LNZ-72. Six SSRs showed variation within the group of heterozygous SSR. SSR VVS3 detected the highest number of variations, with 82 mutated (homozygous) individuals, followed by VChr9b with 23 homozygous individuals. SSRs VChr13c and VChr15b each presented a single homozygous variation in samples LNZ-72 and LNZ-19 (Figure S2a), respectively, while SSRs VMC4F3 (LNZ-76) and VVIb09 (LNZ-73) showed triallelic cases. SSRs VRG2 and VVIp34 were triallelic in this variety, with the latter also presenting a triallelic profile with numerical variation in samples LNZ-16/LNZ-60. Finally, SSR VRG3 was tetraallelic in all Malvasia volcanica samples.

In the only variety from Fuerteventura Island (Listan prieto), this SSR kit showed the full spectrum of allele numbers previously described as well as multiallelic individuals, although only 7 SSRs detected variation among its 18 representatives (Table S5). This variety had 29 homozygous SSRs, 68 heterozygous SSRs, one tetraallelic SSR (VVIp34), and two multiallelic SSRs (VRG2 and VRG3). Among the homozygous SSRs, five showed variation in their MP-SSR profiles: (1) ZAG7 was heterozygous in samples FTV-9/FTV-13/FTV-14/FTV-16 and triallelic in sample FTV-8 (Figure 5 and Figure S2d); (2) VChr9b showed two groups of homozygous accessions with numerical variation (FTV-8 (Figure 5)///FTV-10/FTV-11), two heterozygous samples (FTV-1/FTV-32), and one triallelic individual (FTV-3); (3) VChr15b detected three groups of triallelic samples with different numerical compositions (FTV-1/FTV-2/FTV-13///FTV-4/FTV-12///FTV-7) and one tetraallelic accession (FTV-8, Figure 5); (4) VChr14b appeared as heterozygous with three sample groups showing numerical variation (FTV-7/FTV-16///FTV-12///FTV-8, (Figure 5); and (5) VChr13b showed heterozygous variation in sample FTV-8 (Figure 5). Among the 68 heterozygous SSRs in this variety, only SSRs VVS3 (FTV-2/FTV-9/FTV-13) and VVMD28 (FTV-1/FTV-2/FTV-3/FTV-4/FTV-5/FTV-7/FTV-9/FTV-10/FTV-11/FTV-12/FTV-13/FTV-14/FTV-18/FTV-20) presented homozygous accessions.

The two accessions representing the variety Malvasia Dubrovacka (LNZ-87 and LNZ-88) also showed variation between them (Table S6). Twenty-three SSRs expressed homozygosity in their MP-SSR profiles without any observed variation. Seventy-two SSRs were heterozygous, with five SSRs showing variations: homozygous (VVS3 in LNZ-87/LNZ-88), heterozygous with numerical variation (VChr13b in LNZ-87/LNZ-88), and triallelic (VMC4F3, VVIv33, and VVIn57 in LNZ-87). SSRs VVIb01, VRG2, and VVIt60 showed triallelism in all samples, VRG3 was tetraallelic, and VVIp34 displayed multiallelism. Table S7 shows the performance of this SSR kit for the variety Muscat of Alexandria (LNZ-18 and LNZ-52). Of the 100 SSRs analyzed, 35 were homozygous, 62 were heterozygous, 1 was triallelic (VRG2), and 2 were tetraallelic (VRG3 and VVIv51). Only one homozygous variation was detected for the heterozygous SSR VVMD5 in sample LNZ-18 (Figure S2c). The last three varieties, Listan negro (LNZ-59), Malvasia fina (LNZ-69), and Bermejuela (LNZ-89) were each represented by a single accession. The variety Listan negro (Table S8) had 35% homozygous SSRs, 61% with heterozygous profiles (with the exception of VVS3, which appeared homozygous), 3% triallelic SSRs (VRG2, VVMD14, VVIp34), and one tetraallelic SSR (VRG3). The variety Malvasia fina (Table S9) also had mostly heterozygous SSRs (64 SSRs) and presented one homozygous variation for VVS3. In addition, 33 SSRs were homozygous, 2 were triallelic (VRG2 and VVMD14), and 2 were tetraallelic (VRG3 and VVIp34). Finally, the Canarian variety Bermejuela (Tables S10 and S1) showed 28% homozygous SSRs, 67% heterozygous SSRs, 3% triallelic SSRs (VVIb01, VRG2, VVIv51), and 2% tetraallelic SSRs (VRG2, VVIp34). One SSR likely showed variation in the sample VRG16, which was formally considered homozygous but was placed under the heterozygous column.

3.3. *Pedigree of the Malvasia Volcanica Variety*

The pedigree relationship was satisfied for almost all of the 100 SSRs. Three SSRs were found to be discordant. Specifically, for SSRs VVS3, VRG16, and VChr13b (Figure 6b), the pedigree was not satisfied. As an example, Figure 6a shows the case of pedigree failure for SSR VVS3 as well as for SSR VChr13b (Figure 6b). In cases where one or both parents exhibited triallelism (VVIb01, VMC4F3 (Figure 6c), VVIv33, VVIv51, VVIn57, VVIt60), tetraallelism, or multiallelism (VVIp34) (Figure 6d), the pedigree was satisfied in all cases.

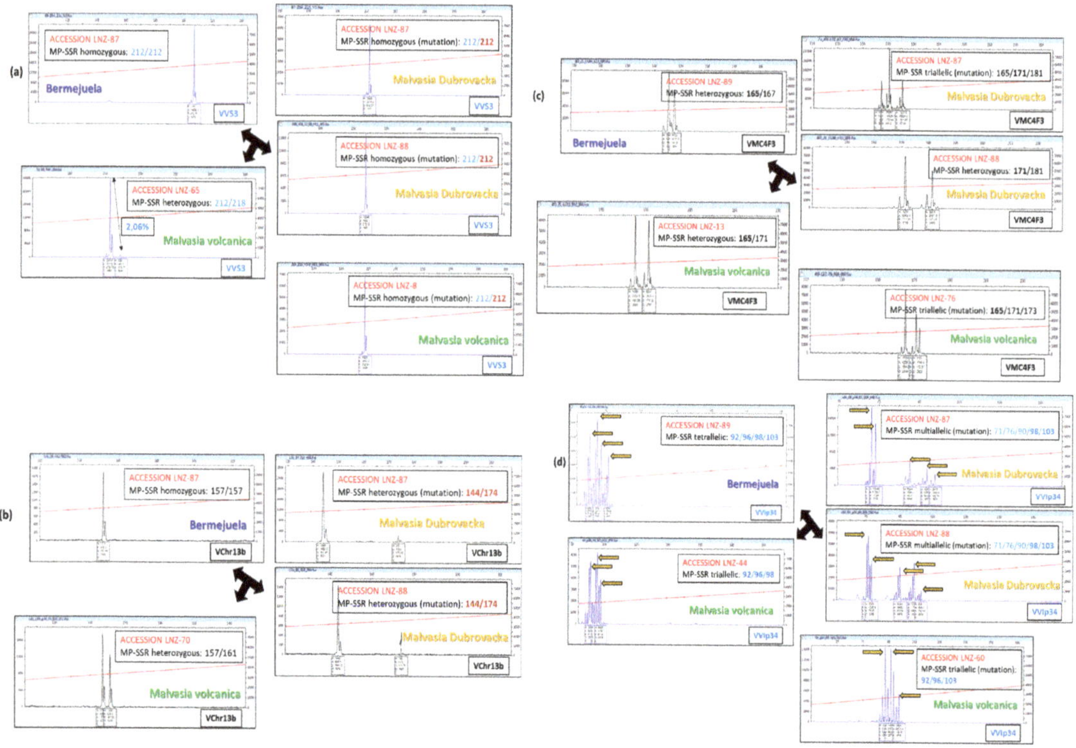

Figure 6. Pedigree of the Malvasia volcanica variety. Examples of four cases in which the parent varieties (Malvasia Dubrovacka and Bermejuela) were mutated. (**a**) The pedigree was not fulfilled. Malvasia volcanica is heterozygous, both parents being homozygous (although it may exhibit homozygous variation). (**b**) The pedigree was not fulfilled. Malvasia volcanica is heterozygous, but its alleles show a numerical change with respect to Malvasia Dubrocacka. In triallelic (**c**), tetrallelic, and multiallelic (**d**) cases, the pedigree was always fulfilled. Number in red: possible allele mutated.

4. Discussion

The primary objective of this study was to investigate variations in MP-SSR markers (intra-varietal variability) within two grapevine varieties cultivated in the Canary Islands: Malvasia volcanica (a local variety from Lanzarote Island) [8,23–26,47,48,50,52,53] and Listan prieto (a very ancient variety from mainland Spain, specifically Castilla) [54,55].

SSR markers were chosen as the genetic analysis tool in response to the historical and biogeographical uniqueness of Canary Islands vineyards [23,56] as well as the successful application of SSRs in studies with samples collected across broad geographic regions [7,21].

In most European wine regions, the phylloxera crisis led to vineyard replanting with new and genetically limited materials, resulting in the homogenization of plant material that often-constrained the ability of SSRs to detect clonal differences [57]. Conversely, vineyards in the Canary Islands, free from phylloxera, have preserved centuries-old plantings (over 500 years), propagated vegetatively, where the natural accumulation of somatic mutations over time has generated greater intra-varietal genetic diversity. As noted by Rade S. Jančić, vineyard aging correlates with an increase in clonal diversity [58].

The interest in Lanzarote Island was to initiate a study on its local and most extensively cultivated variety, Malvasia volcanica. For this purpose, 86 samples were genotyped, with the particularity that four accessions were identified as different varieties: Muscat of Alexandria (LNZ-18 and LNZ-52), Listan negro (LNZ-59), and Malvasia Fina (LNZ-69). This detection of MP-SSR profiles differing from the accession name (inter-varietal variability) and thus sampling errors can be logically explained. The old Canary Islands vineyards do not correspond to uniform plantations of a single variety; rather, they are mixed plantings with a high diversity of grapevine varieties. Additionally, the length of *Vitis vinifera* L. shoots [57] allows for intermixing among adjacent vines planted very closely, both in "hoyos" (planting pits) (Figure 2) and *chabocos* [25]. This explains the appearance of other varieties within this Malvasia volcanica population of 86 accessions. The study continued including these varieties to assess their behavior. On Fuerteventura Island, the focus was on the intra-varietal variability of the Listan prieto variety (18 individuals) as it is highly emblematic on this island. The low sample number reflected the historically residual viticulture on Fuerteventura due to nearly desert-like climatic conditions. Presently, it is the only island without a Protected Designation of Origin (PDO), although viticulture is resurging with the recent European legislation allowing vineyard irrigation [26].

For the reasons outlined above, the results are encouraging, with variability found in 93.46% of the accessions analyzed (100 out of 107 individuals) using 17 out of 100 SSRs employed (Table 2). This overall result demonstrates the following: (1) SSRs were effective in this particular case, and (2) intra-varietal variability was present in the majority of samples. Specifically, for Malvasia volcanica, the variability was 93.9%, detected by 9 SSRs, presenting heterozygous, numerically variable heterozygous, homozygous, triallelic, and numerically variable triallelic variants. SSR VRG3 was tetraallelic across all individuals, thus unable to detect variability, as was triallelic SSR VRG2 (Table 2, Table 3 and Table S4). For Listan prieto, 7 SSRs detected 94.4% variability, with homozygous, numerically variable homozygous, heterozygous, triallelic, and tetraallelic variants. SSR VVIp34 was fully tetraallelic, and SSRs VRG2 (Table 2, Table 3, Table S5 and Figure S2f) and VRG3 (Table 2, Table 3 and Table S5) were fully multiallelic. For Malvasia Dubrovacka, variability reached 100%, with two samples showing heterozygous, numerically variable heterozygous, and triallelic variants across 5 SSRs. These accessions exhibited cases of homozygous variations, heterozygous variations with numerical differences, and triallelism. The following SSR markers showed no capacity for differentiation: (1) VVIb01, VRG2, and VVIt60 as triallelic; (2) VRG3 as tetraallelic; and (3) VVIp34 as multiallelic (Table 2, Table 3 and Table S6). Among the two Muscat of Alexandria samples, only one SSR detected variability in accession LNZ-18, resulting in 50% variability. Here, SSR VRG3 was triallelic, while VVIv51 and VRG2 were tetraallelic (Table 2, Table 3 and Table S7). The other three varieties had only one representative each, making variation detection particularly challenging. For Listan Negro (LNZ-59), a single variation in SSR VVS3 was defined, which is included in the TECNENOL database. Notably, triallelism appeared in SSRs VRG2, VVMD14, and VVIp34, and tetraallelism in VRG3 (Table 2, Table 3 and Table S8). Similarly, Malvasia Fina (LNZ-69) presented a single SSR VVS3 variation known from TECNENOL, with triallelism in VRG2 and VVMD14, and tetraallelism in VRG3 and VVIp34 (Table 2, Table 3 and Table S9). To conclude, Bermejuela showed a single presumed mutation in SSR VRG16, hypothesized based on its involvement in the pedigree of Malvasia volcanica as a progenitor (Section 3.3). SSRs VVIv51 and VRG2 were triallelic, while VRG3 and VVIp34 were tetraallelic for this variety (Table 2, Table 3 and Table S10). Thus, the variety with the highest detected variability was Malvasia Dubrovacka, followed by Listan prieto, Malvasia volcanica, and Muscat of Alexandria showing the least variability. The other three varieties, represented by single individuals, lacked sufficient reference for conclusive analysis.

Accessions with the highest number of variations (Tables 2 and 3) were LNZ-87 (Malvasia Dubrovacka) (Figure 4) and FTV-8 (Listan prieto) (Figure 5), with five variations each. This was followed by FTV-13 with four variations (Figure S3). Three variations were observed in two samples from Lanzarote and five from Fuerteventura. Thirty-six samples showed two variations (five from Fuerteventura), and fifty-four accessions had a single variation (with another five from Fuerteventura). Finally, seven samples showed no MP-SSR variation.

To interpret this classification, one may hypothesize considering that the evolution of grapevines on Lanzarote was shorter than on Fuerteventura. Between 1730 and 1736, the Timanfaya volcanic eruptions destroyed all crops [27], thus grapevine evolution on Lanzarote spans just over 300 years. In contrast, grapevines theoretically never disappeared on Fuerteventura, suggesting over 500 years of continuous evolution. Consequently, vines on Lanzarote had to be reintroduced from neighboring islands, which had uninterrupted evolution for over five centuries [27]. Moreover, natural selection shaped vines imported mainly from mainland Spain to adapt to the harsh desert climate, especially on Fuerteventura, where evapotranspiration remains high due to the lack of protective ground cover like *picón* or *rofe* [25,28].

This explains why the Fuerteventura samples exhibited, both in absolute numbers and proportionally, higher variability than those from Lanzarote. Representative examples include FTV-8, FTV-13, and five additional samples from Fuerteventura with five, four, and three variations, respectively (Tables 2 and 3). Another accession with five variations was a Malvasia Dubrovacka sample from Lanzarote, possibly reflecting an evolution exceeding 500 years from another island before being introduced to Lanzarote. This sample appeared more evolved than any Malvasia volcanica sample, despite being one of its progenitors. Thus, if the crossing occurred on Lanzarote, all Malvasia volcanica samples would be less evolved, having fewer variations as they post-date the Malvasia Dubrovacka sample (LNZ-87). Regardless of the case(s), we assume that the process of incorporating variations into a given genome remains largely unknown in most instances. Although in some cases, it may be possible to reliably estimate the timing of genome fragment introgression [59].

Regarding SSR efficiency for detecting intra-varietal variability in this grapevine collection, the best-performing SSRs were those generating the highest absolute number of MP-SSR variants, and if possible, capable of distinguishing different varieties. In descending order of efficiency (Tables S2 and S3a): (1) VChr15b [44], which distinguished 6 variety groups and 5 variations, totaling 11 MP-SSR detections; (2) VVIp34 [41] with a total of 9 detections (2 variations), distinguishing 7 groups; alongside the international SSR VVMD32, which distinguished 6 groups and 3 variations (9 total); (3) VChr9b [44], with 9 total MP-SSR detections (5 mutations), though distinguishing only 5 groups; (4) the international SSRs VVMD5 and VVMD28 [34]; and (5) fifth in the ranking was the SSR marker designed by the Vitis Microsatellite Consortium, VMC4F3, which demonstrated a total efficiency that enabled the identification of 8 MP-SSRs including 2 variations and 6 varietal groups. The least efficient SSRs were (Tables S2 and S3a,b): (1) VVNTM1 [9,46], which failed to distinguish any MP-SSR; (2) a group of eight SSRs that were only able to identify two MP-SSRs (VChr19a, VRG4, VMC3D8, VChr12a, VChr16a, VRG2, VRG13, VRG3); and (3) a group of eighteen SSRs with similarly low discriminatory power, each distinguishing only three MP-SSRs (VVS3, VVUCH12, VVNTM5, VVIn73, VChr5b, VChr15a, VChr1b, VChr18b, VChr10a, VChr4a, VMD26, VRG1, ZAG67, VRG7, VRG11, VRG15, VChr19b, VVS29), among which only VVS29 was able to detect intra-varietal variation, but it was limited to discriminating just two varietal groups. Therefore, it can be concluded that 27% of the SSRs used in this study showed little to no ability to discriminate MP-SSR. Conversely, if we consider the most efficient group to be those with the highest absolute potential to detect different MP-SSRs in this collection, namely, those capable of distinguishing between 8 and 11 MP-SSRs, this group represents 7% of the SSRs used. If we also include those able to detect 7 distinct MP-SSRs, the proportion of SSRs with good discriminatory capacity rises to 23%.

Vitis vinifera L. is a diploid species characterized by a high level of heterozygosity, which confers a great capacity for inter-varietal variability [60]. This elevated heterozygosity is the result of its evolutionary history (hybridizations between local forms of the subspecies sylvestris and the domesticated subspecies) and its cross-pollinating sexual reproduction (driven by the dioecious nature of sylvestris) [5]. Nevertheless, some varieties exhibit long homozygous regions. Two main mechanisms explain this loss of heterozygosity: (1) cellular displacement in periclinal chimeras [61], and (2) chromosomal replacement or deletion events [62,63]. In our study, the variety that exhibited the highest heterozygosity was Malvasia Dubrovacka, with 72% of heterozygous loci (Table S6), followed by Listan prieto with 68% (Table S5), and Bermejuela with 67% (Table S10). Malvasia Fina presented 64% heterozygous *loci* (Table S9), while Muscat of Alexandria and Listan negro showed 62% and 61%, respectively (Tables S7 and S8). In contrast, the variety with the highest degree of homozygosity was Malvasia volcanica, with 41% of *loci* being homozygous (Table S4). The remaining percentages up to 100% can be explained by the presence of *loci* with more than two alleles. These are cases of triallelism, tetraallelism, and multiallelism (Table 2 and Table S4–Table S10). The emergence of such allelic configurations is primarily attributed to somatic mutations, mainly periclinal chimeras, accumulated over centuries of clonal propagation under anthropogenic selection pressure. However, as Gambino et al. [15] demonstrated, a chimeric plant does not necessarily exhibit multiallelism. Thus, in all of the varieties studied within this population of 107 accessions, regions with three or more alleles were identified (Table 2 and Table S4–Table S10). As an example, one of the most striking cases was observed in the SSR marker VVIp34, where heterozygous, triallelic, tetraallelic, and multiallelic individuals appeared depending on the variety being genotyped (Figure 6d andFigure 7). A frequent issue encountered is whether to consider or disregard a very small peak or family of peaks. This is an important decision, as it may determine whether an individual is classified as heterozygous or homozygous. When an electropherogram displays two families of peaks and one of them is very small or extremely small, the smaller peak is considered valid if the peak ratio (between the smaller and the larger family) falls within or exceeds the range of 1.5 to 2 (Figure 4, Figure 5 and Figure S2). Otherwise, if the ratio is lower, the peak or peak family is disregarded, and the MP-SSR is considered homozygous [64].

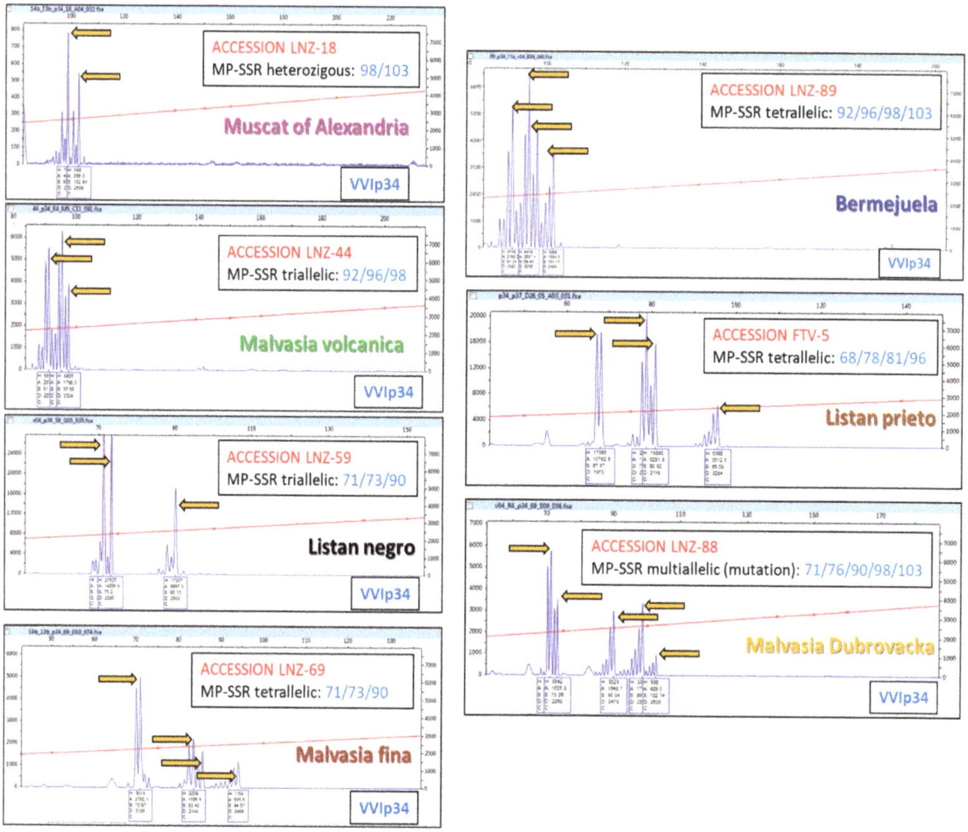

Figure 7. Detail of SSR VVIp34 showing the ability to generate heterozygous MP-SSR profiles in the Muscat of Alexandria variety, triallelic profiles in Malvasia volcanica and Listan negro, tetraallelic profiles in Malvasia fina and Bermejuela, and multiallelic profiles in Malvasia Dubrovacka.

The possibility that Bermejuela and Malvasia Dubrovacka might be the progenitors of Malvasia volcanica was first proposed in 2006 in the book titled "*Variedades de Vid de Cultivo Tradicional en Canarias*", in which the authors hypothesized this relationship based on a study involving six SSR markers [53]. It was not until 2018 that Dr. Immaculada Rodríguez-Torres confirmed this possibility in her book "*Variedades de vid cultivadas en Canarias. Descriptores Morfológicos. Caracterización morfológica, molecular, agronómica y enológica*" using 11 SSRs [50]. Subsequently, our research group published a study in the international journal OenoOne titled "Molecular characterization of *Vitis vinifera* L. local cultivars from volcanic areas (Canary Islands and Madeira) using SSR markers", in which this parental relationship was confirmed using 20 SSR markers [23]. As part of the present study on the intra-varietal variability of Malvasia volcanica, we considered it a suitable opportunity to confirm this pedigree using 100 SSR *loci*. However, not all SSR markers supported the proposed pedigree. Markers VVS3, VRG16, and VChr13b did not validate the parental relationships. Figure 6a,b provides two examples corresponding to markers VVS3 and VChr13b. In Figure 6a, the homozygous MP-SSR profiles of the two Malvasia Dubrovacka samples (variants LNZ-87 and LNZ-88) and the homozygous ARP of Bermejuela (LNZ-89) are shown according to the TECNENOL database [8,23–26,47,48]. Also shown are the MP-SSR profiles of two Malvasia volcanica samples: one corresponding to a heterozygous ARP (LNZ-65), and the other to a mutated sample of Malvasia volcanica (LNZ-8), also recorded in the TECNENOL database [8,23–26,47,48]. The SSR VChr13b marker (Figure 6b) showed a Malvasia volcanica genotype that was heterozygous (157/161), while the putative parents were: (1) Bermejuela, which is homozygous (157/157), and (2) the two analyzed Malvasia Dubrovacka individuals, which are heterozygous (144/174) and possibly mutated (numerical variation). For this particular SSR, no information is available in the TECNENOL database [8,23–26,47,48], nor has any bibliographic reference been found for these varieties. For SSR VRG16, the mutated parent is most likely Bermejuela, which is homozygous (250/250), while the two Malvasia Dubrovacka samples are heterozygous (240/250), and Malvasia volcanica is also homozygous (240/240). Assuming that the inconsistencies in these pedigrees are due to mutations in one of the parents, which is plausible given the low number of genotyped parental samples (only two Malvasia Dubrovacka accessions and one Bermejuela accession), it would be reasonable to affirm that the parents of Malvasia Volcánica are Malvasia Dubrovacka and the local Canarian variety Bermejuela. In cases where one or both progenitors exhibited triallelism (e.g., VVIb01, VMC4F3, VVIv33, VVIv51, VVIn57, VVIt60), tetraallelism, or multiallelism (e.g., VVIp34), the pedigree was confirmed in all cases. Figure 6c,d illustrate two examples supporting these parental relationships through loci with more than two alleles.

5. Conclusions

Certainly, the most critical challenge viticulture faces in the 21st century is climate change. For this reason, experts emphasize the need to explore the biodiversity of the species *Vitis vinifera* L. at all levels. Obtaining plant material resilient to water stress, heatwaves (extreme temperatures), and excessive sunlight is essential as a key strategy to mitigate the effects of climate change. In this context, the present study explored the intra-varietal biodiversity of two grapevine varieties widely cultivated in the Canary Islands of Lanzarote and Fuerteventura (Spain). These islands are characterized by a desert climate and are strongly influenced by the trade winds and Saharan dust (Calima). In these volcanic, phylloxera-free areas, grapevines have evolved over three to five centuries, adapting to harsh abiotic conditions through both natural and anthropogenic (asexual reproduction) selection, incorporating relevant somatic mutations that have allowed them to persist to this day.

The aim of this study was to identify individuals exhibiting variation in their molecular profiles (clones) within the varieties Malvasia volcanica (a local variety from Lanzarote) and Listan prieto (a widely cultivated and extended variety in Fuerteventura). Among the 86 Malvasia volcanica accessions, inter-varietal variability was detected, identifying two Muscat of Alexandria accessions (LNZ-18 and LNZ-52), one accession corresponding to the local Canarian variety Listan negro (LNZ-59), and one accession identified as the Portuguese Malvasia Fina (LNZ-69). Additionally, 18 Listan prieto samples from Fuerteventura were analyzed. Overall, intra-varietal variability was found in 93.46% of the analyzed accessions (100 out of 107 individuals). The accessions with the highest number of variations were Malvasia Dubrovacka (LNZ-87) and Listan prieto (FTV-8), each exhibiting five variations. One Listan prieto accession (FTV-13) showed four variations. A group of seven individuals exhibited three variations, two belonging to Malvasia volcanica (LNZ-12, LNZ-72) and five to Listan prieto (FTV-1, FTV-2, FTV-7, FTV-9, FTV-12). A group of 36 samples presented two variations, while 54 accessions showed only one variation. Finally, seven samples exhibited no variation in their MP-SSR profiles. The variety with the highest percentage of variation was Malvasia Dubrovacka (100%, based on only two individuals), followed by Listan prieto (94.4%, with 18 samples), Malvasia volcanica (93.9%, with 82 accessions), and Muscat of Alexandria (50%, with two individuals). For the remaining three varieties, no conclusive reference could be established due to each being represented by a single accession.

To study intra-varietal variability, 100 SSR markers were employed. Of these, 17 (VVS3, VVS29, VVMD5, VVMD28, VMC4F3, VRG16, VVMD32, ZAG7, VChr13c, VVIv33, VVIb09, VChr9b, VChr15b, VVIn57, VChr14b, VVIp34, and VChr13b) were informative in this population of 107 individuals. The most efficient markers were: (1) VChr15b; (2) VVIp34 and VVMD32; (3) VChr9b; (4) VVMD5 and VVMD28; and (5) VMC4F3, detecting 11, 9, and 8 MP-SSRs, respectively. These corresponded to only 7% of the SSRs used. Including those that detected 7 distinct MP-SSRs (a group of 16 SSRs), the percentage of highly discriminating SSRs rises to 23%. The least efficient SSR markers were: (1) VVNTM1; (2) VChr19a, VRG4, VMC3D8, VChr12a, VChr16a, VRG2, VRG13, and VRG3. Thus, 27% of the SSRs used in this study (those detecting 0, 1, 2, or 3 MP-SSRs) showed little to no capacity for MP-SSR discrimination.

The most homozygous variety was Malvasia volcanica, followed by Listan negro, Muscat of Alexandria, Malvasia Fina, Bermejuela, and Listan prieto. The most heterozygous variety was Malvasia Dubrovacka.

The pedigree of Malvasia volcanica is supported by the results from 100 SSR markers, under the assumption that for three SSRs, one of the parents exhibits variation (mutation).

Therefore, under the studied conditions (phylloxera-free, volcanic, and isolated areas), it is possible to detect intra-varietal variability using SSR (microsatellite) markers.

Supplementary Materials: The following supporting information can be downloaded at: https://www.mdpi.com/article/10.3390/horticulturae11070823/s1. Table S1. List of the 107 accessions collected on Lanzarote Island. General information on the resulting analyzed variety (VIVC) and entry registration details; Table S2. List of primers used for the amplification of the selected microsatellite regions. Characteristics; Figure S1. Approximation of the genomic SSR map used in this study. Consensus location of each of the regions selected for molecular characterization; Figure S2. Examples of samples showing variation (mutation) compared with the most widespread profile (ARP) including their replicates. N° %: Examples of peak ratios (percentage indicating the ratio between the smallest and largest peaks), used to disregard extremely small peaks; Figure S3. Lisan prieto profiles (FTV-13) showing variations compared with the most widespread or reference molecular profile (ARP). On the left, ARP electropherograms; Table S3a. List of SSR markers ranked from the most informative, detecting both intra-varietal and inter-varietal variability, to the least informative, which did not detect any variability, meaning it only identified a single MP-SSR; Table S3b. List of SSR markers ranked from the most informative in distinguishing varieties to the least informative, which did not detect any differences, meaning it only identified a single MP-SSR; Table S4. List of SSR markers showing predominant homozygosity, heterozygosity, and multiallelism in the Malvasia volcanica variety. *: Accessions showing variation; Table S5. List of SSR markers showing predominant homozygosity, heterozygosity, and multiallelism in the Listan prieto variety. *: Accessions showing variation; Table S6. List of SSR markers showing predominant homozygosity, heterozygosity, and multiallelism in the Malvasia Dubrovacka variety. *: Accessions showing variation; Table S7. List of SSR markers showing predominant homozygosity, heterozygosity, and multiallelism in the Muscat of Alexandria variety. *: Accessions showing variation; Table S8. List of SSR markers showing predominant homozygosity, heterozygosity, and multiallelism in the Listan prieto variety. *: Accessions showing variation; Table S9. List of SSR markers showing predominant homozygosity, heterozygosity, and multiallelism in the Malvasia fina variety. *: Accessions showing variation; Table S10. List of SSR markers showing predominant homozygosity, heterozygosity, and multiallelism in the Bermejuela variety. *: Accessions showing variation.

Author Contributions: Conceptualization, F.F. and L.R.S.-A.; Methodology, L.R.S.-A., Q.L.-Y. and F.F.; Software, L.R.S.-A., Q.L.-Y. and F.F.; Validation, F.F. and L.R.S.-A.; Formal analysis, F.F., Q.L.-Y. and L.R.S.-A.; Investigation, F.F., Q.L.-Y. and L.R.S.-A.; Resources, F.F. and L.R.S.-A.; Data curation,

F.F. and L.R.S.-A.; Writing—original draft preparation, F.F., L.R.S.-A. and L.D.; Writing—review and editing, L.D. and L.R.S.-A.; Visualization, J.M.C. and F.Z.; Supervision, F.F., Q.L.-Y., L.R.S.-A., L.D., J.M.C. and F.Z. All authors have read and agreed to the published version of the manuscript.

Funding: This study was funded by the Canary Islands Rural Development Fund (Government of the Canary Islands) through the Island Council of Lanzarote and by the Department of Agriculture, Livestock, and Fisheries of the Island Council of Fuerteventura.

Data Availability Statement: Data are not available due to confidentiality.

Acknowledgments: The authors wish to express their gratitude to the Island Council of Lanzarote for their interest in supporting a study of this nature. In particular, they thank Ana Garrido for being the driving force behind the project. They also appreciate the contributions of Darío Pérez, representative of the Regulatory Council of the Protected Designation of Origin "Vinos de Lanzarote". Additionally, the authors thank the Island Council of Fuerteventura and the Majuelo Viticulturists Association for promoting the intra-varietal study of the Listan prieto variety. Finally, sincere thanks are extended to Rosa Pastor, Braulio Esteve-Zarzoso, Laia Fañanás, Cristina Domènech, and Iris Ginés for their invaluable logistical support in the completion of this work.

Conflicts of Interest: The authors declare no conflicts of interest.

References

1. Myles, S.; Boyko, A.R.; Owens, C.L.; Brown, P.J.; Grassi, F.; Aradhya, M.K.; Prins, B.; Reynolds, A.; Chia, J.-M.; Ware, D.; et al. Genetic structure and domestication history of the grape. *Proc. Natl. Acad. Sci. USA* **2011**, *108*, 3457–3462. [CrossRef] [PubMed]
2. Emanuelli, F.; Lorenzi, S.; Grzeskowiak, L.; Catalano, V.; Stefanini, M.; Troggio, M.; Myles, S.; Martinez-Zapater, J.M.; Zyprian, E.; Moreira, F.M.; et al. Genetic diversity and population structure assessed by SSR and SNP markers in a large germplasm collection of grape. *BMC Plant Biol.* **2013**, *13*, 39. [CrossRef]
3. International Organisation of Vine and Wine. *State of the World Vine and Wine Sector, 2023*; OIV: Paris, France, 2023; Available online: https://www.oiv.int/sites/default/files/documents/OIV_State_of_the_world_Vine_and_Wine_sector_in_2022_2.pdf (accessed on 15 May 2025).
4. Wolkovich, E.M.; García de Cortázar-Atauri, I.; Morales-Castilla, I.; Nicholas, K.A.; Lacombe, T. From Pinot to Xinomavro in the world's future wine-growing regions. *Nat. Clim. Change* **2018**, *8*, 29–37. [CrossRef]
5. Pelsy, F.; Dumas, V.; Bévilacqua, L.; Hocquigny, S.; Merdinoglu, D. Chromosome replacement and deletion lead to clonal polymorphism of berry color in grapevine. *PLoS Genet.* **2015**, *11*, e1005081. [CrossRef]
6. Jackson, R.S. *Wine Science: Principles and Applications*, 4th ed.; Elsevier: Amsterdam, The Netherlands, 2014; pp. 1–978.
7. Tympakianakis, S.; Trantas, E.; Avramidou, E.V.; Ververidis, F. *Vitis vinifera* genotyping toolbox to highlight diversity and germplasm identification. *Front. Plant Sci.* **2023**, *14*, 1139647. [CrossRef] [PubMed]
8. Fort, F.; Lin-Yang, Q.; Valls, C.; Sancho-Galán, P.; Canals, J.M.; Zamora, F. Analysis of the diversity presented by *Vitis vinifera* L. in the volcanic island of La Gomera (Canary Archipelago, Spain) using simple sequence repeats (SSRs) as molecular markers. *Horticulturae* **2024**, *10*, 14. [CrossRef]
9. Migliaro, D.; De Nardi, B.; Vezzulli, S.; Crespan, M. An upgraded core set of 11 SSR markers for grapevine cultivar identification: The case of berry-color mutants. *Am. J. Enol. Vitic.* **2017**, *68*, 496–502. [CrossRef]
10. Peiró, R.; Soler, J.X.; Crespo, A.; Jiménez, C.; Cabello, F.; Gisbert, C. Genetic variability assessment in 'Muscat' grapevines including 'Muscat of Alexandria' clones from selection programs. *Span. J. Agric. Res.* **2018**, *16*, e0702. [CrossRef]
11. Carrier, G.; Le Cunff, L.; Dereeper, A.; Legrand, D.; Sabot, F.; Bouchez, O.; Audeguin, L.; Boursiquot, J.M.; This, P. Transposable elements are a major cause of somatic polymorphism in *Vitis vinifera* L. *PLoS ONE* **2012**, *7*, e32973. [CrossRef]
12. Meneghetti, S.; Calò, A.; Bavaresco, L. A strategy to investigate the intravarietal genetic variability in *Vitis vinifera* L. for clones and biotypes identification and to correlate molecular profiles with morphological traits or geographic origins. *Mol. Biotechnol.* **2012**, *52*, 68–81. [CrossRef]
13. De Lorenzis, G.; Squadrito, M.; Rossoni, M.; Di Lorenzo, G.S.; Brancadoro, L.; Scienza, A. Study of intra-varietal diversity in biotypes of Aglianico and Muscat of Alexandria (*Vitis vinifera* L.) cultivars. *Aust. J. Grape Wine Res.* **2017**, *23*, 132–142. [CrossRef]
14. Calderón, L.; Mauri, N.; Muñoz, C.; Carbonell-Bejerano, P.; Bree, L.; Bergamin, D.; Sola, C.; Gomez-Talquenca, S.; Royo, C.; Ibáñez, J.; et al. Whole genome resequencing and custom genotyping unveil clonal lineages in 'Malbec' grapevines (*Vitis vinifera* L.). *Sci. Rep.* **2021**, *11*, 7775. [CrossRef]

15. Gambino, G.; Dal Molin, A.; Boccacci, P.; Minio, A.; Chitarra, W.; Avanzato, C.G.; Tononi, P.; Perrone, I.; Raimondi, S.; Schneider, A.; et al. Whole-genome sequencing and SNV genotyping of 'Nebbiolo' (*Vitis vinifera* L.) clones. *Sci. Rep.* **2017**, *7*, 17294. [CrossRef] [PubMed]
16. Araya-Ortega, D.; Gainza-Cortés, F.; Riadi, G. Exploring genomic differences between a pair of *Vitis vinifera* clones using WGS data: A preliminary study. *Horticulturae* **2024**, *10*, 1026. [CrossRef]
17. Villano, C.; Procino, S.; Blaiotta, G.; Carputo, D.; D'Agostino, N.; Di Serio, E.; Fanelli, V.; La Notte, P.; Miazzi, M.M.; Montemurro, C.; et al. Genetic diversity and signature of divergence in the genome of grapevine clones of Southern Italy varieties. *Front. Plant Sci.* **2023**, *14*, 1201287. [CrossRef]
18. Procino, S.; Miazzi, M.M.; Savino, V.N.; La Notte, P.; Venerito, P.; D'Agostino, N.; Taranto, F.; Montemurro, C. Genome scan analysis for advancing knowledge and conservation strategies of Primitivo clones (*Vitis vinifera* L.). *Plants* **2025**, *14*, 437. [CrossRef]
19. Esteras, C.; Peiró, R.; Soler, J.X.; Gisbert, C. Genetic variability in grapevine clones of 'Muscat of Alexandria'. *Acta Hortic.* **2019**, *1248*, 77–80. [CrossRef]
20. Urra, C.; Sanhueza, D.; Pavez, C.; Tapia, P.; Núñez-Lillo, G.; Minio, A.; Miossec, M.; Blanco-Herrera, F.; Gainza, F.; Castro, A.; et al. Identification of grapevine clones via high-throughput amplicon sequencing: A proof-of-concept study. *G3 Genes Genomes Genet.* **2023**, *13*, jkad145. [CrossRef]
21. Jahnke, G.; Májer, J.; Varga, P.; Szőke, B. Analysis of clones of Pinots grown in Hungary by SSR markers. *Sci. Hortic.* **2011**, *129*, 32–37. [CrossRef]
22. Baltazar, M.; Castro, I.; Gonçalves, B. Adaptation to climate change in viticulture: The role of varietal selection—A review. *Plants* **2025**, *14*, 104. [CrossRef]
23. Marsal, G.; Mendez, J.J.; Mateo-Sanz, J.M.; Ferrer, S.; Canals, J.M.; Zamora, F.; Fort, F. Molecular characterization of *Vitis vinifera* L. local cultivars from volcanic areas (the Canary Islands and Madeira) using SSR markers. *OENO One* **2019**, *53*, 667–680. [CrossRef]
24. Fort, F.; Lin-Yang, Q.; Suárez-Abreu, L.R.; Sancho-Galán, P.; Canals, J.M.; Zamora, F. Study of molecular biodiversity and population structure of *Vitis vinifera* L. ssp. vinifera on the volcanic island of El Hierro (Canary Islands, Spain) by using microsatellite markers. *Horticulturae* **2023**, *9*, 1297. [CrossRef]
25. Fort, F.; Marsal, G.; Mateo-Sanz, J.M.; Pena, V.; Canals, J.M.; Zamora, F. Molecular characterisation of the current cultivars of *Vitis vinifera* L. in Lanzarote (Canary Islands, Spain) reveals nine individuals which correspond to eight new varieties and two new sports. *OENO One* **2022**, *56*, 281–295. [CrossRef]
26. Fort, F.; Lin-Yang, Q.; Valls, C.; Sancho-Galán, P.; Canals, J.M.; Zamora, F. Characterisation and identification of vines from Fuerteventura (Canary Volcanic Archipelago (Spain)) using simple sequence repeat markers. *Horticulturae* **2023**, *9*, 1301. [CrossRef]
27. Mendez, J.J. *Acerca del Canary Wine. Compendio de la Vitivinicultura del Archipiélago Canario*, 2nd ed.; Asociación de Viticultores y Bodegueros de Canarias AVIBO: Santa Cruz de Tenerife, Spain, 2024.
28. González Morales, A. El Cultivo y Producción del vino en la isla de Lanzarote. Territoires du vin [Online], 1 March 2011. Available online: https://preo.ube.fr/territoiresduvin/index.php?id=1407 (accessed on 15 May 2025).
29. Marsal, G.; Boronat, N.; Canals, J.M.; Zamora, F.; Fort, F. Comparison of the efficiency of some of the most usual DNA extraction methods for woody plants in different tissues of *Vitis vinifera* L. *J. Int. Sci. Vigne Vin* **2013**, *47*, 227–237. [CrossRef]
30. Marsal, G.; Baiges, I.; Canals, J.M.; Zamora, F.; Fort, F. A fast, efficient method for extracting DNA from leaves, stems, and seeds of *Vitis vinifera* L. *Am. J. Enol. Vitic.* **2011**, *62*, 376–381. [CrossRef]
31. Fort, F.; Hayoun, L.; Valls, J.; Canals, J.M.; Arola, L.; Zamora, F. A new and simple method for rapid extraction and isolation of high-quality RNA from grape (*Vitis vinifera*) berries. *J. Sci. Food Agric.* **2008**, *88*, 179–184. [CrossRef]
32. Thomas, M.R.; Scott, N.S. Microsatellite repeats in grapevine reveal DNA polymorphisms when analysed as sequence-tagged sites (STSs). *Theor. Appl. Genet.* **1993**, *86*, 985–990. [CrossRef]
33. Bowers, J.E.; Dangl, G.S.; Vignani, R.; Meredith, C.P. Isolation and characterization of new polymorphic simple sequence repeat loci in grape (*Vitis vinifera* L.). *Genome* **1996**, *39*, 628–633. [CrossRef]
34. Bowers, J.E.; Dangl, G.S.; Meredith, C.P. Development and characterization of additional microsatellite DNA markers for grape. *Am. J. Enol. Vitic.* **1999**, *50*, 243–246. [CrossRef]
35. Sefc, K.M.; Regner, F.; Turetschek, E.; Glossl, J.; Steinkellner, H. Identification of microsatellite sequences in *Vitis riparia* and their applicability for genotyping of different *Vitis* species. *Genome* **1999**, *42*, 367–373. [CrossRef] [PubMed]
36. Scott, K.D.; Eggler, P.; Seaton, G.; Rosseto, M.; Ablett, E.M.; Lee, L.S.; Henry, R.J. Analysis of SSRs derived from grape ESTs. *Theor. Appl. Genet.* **2000**, *100*, 723–726. [CrossRef]
37. Di Gaspero, G.; Peterlunger, E.; Testolin, R.; Edwards, K.J.; Cipriani, G. Conservation of microsatellite loci within the genus *Vitis*. *Theor. Appl. Genet.* **2000**, *101*, 301–308. [CrossRef]
38. Lefort, F.; Kyvelos, C.; Zervou, M.; Edwards, K.; Roubelakis-Angelakis, K. Characterization of new microsatellite loci from *Vitis vinifera* and their conservation in some *Vitis* species and hybrids. *Mol. Ecol. Resour.* **2002**, *2*, 20–21. [CrossRef]
39. Crespan, M. The parentage of Muscat of Hamburg. *Vitis* **2003**, *42*, 193–197. [CrossRef]

40. Arroyo-García, R.; Martínez-Zapater, J.M. Development and characterization of new microsatellite markers for grape. *Vitis* **2004**, *43*, 175–178. [CrossRef]
41. Merdinoglu, D.; Butterlin, G.; Bevilacqua, L.; Chiquet, V.; Adam-Blondon, A.-F.; Decroocq, S. Development and characterization of a large set of microsatellite markers in grapevine (*Vitis vinifera* L.) suitable for multiplex PCR. *Mol. Breed.* **2005**, *15*, 349–366. [CrossRef]
42. Doligez, A.; Adam-Blondon, A.F.; Cipriani, G.; Di Gaspero, G.; Laucou, V.; Merdinoglu, D.; Meredith, C.P.; Riaz, S.; Roux, C.; This, P. An integrated SSR map of grapevine based on five mapping populations. *Theor. Appl. Genet.* **2006**, *113*, 369–382. [CrossRef]
43. Regner, F.; Hack, R.; Santiago, J.L. Highly variable *Vitis* microsatellite loci for the identification of Pinot Noir clones. *Vitis* **2006**, *45*, 85–91. [CrossRef]
44. Cipriani, G.; Marrazzo, M.T.; Di Gaspero, G.; Pfeiffer, A.; Morgante, M.; Testolin, R. A set of microsatellite markers with long core repeat optimized for grape (*Vitis* spp.) genotyping. *BMC Plant Biol.* **2008**, *8*, 127. [CrossRef]
45. Fournier-Level, A.; Le Cunff, L.; Gomez, C.; Doligez, A.; Ageorges, A.; Roux, C.; Bertrand, Y.; Souquet, J.M.; Cheynier, V.; This, P. Quantitative genetic bases of anthocyanin variation in grape (*Vitis vinifera* L. ssp. *sativa*) berry: A quantitative trait locus to quantitative trait nucleotide integrated study. *Genetics* **2009**, *183*, 1127–1139. [CrossRef] [PubMed]
46. This, P.; Jung, A.; Boccacci, P.; Borrego, J.; Botta, R.; Costantini, L.; Crespan, M.; Dangl, G.S.; Eisenheld, C.; Ferreira-Monteiro, F.; et al. Development of a standard set of microsatellite reference alleles for identification of grape cultivars. *Theor. Appl. Genet.* **2004**, *109*, 1448–1458. [CrossRef] [PubMed]
47. Marsal, G.; Mateo, J.M.; Canals, J.M.; Zamora, F.; Fort, F. SSR analysis of 338 accessions planted in Penedès (Spain) reveals 28 unreported molecular profiles of *Vitis vinifera* L. *Am. J. Enol. Vitic.* **2016**, *67*, 466–470. [CrossRef]
48. Marsal, G.; Bota, J.; Martorell, A.; Canals, J.M.; Zamora, F.; Fort, F. Local cultivars of *Vitis vinifera* L. in Spanish islands: Balearic Archipelago. *Sci. Hortic.* **2017**, *226*, 122–132. [CrossRef]
49. Maul, E.; Röckel, F. Vitis International Variety Catalogue. Available online: http://www.vivc.de (accessed on 15 May 2025).
50. Rodríguez-Torres, I. *Variedades de Vid Cultivadas en Canarias. Descriptores Morfológicos. Caracterización Morfológica, Molecular, Agronómica y Enológica*; Instituto Canario de Investigaciones Agrarias, Gobierno de Canarias: Santa Cruz de Tenerife, Spain, 2018.
51. El Oualkadi, A.; Ater, M.; Messaoudi, Z.; Laucou, V.; Boursiquot, J.M.; Lacombe, T.; This, P. Molecular characterization of Moroccan grapevine germplasm using SSR markers for the establishment of a reference collection. *OENO One* **2009**, *43*, 135–148. [CrossRef]
52. Fort, F.; Suárez-Abreu, L.R.; Lin-Yang, Q.; Deis, L.; Canals, J.M.; Zamora, F. Origin and Possible Members of the 'Malvasia' Family: The New Fuencaliente de La Palma Hypothesis on the True 'Malvasia'. *Horticulturae* **2025**, *11*, 561. [CrossRef]
53. Zerolo, J.; Cabello, F.; Espino, A.; Borrego, J.; Ibañez, J.; Rodriguez-Torres, I.; Muñoz-Organero, G.; Rubio, C.; Hernández, M. *Variedades de Vid de Cultivo Tradicional en Canarias*, 1st ed.; Instituto Canario de Calidad Agroalimentaria, Gobierno de Canarias: Santa Cruz de Tenerife, Spain, 2006; ISBN 978-84-606-3977-0.
54. Aliquo, G.; Torres, R.; Lacombe, T.; Boursiquot, J.M.; Laucou, V.; Gualpa, J.; Fanzone, M.; Sari, S.; Pérez-Peña, J.; Prieto, J.A. Identity and parentage of some South American grapevine cultivars present in Argentina. *Aust. J. Grape Wine Res.* **2017**, *23*, 452–460. [CrossRef]
55. Galet, P. *Dictionnaire Encyclopédique des Cépages*, 1st ed.; Hachette: Paris, France, 2000.
56. Macías, A. El paisaje vitícola de Canarias. Cinco siglos de historia. *Ería* **2005**, *68*, 351–364. Available online: https://reunido.uniovi.es/index.php/RCG/article/view/1525 (accessed on 15 May 2025).
57. Hidalgo, J.; Hidalgo, L. La Filoxera. In *Tratado de Viticultura*, 2nd ed.; Mundi-Prensa: Madrid, Spain, 2019; Volume 1.
58. Jančić, R.S. Morfološka i Molekularna Karakterizacija Potencijalnih Klonova Sorte Vinove loze Vranac [Morphological and Molecular Characterization of Potential Clones of the Grapevine Variety Vranac]. Ph.D. Thesis, University of Novi Sad, Novi Sad, Serbia, 2022. NaRDuS Repository. Available online: https://nardus.mpn.gov.rs/handle/123456789/21475 (accessed on 15 May 2025).
59. Dong, Y.; Duan, S.; Xia, Q.; Liang, Z.; Dong, X.; Margaryan, K.; Musayev, M.; Goryslavets, S.; Zdunić, G.; Bert, P.-F.; et al. Dual domestications and origin of traits in grapevine evolution. *Science* **2023**, *379*, 892–901. [CrossRef]
60. This, P.; Lacombe, T.; Thomas, M.R. Historical origins and genetic diversity of wine grapes. *Trends Genet.* **2006**, *22*, 511–519. [CrossRef]
61. Hocquigny, S.; Pelsy, F.; Dumas, V.; Kindt, S.; Heloir, M.-C.; Merdinoglu, D. Diversification within grapevine cultivars goes through chimeric states. *Genome* **2004**, *47*, 579–589. [CrossRef] [PubMed]
62. Zhou, Y.; Massonnet, M.; Sanjak, J.S.; Cantu, D.; Gaut, B.S. Evolutionary genomics of grape (*Vitis vinifera* ssp. *vinifera*) domestication. *Proc. Natl. Acad. Sci. USA* **2019**, *116*, 3720–3725. [CrossRef] [PubMed]

63. Li, B.; Gschwend, A.R. *Vitis labrusca* genome assembly reveals diversification between wild and cultivated grapevine genomes. *Front. Plant Sci.* **2023**, *14*, 1234130. [CrossRef] [PubMed]
64. Butler, J.M. *Advanced Topics in Forensic DNA Typing: Interpretation*; Academic Press: London, UK; Oxford, UK; Boston, MA, USA; New York, NY, USA; San Diego, CA, USA, 2014; ISBN 9780124052130.

horticulturae

MDPI

Article

Uncovering the Genetic Identity and Diversity of Grapevine (*Vitis vinifera* L.) in La Palma Island (Canary Archipelago, Spain) Through SSR-Based Varietal Profiling and Population Structure Analysis

Qiying Lin-Yang [1], Leonor Deis [1,2], Joan Miquel Canals [1], Fernando Zamora [1] and Francesca Fort [1,*]

1 Grupo de Tecnología Enológica (TECNENOL), Department of Biochemistry and Biotechnology, Faculty of Oenology, Rovira i Virgili University, Sescelades Campus, C/Marcel·lí Domingo, 1, 43007 Tarragona, Spain; qiying.lin@estudiants.urv.cat (Q.L.-Y.); leonor.deis@urv.cat (L.D.); jmcanals@urv.cat (J.M.C.); fernando.zamora@urv.cat (F.Z.)

2 Plant Physiology, Faculty of Agricultural Sciences, National University of Cuyo, Mendoza CPA M5528AHB, Argentina

* Correspondence: mariafrancesca.fort@urv.cat

Abstract

The primary challenge facing modern agriculture, including viticulture, is the impact of climate change. The scientific community recommends exploring and utilizing both inter-varietal and intra-varietal variability of local grapevines within each region. The goal is to prioritize planting local varieties over international and imported ones to mitigate the effects of climate change. Within this context, La Palma Island has undertaken a comprehensive assessment evaluating its viticultural heritage. A total of 96 individuals were collected and subjected to genotyping utilizing 20 simple sequence repeats (SSRs). This analysis yielded 44 unique molecular profiles, of which 3 represent new varieties reported for the first time (Aromatica Eufrosina, Cagarruta de oveja, and Viñarda rosada). Additionally, fourteen previously unreported mutations were identified, of which two contain triallelic SSRs. Consequently, the present population of local grapevines on La Palma Island comprises seven varieties (Albillo criollo, Aromatica Eufrosina, Bienmesabe tinto, Cagarruta de oveja, Gual Mazo, Sabro, and Viñarda rosada). The Bienmesabe tinto variety is possibly an interspecific cross. The varieties Aromatica Eufrosina and Viñarda rosada also presented somewhat particular behavior. The distinctiveness of this grapevine population from La Palma Island reinforces the notion that the Canary Archipelago represents a significant center of grapevine biodiversity. The volcanic activity of Tajogaite (2021) did not have a significant impact on grapevine biodiversity on the island.

Keywords: *Vitis vinifera* L.; SSR; microsatellite; characterization; identification; volcanic; Canary Islands; La Palma Island

1. Introduction

Vitis vinifera L., a diploid species with 19 chromosome pairs and a genome of approximately 500 Mb [1], is among the oldest and most globally cultivated crops. Over centuries, human selection and adaptation to diverse environments have endowed grapevine populations with remarkable genetic diversity [1,2]. This genetic variability is increasingly important in the context of climate change, as rising global temperatures threaten traditional cultivation patterns, increase vulnerability to pests and diseases, and alter grape and wine quality [3,4]. Traditional or lesser-known grapevine varieties are therefore a critical resource—not only for environmental resilience and breeding but also for producing wines with distinct profiles that can provide commercial differentiation [5].

The Canary Islands (IC), a Spanish volcanic archipelago in the Macaronesia region (which also includes the Azores, Madeira, the Savage Islands, and Cape Verde), are of special interest for viticultural research due to their geological and geographic isolation (Figure 1a). The archipelago includes eight main islands: El Hierro (HI), La Palma (LP), La Gomera (LG), Tenerife, Gran Canaria, Fuerteventura (FT), Lanzarote (LZ), and La Graciosa (Figure 1b). Viticulture in this region began in the 15th century with the arrival of European settlers [6]. The islands' volcanic soils, inhospitable to phylloxera (*Daktulosphaira vitifoliae*), have allowed for the preservation of ungrafted vines and local varieties that have evolved uniquely over centuries [7]. These conditions make the IC, and LP in particular, a unique site for the study of grapevine biodiversity.

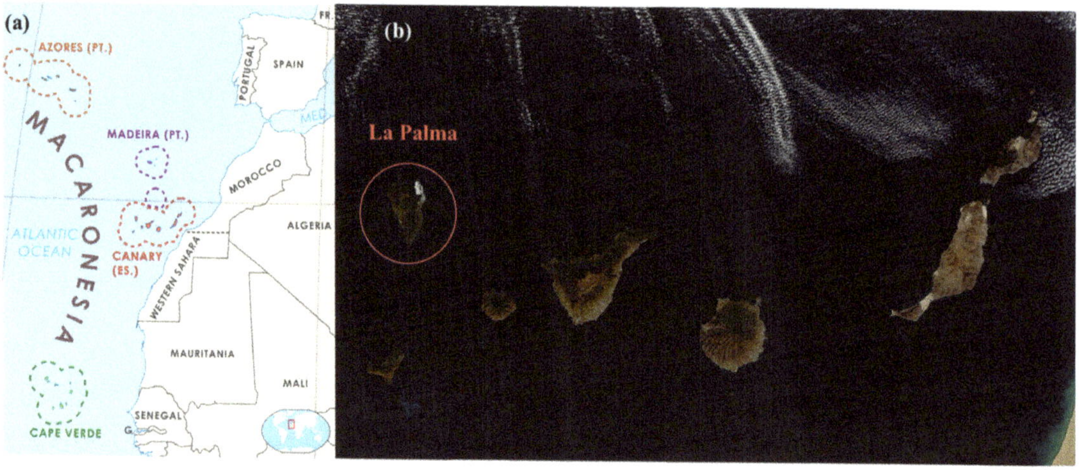

Figure 1. (**a**) Map of Macaronesia [8]; (**b**) Canary Islands Archipelago and La Palma Island [9].

La Palma (28°40′ N, 17°52′ W) is the second highest island in the archipelago, covering 708.32 km^2 and reaching elevations of 2426 m at Roque de los Muchachos [10] (Figure 2a). Its complex topography and climatic variability create diverse viticultural terroirs [11], with influences such as altitude, prevailing trade winds, variable rainfall, and sun exposure contributing to wines with unique organoleptic qualities [12]. The island's Appellation d'Origine Contrôlée/Protégée (AOC/AOP La Palma), established in 1994 [13], organizes viticulture into three subzones based on geographic and microclimatic traits (Figure 2b). The North subzone has fertile, non-sandy soils and uses terrace cultivation on steep slopes, where vines are often grown in goblet form at elevations ranging from 100 to 2000 m [14]. In Hoyo de Mazo (eastern LP), vines are grown on slopes covered with volcanic stones or fine gravel, between 200 and 700 m in elevation [15]. Fuencaliente, the island's hottest and driest region, features vineyards grown on deep volcanic ash layers (often more than two meters thick), using dry-stone windbreaks and terraces at 200–1900 m [16]. It remains a key zone for cultivating Malvasia grapes.

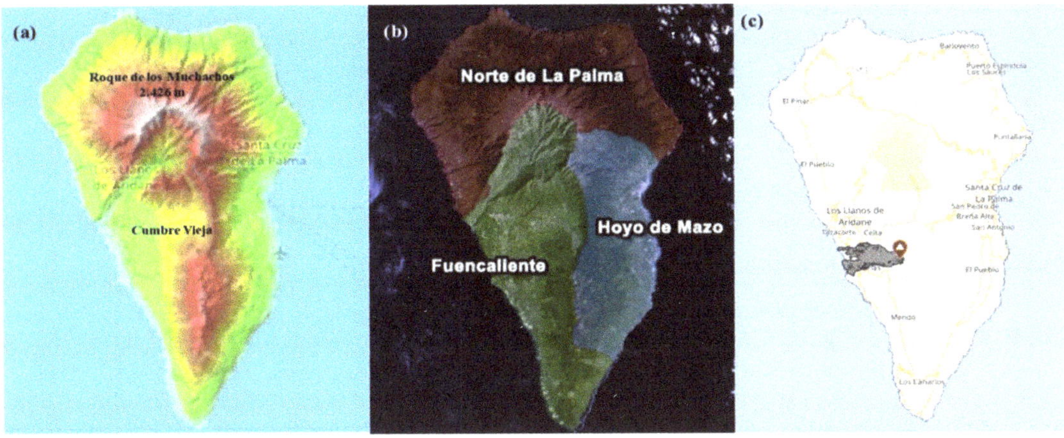

Figure 2. (**a**) Topographic map of La Palma Island [17]; (**b**) viticultural production zones of La Palma [18]; (**c**) mapped map of lava flow and extent [19].

LP's geographic isolation, traditional farming methods, and minimal introduction of external plant material have preserved a rich diversity of local grapevine varieties, many of which are not cataloged in international databases [20]. Among red varieties, Negramoll (Principal name (PN): Mollar cano) is the most widespread, followed by Listan negro (a local Canary variety), Bastardo negro (PN: Trousseau noir), and minor cultivars like Tintilla. For white grapes, prominent varieties include Gual (PN: Malvasia fina), Malvasia (PN: Malvasia Dubrovacka), Verdello (PN: Verdelho branco), and the local Albillo criollo. Other cultivated white varieties include Bastardo blanco (PN: Samarrinho), Bermejuela, Bujariego (or Vijariego blanco), Listan blanco de Canarias (PN: Palomino Fino), and Sabro, a local LP variety. Rarely found cultivars on the island include several pre-phylloxera European and Spanish varieties: red grapes like Alfrocheiro, Flot rouge, Molar, Muscat Hamburg, De Rey, Ferral, and Listan prieto; and white grapes such as Chasselas blanc, Morskoi 94, Muscat of Alexandria, and Beba [21]. These vines, many of them ancient and adapted to LP's microclimates, produce wines with pronounced typicity, strongly influenced by the island's volcanic soil. This soil imparts unique mineral and saline characteristics to the wine, contributing to its distinct sensory profile. Although highly valuable, the genetic documentation of these cultivars remains limited, hindering conservation and adaptation strategies [22].

Volcanic activity has also influenced LP's agricultural and genetic biodiversity. Among the IC, LP has experienced the highest frequency of historical eruptions, all along the Cumbre Vieja Ridge, an active volcanic zone in the island's south [23,24]. The most recent eruption—the Tajogaite volcano in 2021 (Figure 2c)—lasted 85 days and affected over 1200 hectares, including 370 hectares of crops and many vineyards [25–27]. This and other eruptions have likely reduced LP's grapevine genetic diversity, with the island now believed to host fewer genetically distinct varieties compared to its neighbors [21].

Despite these challenges, viticulture remains central to the island's agriculture. The AOC/AOP La Palma unites producers and regulates grape growing and winemaking [28–31]. Wines produced include reds, whites, rosés, and sweet wines. Especially noteworthy are naturally sweet wines from Malvasia aromatica, primarily from Fuencaliente and Villa de Mazo. These are made from overripe grapes with high sugar and alcohol content [32].

In this context, the current study aims to genetically characterize LP's cultivated grapevine population, with particular emphasis on areas such as Fuencaliente, where extreme conditions prevail. These areas may be key to identifying genotypes with high tolerance to heat and drought—traits vital under climate change. The study sets out three main objectives: (1) To document the island's genetic diversity, both inter-varietal and, where possible, intra-varietal. (2) To clarify varietal identities, correcting mislabeling and identifying synonymies or unique genotypes. To analyze genetic structure, evaluating how LP's grapevine population behaves and differentiates from others. Furthermore, this research includes an essential conservation component: it examines plant material collected before the 2021 eruption, especially from areas affected by volcanic ash. This offers a rare opportunity to preserve potentially valuable genotypes that might otherwise be lost or obscured beneath volcanic deposits. The findings will not only help clarify LP's unique genetic contributions to *Vitis vinifera* L. diversity but also provide valuable insights for climate adaptation and the development of distinct wines. In line with Wolkovich et al. [4], this work emphasizes the need to explore and protect grapevine diversity, both to buffer the effects of climate change through intra-varietal resilience and to counteract wine industry homogenization through inter-varietal innovation. Ultimately, the discovery and preservation of unique LP varieties open the door to creating single-varietal wines with unmatched sensory profiles and ecological adaptation.

2. Materials and Methods

2.1. Plant Material Collection

A total of 96 grapevine (*Vitis vinifera* L.) cane samples were collected from different cultivated plants across LP Island. The material was selected using a mass selection strategy in collaboration with various local winegrowers and under the supervision of technicians from the AOC/AOP "Vinos de La Palma" and the Consejería de Agricultura, Ganadería, Pesca y Soberanía Alimentaria del Cabildo Insular de la Isla de La Palma.

Sampling was conducted during vine dormancy, at the time of winter pruning, ensuring that the vines were physiologically mature and free of visible disease symptoms. After collection, the samples were transported under refrigerated conditions to Rovira i Virgili University and subsequently stored at −20 °C until laboratory processing. Specific information for each accession, including geographical location, local name, and preliminary phenotypic observations, is provided in Table S1.

2.2. Sample Preparation and DNA Extraction

The cane samples were processed through selective plant tissue preparation, carefully removing the outer woody bark (rhytidome) and the central pith to minimize the presence of contaminants that interfere with DNA extraction, such as lignin, tannins, polyphenols, and polysaccharides. Only the green cortical zone corresponding to the phloem and young vascular tissues were retained. The selected tissue fragments were immediately frozen and ground into a fine, homogeneous powder in the presence of liquid nitrogen (−196 °C). Aliquots of 0.2 g of powdered tissue were weighed and stored in Eppendorf tubes at −20 °C until use.

Genomic DNA was extracted following a modified protocol based on Fort et al. [33], and later optimized by Marsal et al. [34,35] for woody tissues. The procedure included two chloroform–isoamyl alcohol washes to enhance the removal of protein contaminants. DNA quality and concentration were assessed using spectrophotometry (NanoDrop™ 1000, Thermo Fisher Scientific, Wilmington, DE, USA). Extracts with an A260/A280 ratio between 1.8 and 2.0 and an A260/A230 ratio above 1.8 were considered optimal. Samples not meeting these criteria were either discarded or re-purified.

2.3. Selection of Microsatellite Markers (SSR)

A total of 20 highly polymorphic nuclear SSR markers were employed, selected for their proven efficacy and discriminating power in varietal identification and genetic diversity studies in *Vitis vinifera* L. These markers were chosen based on their high heterozygosity, reproducibility, and wide distribution across the grapevine genome: VVS2, VVS3, and VVS29 [36]; VVMD5, VVMD6, and VVMD7 [37]; VVMD27, VVMD28, and VVMD36 [38]; VrZAG21, VrZAG47, VrZAG62, VrZAG64, VrZAG79, and VrZAG83 [39]; SCU06vv [40]; VvUCH11, VvUCH12, and VvUCH19 [41]; and VChr19a [42].

Among these, VrZAG47 and VVMD27 are not independent markers, as they target the same genomic region using different primer pairs [43]. Additionally, 7 of the 9 SSR reference markers internationally accepted for grapevine varietal identification were included in this study [44]. Detailed information on primer sequences, genomic loci, fluorophores, and technical specifications is provided in Table S2, while their consensus location on the genetic map is illustrated in Figure S1.

2.4. DNA Amplification by Polymerase Chain Reaction (PCR)

The amplification of specific DNA regions was performed using the polymerase chain reaction (PCR) technique, employing the set of 20 previously described nuclear microsatellite (SSR) markers. Reactions were performed using the AmpliTaq DNA polymerase kit (Applied Biosystems, Foster City, CA, USA) in a final volume of 12.5 μL, containing 4 ng of genomic DNA, 1 μM of each primer (Fw and Rv), 1.25 μL of buffer, 2 μL of dNTPs, 0.125 μL of deionized formamide, and 0.0625 μL of Taq polymerase.

Forward primers were labeled with specific fluorochromes for subsequent detection via capillary electrophoresis (6-FAM: VVS3, VVMD7, VVMD28, VVMD36, VrZAG47, VrZAG62, VrZAG83, VvUCH11, and VvUCH19; HEX: VVS2, VVS29, VVMD6, VVMD27, VrZAG21, VrZAG79, and VChr19a; NED: VVMD5, VrZAG64, SCU06vv, and VvUCH12). SSRs were grouped according to their annealing temperature (Ta), which ranged between 50 and 58 °C (Table S2). PCRs were performed using an Applied Biosystems™ 2720 thermal cycler (Foster City, CA, USA) with the following cycling conditions: initial denaturation at 95 °C for 5 min; followed by 40 cycles of 95 °C for 45 s, specific annealing temperature for 30 s, and 72 °C for 90 s; with a final extension step at 72 °C for 7 min.

2.5. Capillary Electrophoresis and Fragment Analysis

PCR products were prepared for capillary electrophoresis by adding 20.5 μL of deionized formamide and 0.25 μL of the internal size standard GeneScan™ ROX 500 (Applied Biosystems, Foster City, CA, USA) to each sample. Different volumes of amplified product were added depending on the fluorophore used: 2 μL for HEX, 4 μL for 6-FAM, and 6 μL for NED. The plates were briefly denatured at 95 °C for 3 min before loading onto the ABI PRISM 3730® Genetic Analyzer (Applied Biosystems, Foster City, CA, USA).

Fragment separation by size and fluorescent signal detection enabled the generation of electropherograms, which were analyzed using Peak Scanner™ Software v.1.0 (Applied Biosystems, Sparta, NJ, USA) to determine precise fragment lengths in base pairs (bp), allowing allele calling at each locus. Each sample was analyzed at least twice from independent DNA extractions to minimize genotyping errors.

2.6. Genotyping and Varietal Identification

Allelic profiles were compared with a local database developed by the TECNENOL Research Group [20,45–50]. Genotypes that did not match any component of our database were compared with entries in a much larger database, the *Vitis* International Variety Catalogue (VIVC) [51]. This approach also allows the detection of synonymous cultivar names and possible allelic mutations.

2.7. Analysis of Genetic Data

The genetic profiles obtained from the SSR markers were analyzed to characterize both intra- and inter-varietal genetic diversity among the *Vitis vinifera* L. accessions, and to infer phylogenetic relationships and population structure. For this purpose, a suite of well-established statistical and bioinformatics tools commonly used in population genetics studies was employed.

First, a multilocus genotype matrix was constructed and formatted for analysis in GenAlEx v6.5 [52,53]. Several genetic diversity parameters were calculated to assess the efficiency and reliability of the 20 SSR markers, including the number of alleles per locus (Na), effective number of alleles (Ne), observed heterozygosity (Ho), expected heterozygosity (He), fixation index (F), and probability of identity (PI). The software also facilitated the identification of genetically identical individuals, permitting the removal of redundant accessions and the detection of somatic mutations.

To investigate the underlying genetic structure of the collection, STRUCTURE 2.3 [54,55] was used. This Bayesian clustering method estimates the probability of individual membership in one or more genetic clusters (K), defined by allele frequencies at each locus under Hardy–Weinberg equilibrium. Simulations were conducted for K values ranging from 1 to 7, using an admixture model with correlated allele frequencies. Each run consisted of a burn-in period of 100,000 iterations followed by 1,000,000 additional MCMC iterations. The optimal number of genetic clusters (K) was determined using the ΔK method described by Evanno et al. [56], which calculates the second-order rate of change in the likelihood function across successive K values. Individuals were assigned to a genetic cluster if their coefficient (q) was $\geq 85\%$; individuals with lower values were considered admixed.

Assignment tests were subsequently performed in GenAlEx v6.5 to verify the distribution of individuals across structure-derived populations based on observed allele frequencies [57]. The log-likelihood probability of each individual's membership to each subpopulation was calculated, and individuals were assigned to the group with the highest log-likelihood value. GenAlEx was also used to calculate the genetic differentiation coefficient (Fst) between populations through an Analysis of Molecular Variance (AMOVA), using 999 permutations and assuming the infinite allele model for SSR genotypes. Additionally, a Principal Coordinates Analysis (PCoA) was conducted based on standardized covariance of genetic distances derived from codominant markers, enabling graphical two-dimensional representations at both the individual and population levels. For three-dimensional visualization of the PCoA, the Python 3.5 programming [58] and the Matplotlib 2.0 library were used to generate spatial representations that enhanced the interpretation of clustering patterns and the identification of relationships among individuals and populations.

Furthermore, phylogenetic relationships and clustering designs were inferred using MEGA v7.0 software [59], applying the Neighbor-Joining algorithm [60], which groups genetic samples based on profile similarity. This approach enabled the construction of phylogenetic trees and circular dendrograms to detect consistent groupings and potential kinship relationships among the accessions, thereby providing deeper insights into their genetic structure and interrelationships.

3. Results

A total of 96 grapevine accessions from LP Island were genotyped (Table S1) using the set of 20 SSR markers that the TECNENOL Research Group has been applying [20,45–50] (Table S2 and Figure S1). This study evaluated the performance of the SSR kit for this population and investigated the presence of synonyms, homonyms, and identification errors among the samples. Understanding the extent of inter-varietal variability was one of the primary objectives, with the goal of assessing the grapevine biodiversity on this island. Finally, studying the singularity of the local variety population would highlight the significance of the viticultural heritage of LP Island.

3.1. SSR Polymorphism

The molecular profile (MP-SSR) found for each of the 96 individuals analyzed allowed the detection of identical accessions. Thus, the first normalization of the data obtained for this grapevine population consisted of eliminating redundant MP-SSRs. Consequently, 52 accessions were removed. Table S3 presents the 44 unique profiles found in this population, as well as two sport samples (same MP-SSR but with a change in berry color) that are also exceptionally included in this table. With the population of the 44 unique MP-SSRs, the study of the statistical parameters began, which would demonstrate the efficacy of the selected SSR kit for carrying out the present work.

Table S4 shows the main evaluated statistical parameters. The average number of alleles (Na) present in this grapevine population from LP Island was 11.2 alleles. The SSR markers that exhibited the highest number of alleles (Na) and the greatest capacity to discriminate among genotypes were VVMD36 (18 alleles), ZAG79 (17 alleles), and VVMD27 (16 alleles), while the SSRs with the lowest Na were VVS3 and VVS29 (5 alleles), ZAG62 (7 alleles), and UCH19 (8 alleles). The average effective number of alleles (Ne) for this population was 5.2, with the best-performing SSRs being VVMD27 (9.410) and ZAG79 (8.620), while the poorest performers were VVS29 (1.371) and VVS3 (1.739). The averages for observed heterozygosity (Ho) and expected heterozygosity (gene diversity index) (He) were 0.783 and 0.746, respectively. The SSR with the highest Ho was ZAG79 (0.953, followed by VChr19a (0.930), and VVS2, VVMD7, and VVMD27, all with values of 0.907. At the opposite extreme, the lowest values were observed in VVS29 (0.256), VVS3 (0.442), and UCH19 (0.488). The SSR with the highest He was VVMD27 (0.894) followed by ZAG79 (0.884) and VVMD36 (0.874), while the lowest results were observed for VVS29 (0.270), VVS3 (0.425), and UCH19 (0.555). The SSRs with the worst fixation index (or probability of null alleles) were UCH19 (0.120), ZAG83 (0.106), and VVS29 (0.054), while the others showed very acceptable performance. Overall, the SSR kit showed a cumulative probability of identity (PI) of 2.0×10^{-23} for this LP population. The SSRs with the highest PI values were VVMD27 (2.0×10^{-2}), ZAG79 (2.3×10^{-2}), and VVMD36 (2.6×10^{-2}), while the worst values corresponded to VVS29 (5.5×10^{-1}), VVS3 (3.7×10^{-1}), and UCH19 (2.3×10^{-1}).

3.2. Grapevine Variety Analysis

Table S1 presents all the original information as well as the data obtained in this study for each of the 96 genotyped accessions. A total of 31 varieties and 2 color sports were identified using the TECNENOL database [20,45–50] and the VIVC database [51]. Table S3 details the 44 individuals with unique MP-SSR profiles, including their genetic profile based on the seven international SSR markers. Exceptionally, MP-SSR profiles for the two sports in this collection were also included.

The population of local peninsular varieties from Spain comprised 34 accessions corresponding to seven varieties registered in the VIVC. Of these, De Rey [2] and Ferral were each represented by a single accession with no observed allelic differences. The variety Beba included three samples: two consistent with known profiles and one displaying a novel mutation at the second allele of SSR marker VVS3 (designated VVS3-2). All five Listan prieto accessions exhibited known variations, VVS3-2 and VVMD28-1 [49,61], while all eight Mollar cano samples shared the VVS3-2 mutation, previously reported in other IC [47–49]. Among the 12 Palomino Fino accessions, the majority matched the common TECNENOL profile, while three exhibited a novel VVMD27-2 variation. Notably, accession P-69 showed four variations: VVS3-2, VVMD27-2, and triallelic profiles at markers ZAG64 and VChr19a. The variety Vijariego blanco, represented by four accessions, showed considerable genetic variation. Individual P-08 aligned with a previously reported LZ profile (VVS3-2 and ZAG64-2) [47,61], whereas P-60 exhibited a novel combination (VVMD28-1 and ZAG64-2). Accessions P-57 and P-68 shared a new composite profile (VVS3-2, VVMD28-1, and ZAG64-2).

The Portuguese group included 13 individuals from the varieties Alfrocheiro, Malvasia fina, Molar, Samarrinho, and Verdelho branco. The single Alfrocheiro accession exhibited novel variations at VVS3-2 and VVMD7-1. Malvasia fina had two individuals with variations (P-12 at VVS3-2, and P-13 at VVS3-2 and VVMD36-2), and its rosé sport (Malvasia fina roxa) included one individual with a variation at VVS3-2. Among the four Molar accessions, one (P-75) matched the reference profile, while three carried a novel mutation at VVS3-2. The three Samarrinho samples showed the most common profile, and therefore no variations. Lastly, Verdelho branco had one non-mutated individual and another with a newly described variation at VVS3-2.

Among non-Iberian varieties, French cultivars Flot rouge and Trousseau noir, both with two accessions, exhibited no allelic variation. Similarly, Muscat of Alexandria (Greece), Muscat d'Hamburg (UK), and Morskoi 94 (Ukraine) displayed no mutations.

Two varieties of uncertain origin were also studied. Chasselas blanc showed a triallelic pattern at VVMD36, suggesting somatic variation or hybrid ancestry. Malvasia Dubrovacka, erroneously listed as Spanish in the VIVC [51], was represented by four accessions with three MP-SSR profiles. P-81 matched the reference, while P-16 and P-35 shared the VVS3-2 mutation known from LG [50]. A fourth accession, P-48, exhibited a new combination of three mutations (VVS3-2, UCH12, SCU06-2).

The collection also included five local varieties to another IC: Bermejuela, Forastera gomerae, Listan negro, Malvasia volcanica, and Torrontes volcanico. While Bermejuela, Forastera gomerae, and Torrontes volcanico showed no variation, Listan negro had three accessions, two of which carried the VVS3-2 mutation, consistent with previous findings [47,49,50,61]. Among four Malvasia volcanica accessions, three exhibited the VVS3-2 mutation noted in LZ [46,61]; one aligned with the reference. Additionally, a sport variety, Malvasia di Sardegna rose, included two unmutated individuals and one with a novel dual-mutation profile (VVS3-2 and SCU06-2).

Seven local varieties were unique to LP Island. Three, Albillo criollo, Bienmesabe tinto, and Gual Mazo, were each represented by a single, no variation accession [20,62,63], while Sabro, with eight individuals, also showed no variation. Three newly described local varieties, Aromatica Eufrosina, Cagarruta de oveja, and Viñarda rosada, were each represented by two genetically distinct individuals, constituting new varietal entities.

At the lexicographic level, the registration status of accession names (Table S1) was as follows: 12 accessions matched the PN in VIVC (in black); 31 were registered under synonyms recorded in VIVC (black and bold); 14 samples with first-time described variations were assigned new names proposed by the authors (green and bold); 14 new synonyms for varieties already registered in VIVC (brown and bold); 8 had erroneous names (red); 1 accession was registered under a synonym belonging to another variety but was considered to also describe the variety it was registered under (fuchsia); 14 were labeled as "unknown" or with generic names indicating lack of knowledge (light green), such as Tintilla; and 1 individual, corresponding to one of the new varieties, presented a new registration name, proposed as a new PN (turquoise blue). Therefore, this study contributes at the lexicographic level: 3 new PN for the new local varieties (turquoise blue), 14 new names defining individuals with first-time described variations (green and bold), 9 new synonyms for varieties already registered in VIVC (brown and bold), 1 existing synonym name for a given variety proposed also for another variety by the authors (fuchsia), 8 MP-SSR profiles registered erroneously, and 4 accessions with names indicating lack of knowledge about the corresponding variety. Table S3 summarizes this information exclusively for the unique MP-SSR profiles.

3.3. La Palma Grapevine Population Genetic Structure

Before proceeding with the study of the genetic structure of the samples that would define the final population of LP Island, a new data normalization step was performed. From the 44 unique MP-SSR profiles, accessions corresponding to sport mutations and to variations relative to the principal or most widespread MP-SSR profiles in the databases were removed. Thus, the final collection of cultivated grapevines on LP Island consisted of representatives from 31 varieties, 28 previously known and 3 described here for the first time. It should be noted that among the representatives of these 31 varieties, 5 corresponded to individuals with variations. Specifically, the varieties Mollar cano and Malvasia fina were represented by individuals who presented a variability of 97.5% in our MP-SSR, while Vijariego blanco, Listan prieto, and Alfrocheiro were represented by individuals exhibiting a 95% variation.

The program STRUCTURE 2.3 allowed us to classify the 31 varieties into 3 populations (K = 3) after testing combinations from K = 1 to K = 7, as shown in Figure S2. Each population was assigned a representative color, and its members were classified based on the q parameter. In our study, we adopted the arbitrary threshold of 85% [64], so that q values $\geq$ 85% corresponded to pure individuals of a given population, while q values < 85% indicated admixed individuals within the same population. Figure 3 shows the population structure diagram for LP Island, with an assignment accuracy of 89%, and Figure S3 presents all individuals classified according to their q membership values.

POP1 was composed of 15 varieties (48%), of which 9 (60%) were considered pure (Mollar cano-97.5%, Beba, Vijariego blanco-95%, Ferral, Morskoi 94, Torrontes volcanico, De Rey, Sabro, and Gual Mazo), while 6 (40%) were considered admixed (Cagarruta de oveja, Listan negro, Malvasia volcanica, Malvasia Dubrovacka, Listan prieto-95%, and Palomino fino). Among pure varieties, 55.5% were Spanish, 22.2% local to LP, 11.1% from the IC, and 11.1% Ukrainian. Most pure varieties (77.8%) were white, with 55.6% used for dual purposes (table and winemaking) and 44.4% exclusively for winemaking. Among admixed varieties, 33.3% were Spanish, 33.3% from IC, 16.7% from LP, and 16.7% of unknown origin. White varieties constituted 66.6% of this group, used for wine (50%), dual-purpose (33.3%), and as table grapes (16.7%).

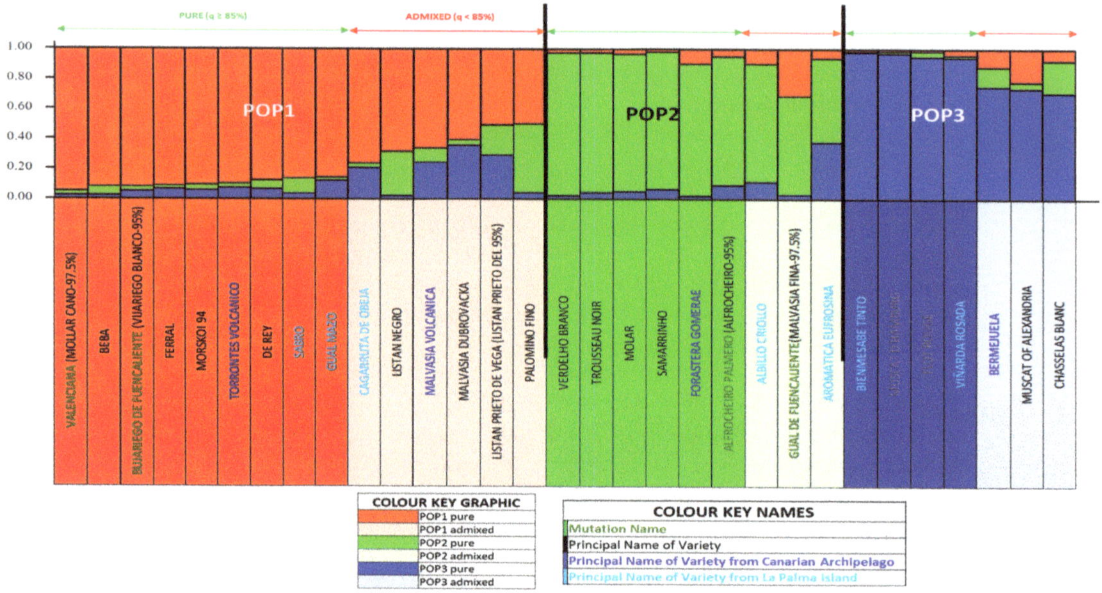

Figure 3. La Palma grapevine varieties population (unique molecular profiles). Structure 3.2. diagram: K = 3 distribution for pure and admixed individuals.

POP2 comprised nine varieties (29%), of which six (67%) were classified as pure (Verdelho branco, Trousseau noir, Molar, Samarrinho, Forastera gomerae, and Alfrocheiro-95%) and three (33%) as admixed (Albillo criollo, Malvasia fina-97.5%, and Aromatica Eufrosina). Among pure varieties, 66.7% were Portuguese, 16.7% French, and 16.7% from IC. Half were white, primarily used for winemaking (66.7%) and dual-purpose (33.3%). The admixed varieties in this population originated from LP (66.7%, including a new variety) and Portugal (33.3%). All were white and used for winemaking.

POP3 included seven varieties (23%), with four (57%) pure (Bienmesabe tinto, Muscat d'Hamburg, Flot rouge, and Viñarda rosada) and three (43%) admixed (Bermejuela, Muscat of Alexandria, and Chasselas blanc). Among pure varieties, 75% were red and 25% rosé, and were used for winemaking (50%), one for dual-purpose (25%), and one with an undetermined use (25%). All admixed varieties were white; 66.7% were used for winemaking, and 33.3% for triple use (table, raisin, and wine production).

Figure 4 shows the population structure analysis of the three cultivated grapevine populations in LP based on their Nei's Genetic Distances. For this analysis, a third normalization step was applied, excluding admixed individuals, reducing the dataset to 19 varieties (Figures 3 and S3). Figure 4a (bidimensional PCoA by populations) displays the distribution of these three populations. Coordinate 1, which accounts for 56.40% of the variation, clearly separates POP1 (right quadrants) from POP2 and POP3 (both in the left quadrants). Coordinate 2 (43.60% goodness), distinguishes POP1 and POP3 (upper quadrants) from POP2 (lower quadrant). According to the *F*st parameter, POP1 was the most genetically distinct population. The distribution of individuals shown in Figure 4b (two-dimensional) and Figure 4c (three-dimensional) show the same pattern observed in Figure 4a.

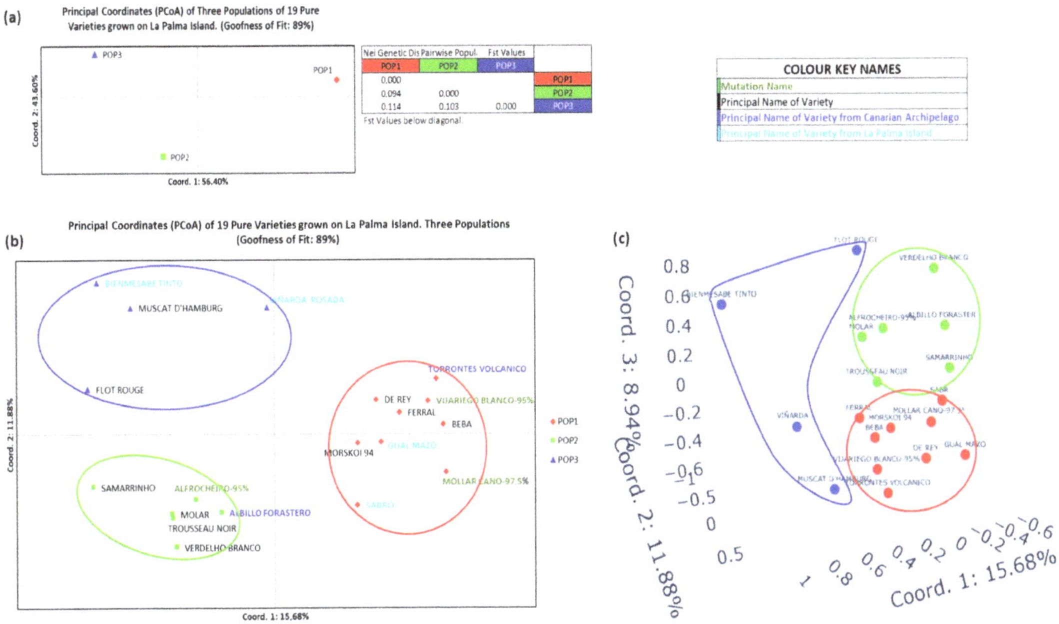

Figure 4. PCoA representation of the grapevine varieties population from La Palma Island normalized for K = 3. (**a**) Two-dimensional representation of the 3 populations. Values of the *F*st statistic for each population. (**b**) Two-dimensional representation of the 3 populations by individuals. (**c**) Three-dimensional representation of the 3 populations by individuals.

3.4. Relation of La Palma Grapevine Population to the Canary Archipelago Population

To explore the genetic relationship between the grapevine population of LP and the rest of the IC, the seven local LP varieties identified in this study were analyzed alongside 33 varieties previously described as local to the archipelago [20,47–50,62,63], obtaining a total of 40 varieties. These were grouped into six insular populations. The IC group comprised four varieties (10%): Vallera (Tenerife), Albillo Monte Lentiscal (Gran Canaria), and two widespread varieties across the archipelago (Listan negro and Bermejuela). The LZ group included 12 varieties (30%): Blanca de la granja del Cabildo, Burra chinija, Diego chinija, Lemes de El Cabezo, Malvasia alistanada fina, Sinforiano chano, Uvillón negro, Vijariego blanco de la granja, Uva de año, Malvasia volcanica, Torrontes volcanico, and Breval negro. HI included 11 varieties (27%): Burra volcanica, Uvalero volcanico, Huevo de gallo, Verijadiego, Verijadiego negro, Pinar negro, Uval negro, Seis de Carlos, Verdello de El Hierro, Uval piñero, and Tesoro blanco. The sole representative of FT was Majorera (3%). LG was represented by five varieties (13%): Forastera gomerae, Barrerita negra, Coello blanca, Malvasia periquin gomerae, and Verdello gomerae. Lastly, the LP population (17%) included four previously described varieties (Bienmesabe tinto, Albillo criollo, Sabro, and Gual Mazo) and three newly identified ones (Cagarruta de oveja, Aromatica Eufrosina, and Viñarda rosada).

Population structure was assessed using STRUCTURE 2.3, testing K values from 1 to 7. Based on the Evanno method [56], the optimal structure was achieved with K = 5, corresponding to 5 ancestral populations for the 40 individuals studied (Figure S4). As in the previous case, each individual was assigned to a specific population and position within that population based on the q parameter. This parameter allowed us to identify pure individuals ($q \geq 85\%$) and admixed individuals ($q < 85\%$).

Figure 5 presents the graphical representation of the optimal distribution for the 40 varieties of the Canary Archipelago, corresponding to five ancestral populations. In addition, Figure S5 shows the distribution of these 40 varieties among the five populations according to their q values.

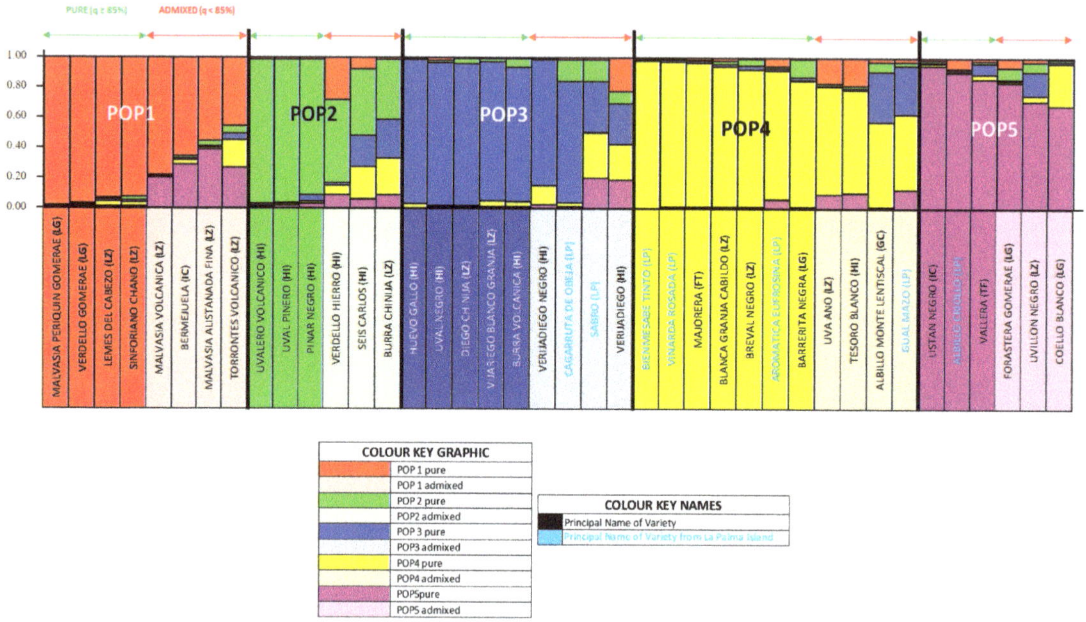

Figure 5. Canary grapevine varieties population (unique molecular profiles). Structure 2.3 Diagram: K = 5 distribution for pure and admixed individuals.

In this analysis, POP1 comprised eight varieties (20%), equally split between pure and admixed. Pure varieties included Malvasia periquin gomerae and Verdello gomerae (LG), and Lemes de El Cabezo and Sinforiano Chano (LZ). The admixed group were Malvasia volcanica, Malvasia alistanada fina, and Torrontes volcanico (LZ), and Bermejuela (IC). POP2 included six varieties (15%), also evenly split between pure and admixed. Pure individuals were Uvalero volcanico, Uval piñero, and Pinar negro (HI), while admixed ones included Verdello de El Hierro and Seis de Carlos (HI), and Burra chinija (LZ). POP3 contained nine varieties (22%), with 56% pure individuals (Huevo de gallo, Uval negro, and Burra volcanica (HI), and Diego chinija and Vijariego blanco de la granja (LZ)) and 44% admixed (Verijadiego and Verijadiego Negro (HI), and Cagarruta de oveja and Sabro (LP)). POP4 was the largest group, with 11 varieties (28%). Pure individuals (64%) included Bienmesabe tinto, Viñarda rosada, and Aromatica Eufrosina (LP), Blanca de la granja del Cabildo and Breval negro (LZ), Majorera (FT), and Barrerita negra (LG). The admixed varieties (36%) were Uva de año (LZ), Tesoro blanco (HI), Albillo monte Lentiscal (IC), and Gual Mazo (LP). POP5 included six varieties (15%), again split evenly. The pure group included Listan negro and Vallera (IC), and Albillo criollo (LP). Admixed individuals were Forastera gomerae and Coello blanco (LG), and Uvillón negro (LZ).

For the population structure analysis, 18 admixed individuals were removed, resulting in a set of 22 pure varieties (Figures 5 and S5). The initial assignment quality (86%) was improved to 100% by reassigning three individuals from POP4 to more appropriate populations (POP1, POP2, and POP3). In the two-dimensional population PCoA (Figure 6a), Coordinate 1 (32.12% variance) clearly separated POP5 (upper right quadrant) from POP1 (upper left quadrant). The remaining populations were more closely clustered: POP3 and POP4 appeared near the origin in the lower right quadrant, with POP2 slightly apart in the lower left quadrant. *Fst* values confirmed that POP1 and POP5 were the most genetically distant, while POP2, POP3, and POP4 were more closely related. The individual level PCoA (Figure 6b,c) showed lower resolution and was less conclusive. Only POP1 remained clearly distinct and consistently separated. POP2, POP3, POP4, and to a lesser extent POP5, showed some degree of overlap, with several outliers. Notably, Bienmesabe tinto from LP appeared clearly differentiated from all other individuals.

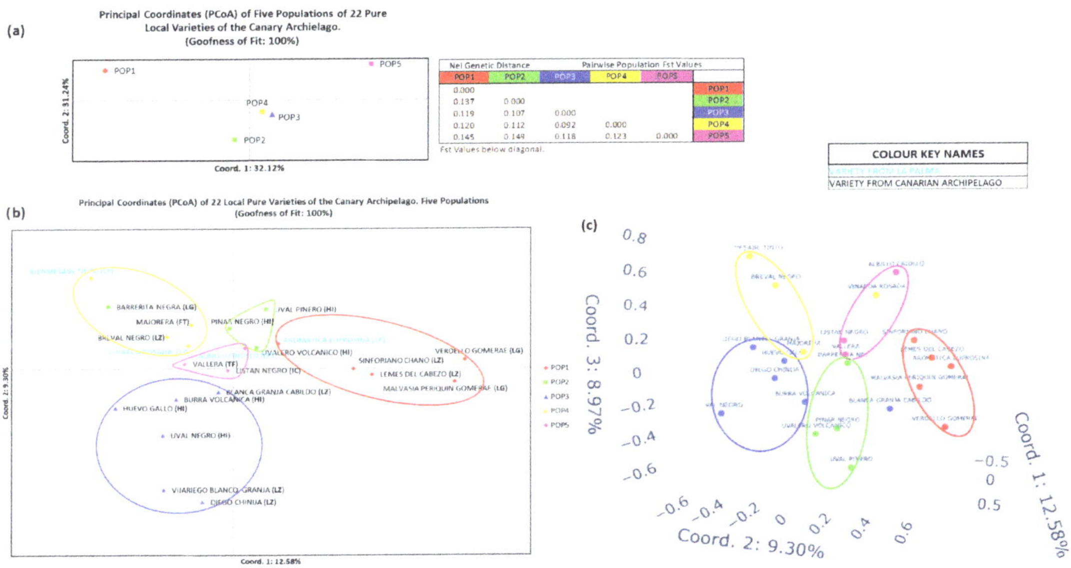

Figure 6. Population of varieties of the Canary Islands (22 pure varieties). (**a**) Two-dimensional representation of the 5 populations of the Canary Islands by population. Values of the *Fst* statistic for each population. (**b**) Two-dimensional representation of the 5 populations by individuals. (**c**) Three-dimensional representation of the 5 populations by individuals.

3.5. Relation of La Palma Grapevine Population to the World Population

To study the seven grapevine genotypes from LP in the context of global diversity, two complementary approaches were employed: a genetic and a geographic strategy.

3.5.1. Genetic Strategy

This approach involved compiling a worldwide dataset of *Vitis vinifera* varieties from natural crosses (from the TECNENOL database [20,45–50]) into a single dataset, regardless of known parentage, totaling 319 genotypes from approximately 23 countries, including those from LP. Using STRUCTURE 2.3, we tested K values from 1 to 7, with the optimal solution revealing two ancestral groups (K = 2) (Figures 7a and S6), based on the q coefficient (Figure S7), differentiating between pure ($q \geq 85\%$) and admixed ($q < 85\%$).

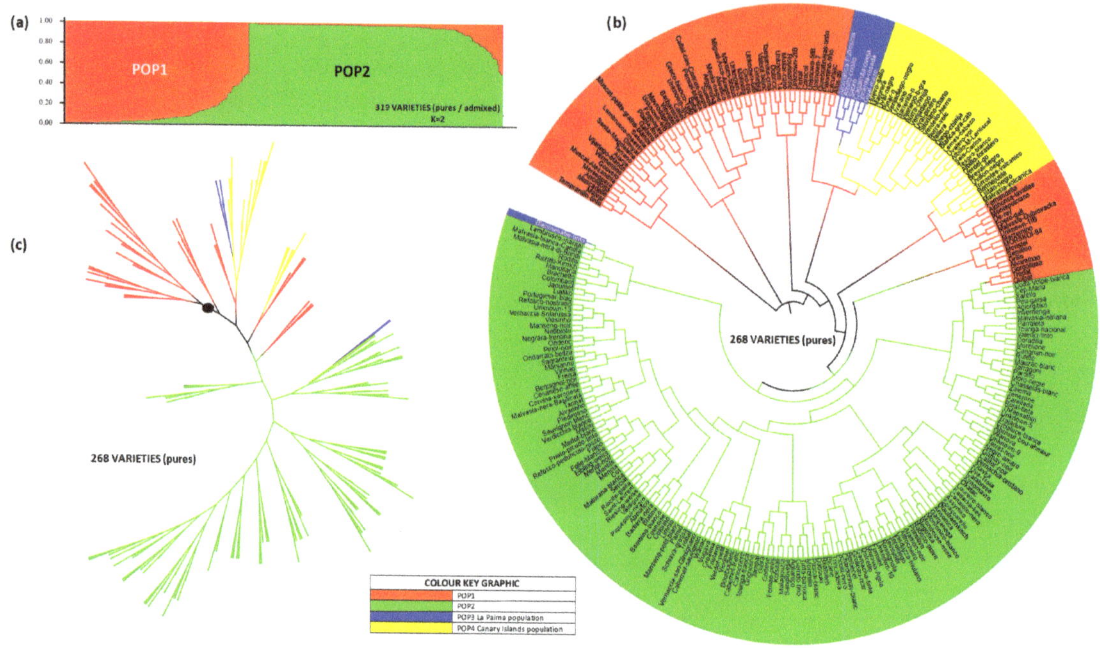

Figure 7. World population (319 individuals) distributed in 2 populations. (**a**) Graphical representation of K = 2 according to Structure 2.3 (with pure and admixed individuals) (**b**) Circular Neighbor-Joining dendrogram of the world population 268 pure individuals, (**c**) Pure individuals world population phylogenetic tree.

POP1 contained 135 varieties (42%), and POP2 had 184 (58%). All IC varieties clustered in POP1. A total of 53 admixed varieties (16.6%) were removed to normalize the dataset: 31 from POP1 and 22 from POP2. Among these, three were from IC (Malvasia periquin gomerae (LG), Malvasia alistanada fina (LZ), and Coello blanco (LG)) and two from LP (Albillo criollo and Aromatica Eufrosina). After normalization, POP1 comprised 104 varieties: 71 Spanish, 10 of unknown origin (including Malvasia Dubrovacka [65]), 9 Italian, 4 French, 3 Portuguese, 3 Greek, and 1 each from Bulgaria, Croatia, Lebanon, and Serbia. POP2 included 162 varieties: 51 Italian, 39 French, 24 Spanish, 12 Portuguese, 11 Greek, 4 Croatian, 4 unknown, 2 each from Slovenia, Germany, Bulgaria, Hungary, and Bosnia and Herzegovina, and 1 each from Turkey, Montenegro, Switzerland, Israel, Ukraine, Georgia, and Algeria. Assignment precision reached 99%, and after reassigning one misclassified variety, accuracy was 100% with a final normalized dataset of 266 pure varieties.

To evaluate the LP and IC populations more specifically, these individuals were extracted from POP1 and reclassified into two new populations: LP (POP3) and IC (POP4). IC remained normalized, while the two previously excluded admixed LP varieties (Albillo criollo and Aromatica Eufrosina) were reinstated. This slightly reduced the overall assignment accuracy to 88% (compared to 89% if LP had remained fully normalized).

The circular dendrogram (Figure 7b) illustrates the distribution of the 268 pure varieties (including the two admixed LP varieties) across four clusters: POP1 (70), POP2 (161), IC (30), and LP (seven: five pure and two admixed). The IC population, nested within POP1, formed two sub-branches (in yellow). Within one of these, a further differentiation emerges, forming two smaller sub-branches, one of which comprised six LP varieties (in blue). The exception was Bienmesabe tinto, which clustered separately with POP2. This pattern also appeared in the phylogenetic tree (Figure 7c).

Figure 8 presents the PCoA analysis. In the population-level plot (Figure 8a), Coordinate 1 (76.79% of variance) separated POP2 (right quadrants) from all others (left), while Coordinate 2 (21.06%) distinguished POP1 (upper quadrants) from IC and LP (lower quadrants). Thus, POP2 appeared isolated (upper right), POP1 in the upper left, and IC and LP in the lower left. *F*st values supported these patterns. At the individual level (Figure 8b), most POP2 varieties grouped on the left, while other populations clustered on the right. Coordinate 1 explained 5.64% of variance and Coordinate 2 explained 3.57%, providing limited resolution. However, LP varieties consistently appeared in the lower right quadrant, close to the Coordinate 1 axis. The three-dimensional PCoA (Figure 8c) did not provide additional insights beyond those revealed in the two-dimensional representation, though Bienmesabe tinto from LP remained clearly separated from all others, and Aromatica Eufrosina appeared in the upper region of POP2.

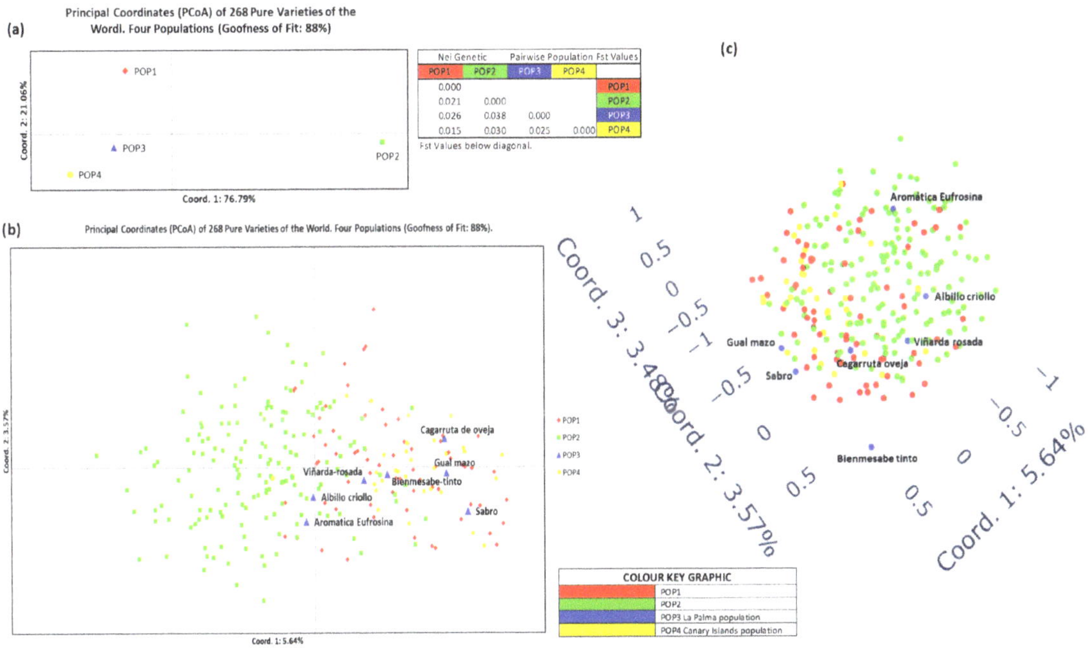

Figure 8. PCoA representation of La Palma Island, Canary Islands, and worldwide population, normalized for K = 2. (**a**) Two-dimensional representation of the 4 populations per population, and values of the *F*st statistic for each population, (**b**) two-dimensional representation of the 4 populations per individual. (**c**) three-dimensional representation of the 4 populations per individual.

3.5.2. Geographic Strategy

To validate the previous results, a geographic criterion was applied to the global grapevine dataset (319 varieties), grouping varieties into seven regions based on their countries of origin according to the VIVC database [51]. The decision to group by areas rather than individual countries was made because some countries were represented by only a single variety. The regions defined (Figure S8) were EASTMED-CAU (Algeria, Cyprus, Georgia, Israel, Lebanon, Tunisia, and Turkey), BP (Bosnia and Herzegovina, Bulgaria, Croatia, Greece, Serbia, Slovenia, and Montenegro), ITA (Italy), FRA-CE (Austria, France, Germany, Hungary, Ukraine, and Switzerland), IP (Spain and Portugal), IC, and LP.

An assignment test in GenAlEx 6.5 yielded a 60% accuracy rate, allowing for the identification of misassigned and admixed individuals. The group compositions were as follows: EASTMED-CAU (12 varieties, 4%) contained 42% pure (5) and 58% admixed (7); BP (28 varieties, 9%) contained 50% pure (14) and 50% admixed (14); ITA (73 varieties, 23%) contained 53% pure (39) and 47% admixed (34); FRA-CE (61 varieties, 19%) contained 69% pure (42) and 31% admixed (19); IP (105 varieties, 33%) contained 65% pure (68) and 35% admixed (37); IC (33 varieties, 10%) contained 70% pure (23) and 30% admixed (10); and LP (7 varieties, 2%), where all individuals were retained including admixed varieties.

After removing 122 admixed or misassigned individuals (except from LP), the normalized dataset comprised 197 well-assigned or pure varieties. A second assignment test improved accuracy to 87%.

Figure 9 shows a circular dendrogram (Figure 9a) and phylogenetic tree (Figure 9b) based on this geographic grouping. In both representations, IC and LP stood out clearly, always clustering within the IP domain and forming a major sub-branch. Notably, LP also formed a distinct sub-branch within IC. Under these conditions, Bienmesabe tinto was fully integrated into the LP group.

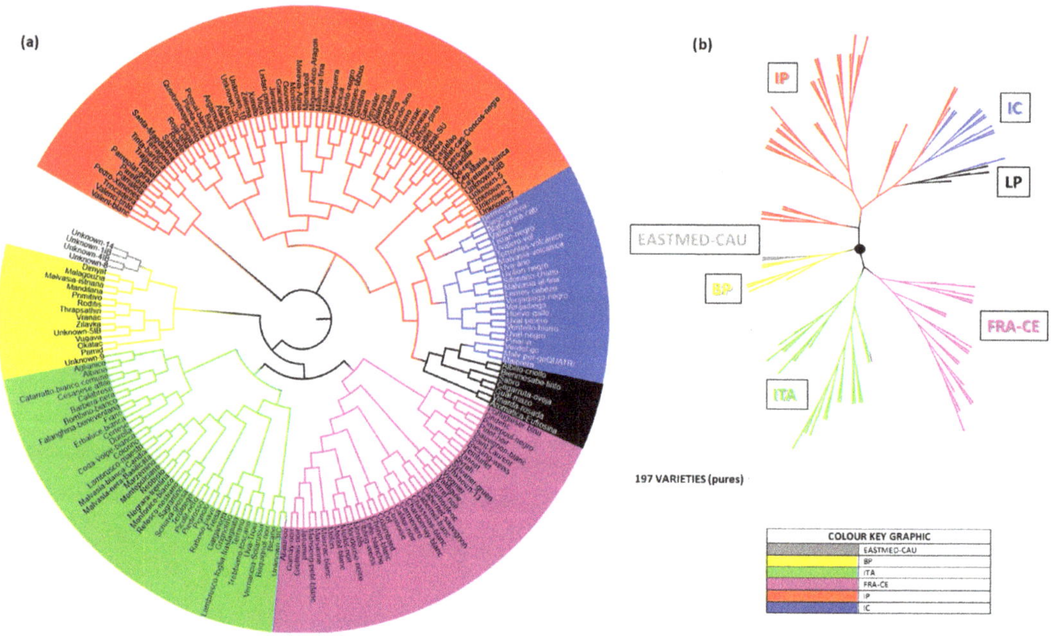

Figure 9. World population (197 individuals) distributed in populations corresponding to 7 geographical areas. (**a**) Circular Neighbor-Joining dendrogram of the 197 pure individuals of the world population and La Palma, (**b**) phylogenetic tree of 7 populations distribution with all their individuals.

Figure 10 presents the graphical representation of the PCoA applied to the global population (197 individuals), distributed across seven arbitrary regions based on geographic criteria.

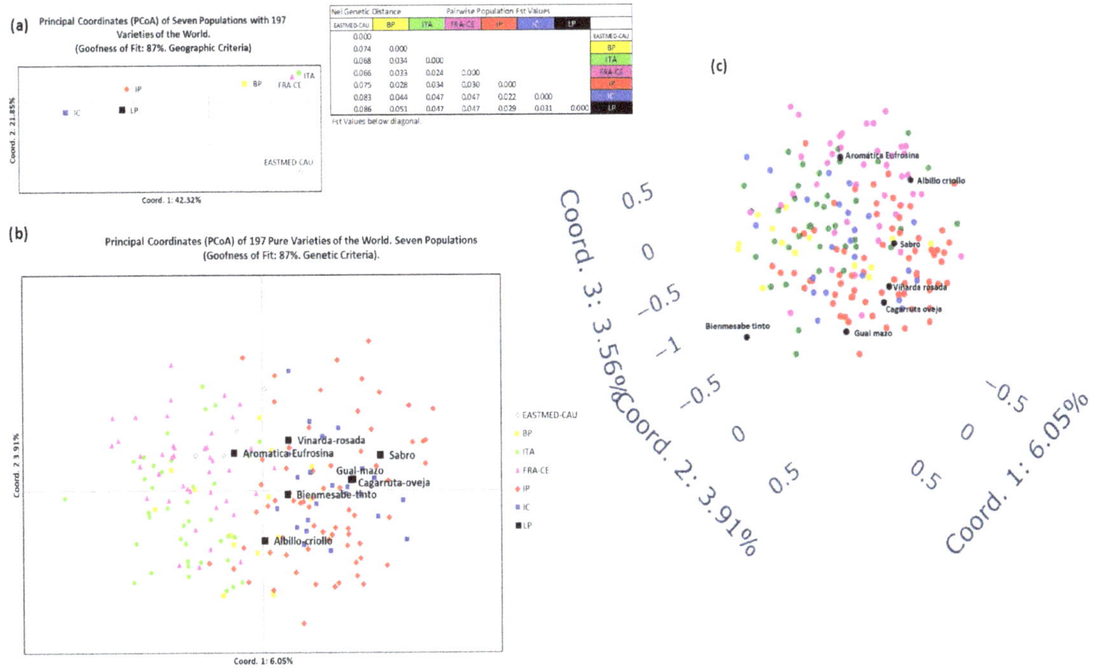

Figure 10. PCoA representation of La Palma Island, Canary Islands, and world population for the geographical criterion. (**a**) Two-dimensional representation of the 7 populations per population. Values of the *F*st statistic for each population, (**b**) two-dimensional representation of the 7 populations per individual. (**c**) Three-dimensional representation of the 7 populations per individual.

In the two-dimensional population-level plot (Figure 10a), Coordinate 1 (42.32% of variance) separated IP-related groups (IP, IC, LP) into the left quadrants. Coordinate 2 (21.85%) slightly differentiated IP and separated BP, ITA, and FRA-CE from IC, LP, and EASTMED-CAU. The most distant cluster was EASTMED-CAU (lower right quadrant). Furthermore, two main clusters can be identified: (1) BP, ITA, FRA-CE (upper right), and (2) IP, LP, and, to a lesser extent, IC (central left). These groupings were supported by *F*st values.

The two-dimensional individual-level plot (Figure 10b) reflected the same structure, with a rotation of the first coordinate axis. Coordinate 1 (6.05% of variance) separated EASTMED-CAU, BP, ITA, and FRA-CE (left) from IP, IC, and LP (right). Coordinate 2 (3.91%) does not reveal any discernible structure. Within LP, Albillo criollo and Aromatica Eufrosina were slightly divergent, while Bienmesabe tinto clustered with the rest.

The three-dimensional PCoA plot (Figure 10c) supports these observations. EASTMED-CAU, BP, ITA, and FRA-CE occupied the lower/internal space, while IP, IC, and LP appeared toward the front/upper space. Within LP, Albillo criollo, Aromatica Eufrosina, and Bienmesabe tinto were clearly separated, with the latter showing the greatest degree of differentiation from the other varieties included in the overall study.

4. Discussion

This study represents the first exhaustive genetic characterization of *Vitis vinifera* ssp. *vinifera* in LP using SSR markers. The results confirm that, despite the island's small size and a history marked by frequent volcanic disturbances, LP preserves a genetically diverse and unique grapevine germplasm. This genetic resource should be considered of high value for both the conservation of viticulture heritage and for its potential use in adaptation strategies against climate change.

4.1. Analysis of Grapevine Varieties

The study conducted on LP Island represents a comprehensive genetic survey of its cultivated grapevine population. A total of 96 vine accessions were collected across the island's principal viticultural zones: Northern, Hoyo de Mazo, and Fuencaliente. After genotyping using SSR markers and comparing against the TECNENOL database, 52 redundant profiles were removed. This level of redundancy (54.2%) is consistent with that of another IC: similar to HI (52.9%), slightly below LG (56.67%) and LZ (55%), but notably higher than FT (37.5%). Ultimately, 44 unique multilocus profiles (MP-SSR) were identified (Table S3). Among these, 28 were known varieties referenced in the TECNENOL and/or VIVC databases, while three remained unidentified. Notably, several profiles exhibited minor allelic variations when compared to their standard references. To distinguish between mutants and new varieties, the study adopted a similarity threshold of 87.5%, i.e., up to five allele differences across 20 SSR loci were considered mutations, whereas greater divergence indicated a new variety. This follows strategies previously outlined by Ibáñez et al. [66], Vélez [67], and Cabezas et al. [68].

The grapevine varieties on LP Island were found to originate from nine distinct regions. Spanish varieties, most genetically related to Heben, included Beba, De Rey, Ferral, Listán prieto (95% similarity), Mollar cano (97.5%), Palomino fino, and Vijariego blanco. Portuguese varieties, primarily related to Savagnin blanc, included Alfrocheiro (95%), Malvasia fina (97.5%), Molar, Samarrinho, and Verdelho branco. Other international varieties found included Trousseau noir and Flot rouge (France), Chasselas blanc (possibly Central European [69]), Malvasia Dubrovacka (likely Balkan [65]), Muscat of Alexandria (Greece), Muscat d'Hamburg (England), and Morskoi 94 (Ukraine), which corresponds to "Unknown No. 5" from Rodríguez-Torres [62]. The local IC varieties included Bermejuela, Forastera gomerae, Listan negro, Malvasia volcanica, and Torrontes volcanico. In addition, LP is home to local varieties such as Albillo criollo, Bienmesabe tinto (likely an interspecific hybrid [20]), Gual Mazo (erroneously listed as Italian in VIVC [51,63]), and Sabro. Two previously known color sports were also recorded: Malvasia di Sardegna rosada (Canary Island) and Malvasia fina roxa (Portugal). Three entirely new varieties were identified—Aromatica Eufrosina, Cagarruta de oveja, and Viñarda Rosada—bringing the total to 31 distinct varieties cultivated on LP.

Significant intra-varietal diversity was detected (Tables S1 and S3). Forty accessions showed such variation, leading to 20 unique MP-SSR profiles. Several variants corresponded to those found on other islands: Listán negro santanero (variation in VVS3-2, first described in LZ [47,49,50,61]), Listán prieto de Vega (VVS3-2 and VVMD28-1, from FT [49,61]), Malvasia blanca de Agulo (LG [50]), Malvasia volcánica cabezuda (LZ [47,61]), Mollar Bonilla (LZ [47,49]), and Diego de El Raso (VVS3-2 and ZAG64-2, LZ [47,61]). The study adheres to the VIVC [51] practice of naming somatic mutations (e.g., Pinot meunier from Pinot noir). Thus, 14 new names were proposed for detected mutants: Alfrocheiro palmero, Beba de Bienes, Malvasia de Mazo, Malvasia rosada de Breña, Gual de Fuencaliente, Gual Jeremias, Gual rosado de Armas, Molar de Bienes, Listan blanco menudo, Verdello palmero, Bujariego de Fuencaliente, Bujariego palmero, Chasselas palmero (a triallelic variant), and Albillo baboso palmero (which showed two SSR mutations and two cases of triallelism). Additionally, 8 misidentified accessions (marked in red) and 14 unknown to local growers (light green) were identified. One notable case was P-35, locally called Malvasia blanca (fuchsia in Table S1), a VIVC synonym for Alarije, commonly mistaken for

"Malvasia", although it matched Malvasia Dubrovacka genetically. The authors argue the synonym is better reserved for a true "Malvasia" [65].

The initial sampling was conducted shortly before the 2021 eruption of the Tajogaite volcano. Fourteen accessions were buried by lava or ash (Table S1), including representatives of De Rey, Flot rouge, Forastera gomerae, Muscat of Alexandria, Samarrinho, Trousseau noir, and Sabro, with no variants among them. Of the four variants affected, Malvasia volcanica cabezuda and Mollar Bonilla were collected and preserved previously on the LZ island. Only the sport known as Malvasia fina roxa was lost. Therefore, the impact of the eruption on grapevine biodiversity was much smaller than anticipated.

The authors propose several additions to the VIVC ampelographic catalog: (1) the three new variety names (in turquoise): Aromatica Eufrosina, Cagarruta de oveja, and Viñarda rosada; (2) the 14 newly named somatic variants (dark green); and (3) twelve new synonyms (brown), including Gallo (PN: Beba), Tintilla palmera (PN: Flot rouge), Almuñeco blanco (PN: Forastera gomera), Almuñeco negro and Muñeco (PN: Listan negro/mutation: Listán negro santanero), Tinta milrera (PN: Listan prieto/mutation: Listan prieto de Vega), Valenciana (PN: Mollar cano/mutation: Mollar Bonilla), Dulzal (PN: Morskoi 94), Moscatel antiguo (PN: Muscat of Alexandria), Listan blanco alto (PN: Palomino fino), Verdello grande (PN: Sabro), and Baboso blanco (PN: Samarrinho). These names were submitted to VIVC for inclusion in 2023 [48]. Hence LP Island's grapevine population demonstrates remarkable genetic diversity, with 31 distinct varieties and numerous somatic variants. This richness is of great relevance for both conservation and viticultural innovation, particularly considering climate stress and volcanic challenges. The authors emphasize the need to preserve and study these genotypes further, recognizing LP as a valuable genetic reservoir within the global *Vitis vinifera* L. landscape.

4.2. Genetic Structure of the Grapevine Population in La Palma

Before delving into the analysis of population genetic structure, three methodological clarifications are necessary. First, the program Structure 2.3 was used to assess population structure. This involved determining the most accurate distribution of grapevine individuals, with membership evaluated via the q statistic, which estimates the degree of genetic affiliation to each population. Based on q, individuals were classified as "pure" or "admixed." To ensure consistency and reduce analytical noise, admixed were excluded from downstream analyses (Principal Coordinates Analysis (PCoA), dendrograms, and phylogenetic trees) so as to better understand the genetic identity of LP's cultivated grapevines. Second, the use of 20 SSR markers generated up to 40 allelic data points per diploid sample. Although this allows for highly accurate classification, full precision in PCoA would require 40 dimensions, which is not feasible. Since graphical representations are limited to two or three dimensions, some data distortion is inevitable, particularly as sample size increases. This limitation, however, is widely recognized in the literature, and PCoA is typically used to interpret overall patterns rather than precise placement [70,71]. Third, pedigree and origin data for known varieties and their crosses were sourced primarily from the VIVC database [51].

Analysis using Structure 2.3 (Figure 3, Figures S2 and S3) on 31 non-redundant *Vitis vinifera* ssp. *vinifera* varieties revealed three ancestral populations with 89% confidence. Population 1 (POP1) included 15 varieties, predominantly of Spanish origin. Notable members include Beba, Mollar cano (97.5%), Vijariego blanco (95%), Ferral, and Sabro, all presumed descendants of the Spanish variety Heben [69]. Also grouped here were De Rey, Torrontes volcanico (IC), Gual Mazo (LP), Malvasia Dubrovacka (likely Balkan [65]), and the Ukrainian Morskoi 94. Several admixed with Spanish heritage also clustered here: Listan negro, Listan prieto (95%), Malvasia volcanica, and local LP varieties like

Cagarruta de oveja. This distribution underscores a dominant Iberian lineage in LP's grapevines. Population 2 (POP2) encompassed mostly Portuguese varieties or those with a strong Savagnin blanc ancestry. Pure accessions included Verdelho branco, Trousseau noir, Samarrinho, Molar, Alfrocheiro (95%), and the Canary Island Forastera gomerae (a Palomino fino × Verdelho branco cross). LP's Albillo criollo, Malvasia fina (97.5%), and the new variety Aromatica Eufrosina were also grouped here, suggesting Portuguese or broader European ancestry. The influence of Savagnin blanc, believed to have been introduced to northern Iberia via the Camino de Santiago [72], is evident throughout this cluster. Population 3 (POP3) brought together genetically divergent or ambiguous varieties. These included Flot rouge, a complex interspecific cross, Muscat Hamburg (a breeder-developed cross between Schiava grossa and Muscat of Alexandria), and LP's Bienmesabe tinto, which likely has interspecific origins. New variety Viñarda rosada also clustered here, lacking a known pedigree. Admixed varieties in this group include Chasselas blanc (likely Central European), Bermejuela (parent of Malvasia volcanica), and Muscat of Alexandria. These population assignments suggest a nuanced picture of LP's grapevine heritage. Varieties like Sabro, Cagarruta de oveja, and Gual Mazo are more strongly tied to Spanish ancestry. In contrast, Albillo criollo and Aromatica Eufrosina reflect connections to Portuguese or continental lineages. The positions of Bienmesabe tinto and Viñarda rosada remain speculative, pending further genetic and historical investigation.

The PCoA results, shown in Figure 4, further support this structure. Figure 4a, based on Nei's Genetic Distance, displays the clear separation of the three populations, each occupying distinct quadrants with 100% confidence. POP3, the most genetically distinct, showed higher internal diversity. This trend continues in Figure 4b (27.56% confidence) and Figure 4c (36.5%), where POP3 individuals are more widely dispersed. The greatest outliers included Flot rouge, Bienmesabe tinto, Muscat d'Hamburg, and Viñarda rosada, all of which exhibit complex or interspecific backgrounds.

In this way it can be hypothesized that LP's cultivated grapevine population appears to derive from two primary historical sources. The first is linked to the Iberian Peninsula (mainly Spain), likely introduced during the island's colonization under the Castilian Crown. The second reflects Portuguese influence, possibly through settlers from Madeira [73] or the historical introduction of Savagnin blanc-related cultivars. A third, less defined genetic input may stem from the introduction of Direct Producer Hybrids (DPH) during the phylloxera crisis or more recent introductions by professional breeders. Together, these findings highlight the unique and diverse nature of LP's grapevines and support their relevance for conservation, breeding, and further genetic study.

4.3. Relationship of the La Palma Grapevine Population with the Canary Islands

To explore the genetic relationship between the seven local grapevine varieties from LP and those from the rest of the IC, a reference group of 33 individuals from across the archipelago was analyzed. This included 12 from LZ, 11 from HI, 5 from LG, 4 generic IC types, and 1 Majorera variety from FT, along with the seven LP varieties under investigation. All individuals were assigned to specific population groups using a consistent methodology (Figures S4 and S5).

The genetic structure revealed five main populations (POP1–POP5), with Figure 5 displaying the distribution of 40 varieties. However, limited historical and genetic data on many of these cultivars, some recently described, restricted the depth of analysis. POP1 includes eight varieties largely associated with the "Malvasia" group. This cluster is mainly composed of individuals from LZ, including Malvasia volcanica, a known cross of Malvasia Dubrovacka and Bermejuela. Two other varieties in this group also contain "Malvasia" in their names, reflecting their genetic closeness [47,50]. Some accessions from LG were also present. POP2 is composed almost exclusively of HI varieties, with Verdello de El Hierro being the only one with documented parentage (Verijadiego × Alfrocheiro). The genetic data suggests a potential Portuguese or Central European connection in this group. POP3 holds nine varieties, most of Spanish origin. Many in this group are closely related to Vijariego blanco (also known as Diego in IC), which has Heben as a parent. Sabro, another variety with Heben lineage, appears as an admixed member. Two LP varieties, Cagarruta de oveja and Sabro, are also included here as admixed individuals indicating shared ancestry. POP4, the largest group with 11 accessions, includes varieties derived from unusual parentages or interspecific hybridizations. Notably, three pure LP varieties, Bienmesabe tinto, Viñarda rosada, and Aromatica Eufrosina, are included here. Other members such as Gual Mazo, Majorera (FT), Blanca de la granja del Cabildo (LZ), and Tesoro blanco (HI) have been characterized as genetic outliers in earlier studies [48–50]. POP5, comprising six members, cluster varieties are genetically related to Palomino fino, a Spanish cultivar. This group includes Listan negro (IC), Albillo criollo (LP), and Forastera gomerae (LG), all genetically linked through shared parentage. The study confirms a genetic connection between the Spanish variety Heben and the LP cultivars Cagarruta de oveja and Sabro. It also reinforces the close relationship among Bienmesabe tinto, Viñarda rosada, and Aromatica Eufrosina.

To enhance structural analysis, admixed individuals were excluded, yielding a refined dataset of 22 IC varieties. This initially resulted in an 86% assignment accuracy, which was improved to 100% following reassignment of misclassified varieties. Figure 6 presents Principal Coordinate Analysis (PCoA) outcomes. The 2D distribution (Figure 6a) reveals POP1 and POP5 as the most genetically distinct, supported by *F*st statistics and a 63.36% explained variance. POP2 remains distinct, while POP3 and POP4 form a more central, overlapping cluster. Figure 6b,c, illustrating two- and three-dimensional individual-level PCoA plots (21.88% and 30.78% goodness, respectively), confirm these trends. POP1, POP2, and POP5 appear more genetically cohesive, while POP3 and particularly POP4 show greater diversity. Bienmesabe tinto and Viñarda rosada stand out for their unique genetic profiles, likely due to distinct hybridization events.

4.4. Relationship of La Palma Population with the Global Diversity

This final section evaluates the distinctiveness of the LP grapevine population in relation to globally distributed varieties. The primary objective was to clarify the genetic identity and potential affinities of the seven local LP varieties. Two complementary strategies were employed: one based purely on genetic analysis and another incorporating a geographic component. In both approaches, the LP population was kept intact (unnormalized) to preserve individual variety behaviors, despite a slight decrease in assignment accuracy. The dataset used included 319 MP-SSR profiles from varieties originating in 23 countries, all resulting from natural crosses, regardless of known parentage.

In the genetic approach, the varieties were initially grouped into two ancestral populations (Figures S6, S7 and 7a). Population 1 (POP1), largely composed of Iberian Peninsula (IP) varieties, mostly Spanish, contained all LP accessions and members of the IC group. After removing admixed individuals (except Albillo criollo and Aromatica Eufrosina), POP1 had 106 accessions, while POP2 had 162, for a total of 268 varieties with 99% assignment accuracy. Reclassifying one misassigned variety raised accuracy to 100%. For more precise analysis, POP1 was subdivided into IC and LP clusters, which slightly lowered assignment accuracy to 88%. The geo-genetic approach normalized the dataset by removing misallocated varieties in the VIVC database, resulting in 197 total profiles: 190 from global sources and the 7 LP accessions. These were divided into seven populations: EASTMED-CAU, BP,

ITA, FRA-CE, IP, IC, and LP. Assignment accuracy improved significantly from 60% to 87% following this normalization.

Circular dendrograms (Figures 7b and 9a) illustrated that LP varieties formed a distinct group within the broader IC and IP populations. One notable exception was Bienmesabe tinto, which appeared as an outlier in the genetic dendrogram (Figure 7b), but clustered with other LP accessions in the geographic dendrogram (Figure 9a). These trends were further supported by phylogenetic trees (Figures 7c and 9b), where LP accessions consistently retained a distinct genetic position. Principal Coordinate Analysis (PCoA) plots by population (Figures 8a and 10a) showed coherent clustering. In the genetic approach, 97.85% of goodness was explained, while in the geographic method, the explained goodness was lower at 64.17%, likely due to the presence of more population groups. Spanish origin populations clustered on the left of the PCoA space, while non-Spanish groups (POP2) appeared on the right. The geographic analysis further split POP2 into four subpopulations, with EASTMED-CAU emerging as the most genetically distant. At the individual level (Figures 8b,c and 10b,c), clustering patterns were similar. While Bienmesabe tinto grouped with LP accessions in 2D space, its separation in 3D suggested a unique origin. The remaining six LP varieties followed two trends: Cagarruta de oveja, Gual Mazo, and Sabro clustered with Spanish varieties (implying Iberian ancestry), whereas Viñarda rosada, Albillo criollo, and Aromatica Eufrosina aligned more closely with non-Iberian groups, and showing Aromatica Eufrosina a strong divergence.

In conclusion, (1) LP varieties are genetically distinct from IC, IP, and global varieties; (2) Bienmesabe tinto is likely an interspecific hybrid; (3) some LP varieties show connections to Central European lineages; and (4) others have probable Spanish ancestry, which could be from the progenitor Heben or not.

5. Conclusions

This study represents the first comprehensive genetic characterization of the cultivated grapevine population on LP island, using a robust SSR-based methodology. The SSR kit previously used on other IC has again proven efficient and effective.

Among the 96 accessions collected across the island, 44 unique MP-SSRs were identified, including 3 entirely new varieties previously unknown to science: Cagarruta de oveja, Viñarda rosada, and Aromatica Eufrosina. In addition, 14 new mutations exclusive to this island were identified, including 3 cases of triallelism. Additionally, 8 accessions were determined to be labeling or sampling errors, and 14 accessions unknown by either growers or technicians could be identified.

LP's contribution also holds lexicographic value. Thus, inclusion in the VIVC is proposed for (1) the three new varieties PN (Aromatica Eufrosina, Cagarru-ta de oveja, Viñarda rosada), along with their SSR profiles; (2) the 14 novel mutant names listed above; and (3) 12 new synonyms drawn from local viticultural tradition, referring either to existing varieties or specific mutations.

Of the 14 accessions affected by lava or ash during the Tajogaite volcanic eruption, only 1 was permanently lost, a mutated individual of Malvasia fina roxa, making the overall biodiversity loss far lower than initially feared.

Regarding LP's genetic structure, results support the existence of seven representative varieties. Four had already been described while three are new, representing a major finding. These varieties likely originate from Spanish or broader European lineages, with genetic traits suggesting unique evolutionary paths. These genotypes represent an irreplaceable component of LP's viticultural heritage and should be prioritized for conservation, morphological and oenological characterization, and integration into principal germplasm collections. Their genetic and adaptive uniqueness highlights the need to incorporate them into regional and national strategies for plant genetic resource conservation.

Supplementary Materials: The following supporting information can be downloaded at: https://www.mdpi.com/article/10.3390/horticulturae11080983/s1, Table S1. Information on 96 accessions from La Palma (Original and Conclusive); Table S2. List of 20 primers used for the amplification of the selected microsatellite regions. Characteristics; Figure S1. Approximation of the genomic SSR map used in this study. Consensus location of each of the regions selected for molecular characterization; Table S3. Unique molecular profile of 44 accessions (and 2 sport) corresponding to 31 varieties and 2 sport (color mutations) collected during the La Palma Island prospection. International SSR coincides with the SSR of TECNENOL; Table S4. Statistical characterization of the twenty microsatellite markers used in this study; Figure S2. The four steps of the graphical method of Evanno et al. [56], allowing the estimation of the true number of ancestral K groups for a population with 31 individuals from La Palma Island; Figure S3. Genetic structure of La Palma Island population. Distribution K = 3 (Individuals belonging to each group or population). Details of the ratio of pure and admixed individuals according to the value of q (pure ($q \geq 85\%$) and admixed ($q < 85\%$)); Figure S4. The four steps of the graphical method of Evanno et al. [56], allowing the estimation of the true number of ancestral K groups for a population with 40 individuals from Canary Islands collection (IC including La Palma Island); Figure S5. Genetic structure of the Canary Islands population (40 varieties). Distribution K = 5 (Individuals belonging to each group or population). Details of the ratio of pure and admixed individuals according to the value of q (pure ($q \geq 85\%$) and admixed ($q < 85\%$)). Population structure; Figure S6. The four steps of the graphical method of Evanno et al. [56], allowing the estimation of the true number of ancestral K groups for a population with 319 individuals from the TECNENOL database (including La Palma Island); Figure S7. Genetic structure of the world population. Distribution K = 2 (Individuals belonging to each group or population). Detail the proportion of pure and admixed individuals as a function of q value. Nationalities that make up pure and admixed groups; Figure S8. Genetic structure of the world population (319 individuals). Distribution in 7 geographical areas. Detail of the ratio of well-assigned (pure) and misassigned (admixed) individuals. Nationalities that make up each of the groups: EASTMED-CAU (Algeria, Cyprus, Georgia, Israel, Lebanon, Tunisia and Turkey), BP (Bosnia and Herzegovina, Bulgaria, Croatia, Greece, Serbia, Slovenia and Montenegro), ITA (Italy), FRA-CEU (Austria, France, Germany, Hungary, Switzerland and Ukraine), IP (Spain and Portugal), IC (Canary archipelago) and LP (La Palma island).

Author Contributions: Contributions: Conceptualization, Q.L.-Y., F.F. and L.D. methodology, Q.L.-Y., L.D. and F.F.; software, Q.L.-Y., L.D. and F.F.; validation, Q.L.-Y. and L.D.; formal analysis, Q.L.-Y. investigation, Q.L.-Y.; resources, Q.L.-Y. and L.D.; data curation, Q.L.-Y. and L.D.; writing—original draft preparation, Q.L.-Y. and F.F.; writing—review and editing, L.D. and Q.L.-Y.; visualization, J.M.C. and F.Z.; supervision, F.F., Q.L.-Y., L.D., J.M.C. and F.Z. All authors have read and agreed to the published version of the manuscript.

Funding: This research was funded by Consejería de Agricultura, Ganadería, Pesca y Soberanía Alimentaria del Cabildo Insular de la Isla de La Palma at the request of Consejo Regulador de la Denominación de Origen Vinos de La Palma.

Data Availability Statement: Data not available because it is confidential.

Acknowledgments: The authors wish to express their gratitude to the Consejería de Agricultura, Ganadería, Pesca y Soberanía Alimentaria del Cabildo Insular de la Isla de La Palma for their interest in supporting a study of this nature. In particular, they thank Elías Bienes, Eva Hernández, José Adrían Hernández and Elías Bienes for being the driving force behind the project. They also appreciate the contributions of Adalberto Martín representative of the Regulatory Council of the Protected Designation of Origin "Vinos de La Palma." Additionally, the authors thank Carlos Lozano, Asociación Técnica Canaria de Enología president. Finally, sincere thanks are extended to Rosa

Pastor, Braulio Esteve-Zarzoso, Laia Fañanás, Cristina Domènech, and Iris Ginés for their invaluable logistical support in the completion of this work.

Conflicts of Interest: The authors declare no conflicts of interest.

References

1. Sancho-Galán, P.; Amores-Arrocha, A.; Palacios, V.; Jiménez-Cantizano, A. Preliminary Study of Somatic Variants of Palomino Fino (*Vitis vinifera* L.) Grown in a Warm Climate Region (Andalusia, Spain). *Agronomy* **2020**, *10*, 654. [CrossRef]
2. Sancho-Galán, P.; Amores-Arrocha, A.; Palacios, V.; Jiménez-Cantizano, A. Genetical, Morphological and Physicochemical Characterization of the Autochthonous Cultivar 'Uva Rey' (*Vitis vinifera* L.). *Agronomy* **2019**, *9*, 563. [CrossRef]
3. van Leeuwen, C.; Destrac-Irvine, A.; Dubernet, M.; Duchêne, E.; Gowdy, M.; Marguerit, E.; Pieri, P.; Parker, A.; de Rességuier, L.; Ollat, N. An Update on the Impact of Climate Change in Viticulture and Potential Adaptations. *Agronomy* **2019**, *9*, 514. [CrossRef]
4. Wolkovich, E.M.; García de Cortázar-Atauri, I.; Morales-Castilla, I.; Nicholas, K.A.; Lacombe, T. From Pinot to Xinomavro in the world's future wine-growing regions. *Nat. Clim. Change* **2018**, *8*, 29–37. [CrossRef]
5. Sancho-Galán, P.; Amores-Arrocha, A.; Palacios, V.; Jiménez-Cantizano, A. Identification and characterization of white grape varieties autochthonous of a warm climate region (Andalusia, Spain). *Agronomy* **2020**, *10*, 205. [CrossRef]
6. Arco, M.J.; Atiénzar, E.; Rosario, M.C.; Arco, M.M.; González, C.; Arco, M.C. El Menceyato de Icod en el poblamiento de Tenerife: D.; Gaspar, Las Palomas y Los Guanches. Sobre el poblamiento y las estrategias de alimentación vegetal entre los Guanches. *Eres* **2000**, *9*, 67–129.
7. This, P.; Lacombe, T.; Thomas, M.R. Historical origins and genetic diversity of wine grapes. *Trends Genet.* **2006**, *22*, 511–519. [CrossRef] [PubMed]
8. Wikimedia Commons. Macaronesia Location. Available online: https://commons.wikimedia.org/w/index.php?curid=33566121 (accessed on 6 May 2025).
9. NASA Earth Observatory. Canary Islands. Available online: https://www.flickr.com/photos/gsfc/6630087415/ (accessed on 9 May 2025).
10. GEVIC (Gran Enciclopedia Virtual Islas Canarias). Fisiografía de Canarias. Relieve. Available online: http://www.gevic.net/info/contenidos/mostrar_contenidos.php?idcat=22&idcap=91&idcon=529 (accessed on 5 May 2025).
11. Canary Wine. Isla de La Palma. Available online: https://www.canarywine.com/isla-de-la-palma/ (accessed on 5 May 2025).
12. Climate Data. Clima La Palma (España). Available online: https://es.climate-data.org/europe/espana/la-palma-10272/ (accessed on 6 May 2025).
13. Boletín Oficial del Estado (BOE). Ley 25/1994, de 12 de Julio, Por la Que Se Incorpora al Derecho Español la Directiva 91/680/CEE, Sobre el Sistema Común Del Impuesto Sobre el Valor Añadido. BOE núm. 161, pp. 21568–21576. Available online: https://www.boe.es/boe/dias/1994/07/05/pdfs/A21568-21576.pdf (accessed on 5 May 2025).
14. Vinos La Palma. Subzona Norte. Available online: https://www.vinoslapalma.com/viticultura/zonas-de-produccion/subzona-norte.html (accessed on 5 May 2025).
15. Vinos La Palma. Subzona Hoyo de Mazo. Available online: https://www.vinoslapalma.com/viticultura/zonas-de-produccion/subzona-hoyo-de-mazo.html (accessed on 5 May 2025).
16. Vinos La Palma. Subzona Fuencaliente. Available online: https://www.vinoslapalma.com/viticultura/zonas-de-produccion/subzona-fuencaliente.html (accessed on 5 May 2025).
17. Topographic-map.com. Mapa topográfico de La Palma. Available online: https://es-ar.topographic-map.com/map-c3c3cz/La-Palma/ (accessed on 12 May 2025).
18. Vinos La Palma. Zonas de Producción Vitícola. Available online: https://www.vinoslapalma.com/viticultura/zonas-de-produccion.html (accessed on 12 May 2025).
19. Wikipedia. Erupción Volcánica de La Palma de. 2021. Available online: https://es.wikipedia.org/wiki/ErupciÃşn_volcÃąnica_de_La_Palma_de_2021 (accessed on 12 May 2025).
20. Marsal, G.; Mendez, J.J.; Mateo-Sanz, J.M.; Ferrer, S.; Canals, J.M.; Zamora, F.; Fort, F. Molecular characterization of *Vitis vinifera* L. local cultivars from volcanic areas (the Canary Islands and Madeira) using SSR markers. *OENO One* **2019**, *53*, 667–680. [CrossRef]
21. Vinos La Palma. Variedades de Uva. Available online: https://www.vinoslapalma.com/viticultura/variedades-de-uva.html (accessed on 24 May 2025).
22. Mendez, J.J. Acerca del Canary Wine. In *Compendio de la Vitivinicultura del Archipiélago Canario*, 2nd ed.; Asociación de Viticultores y Bodegueros de Canarias AVIBO, Ed.; Centro de la Cultura Popular Canaria (CCPC): Canary Islands, Spain, 2024.
23. GEVIC (Gran Enciclopedia Virtual Islas Canarias). Fisiografía de Canarias. Erupciones Históricas en Canarias. Available online: https://www.gevic.net/info/contenidos/mostrar_contenidos.php?idcomarca=-1&idcon=716&idcap=91&idcat=22 (accessed on 12 May 2025).

24. Instituto Geográfico Nacional (IGN). Descripción Geológica de La Palma. Available online: https://www.ign.es/web/resources/sismologia/tproximos/sismotectonica/pag_sismotectonicas/can_la_palma_en.html (accessed on 12 May 2025).
25. Carracedo, J.C.; Troll, V.R.; Day, J.M.; Junca, M.A.; Soler, V.; Deegan, F.M.; Pérez-Torrado, F.J.; Gisbert, G.; Gazel, E.; Rodríguez-González, A.; et al. The 2021 eruption of the Cumbre Vieja volcanic ridge on La Palma, Canary Islands. *Geol. Today* **2022**, *38*, 94–107. [CrossRef]
26. Marca Canaria. Erupciones Históricas en Canarias. Available online: https://marcacanaria.com/erupciones-historicas-en-canarias/ (accessed on 12 May 2025).
27. RTVE. Erupción Volcánica en La Palma. Finaliza La Erupción del Volcán de La Palma tras 85 Días de Actividad. RTVE Noticias. Available online: https://www.rtve.es/noticias/20211225/finaliza-erupcion-volcan-palma-tras-85-dias-actividad/2244006.shtml (accessed on 12 May 2025).
28. Vinos La Palma. Estadísticas. Available online: https://www.vinoslapalma.com/bodegas/estadisticas.html#:~:text=2019%20561%20937%2019%20645,982%20Buena (accessed on 24 May 2025).
29. Vinos La Palma. La Vendimia 2024 en la Denominación de Origen Protegida Vinos La Palma Avanza Con Bajos Rendimientos y Marcada Por Condiciones Climáticas Adversas. Available online: https://vinoslapalma.com/noticias-vinos-la-palma/93-actualidad/420-la-vendimia-2024-en-la-denominacion-de-origen-protegida-vinos-la-palma-avanza-con-bajos-rendimientos-y-marcada-por-condiciones-climaticas-adversas.html#:~:text=En%20cuanto%20a%20la%20tipologÃ■a,de%20uva%20tinta (accessed on 24 May 2025).
30. Vinos La Palma. Informe de Daños Viñedos y Vendimia 2021. Available online: https://www.vinoslapalma.com/bodegas/vendimia/ano-2021.html (accessed on 24 May 2025).
31. Vinos La Palma. Informe Valoración Daños Erupción Volcánica y Vendimia. Enero 2022. Available online: https://www.vinoslapalma.com/files/noticias/danos-vendimia-2021.pdf (accessed on 24 May 2025).
32. Vinos La Palma. Tipos de Vino. Available online: https://www.vinoslapalma.com/bodegas/tipos-de-vinos.html#:~:text=Vinos%20de%20tea (accessed on 24 May 2025).
33. Fort, F.; Hayoun, L.; Valls, J.; Canals, J.M.; Arola, L.; Zamora, F. A new and simple method for rapid extraction and isolation of high-quality RNA from grape (*Vitis vinifera*) berries. *J. Sci. Food Agric.* **2008**, *88*, 179–184. [CrossRef]
34. Marsal, G.; Baiges, I.; Canals, J.M.; Zamora, F.; Fort, F. A fast, efficient method for extracting DNA from leaves, stems, and seeds of *Vitis vinifera* L. *Am. J. Enol. Vitic.* **2011**, *62*, 376–381. [CrossRef]
35. Marsal, G.; Boronat, N.; Canals, J.M.; Zamora, F.; Fort, F. Comparison of the efficiency of some of the most usual DNA extraction methods for woody plants in different tissues of *Vitis vinifera* L. *J. Int. Sci. Vigne Vin.* **2013**, *47*, 227–237. [CrossRef]
36. Thomas, M.R.; Scott, N.S. Microsatellite repeats in grapevine reveal DNA polymorphisms when analyzed as sequence-tagged sites (STSs). *Theor. Appl. Genet.* **1993**, *86*, 985–990. [CrossRef]
37. Bowers, J.E.; Dangl, G.S.; Vignani, R.; Meredith, C.P. Isolation and characterization of new polymorphic simple sequence repeat loci in grape (*Vitis vinifera* L.). *Genome* **1996**, *39*, 628–633. [CrossRef]
38. Bowers, J.E.; Dangl, G.S.; Meredith, C.P. Development and characterization of additional microsatellite DNA markers for grape. *Am. J. Enol. Vitic.* **1999**, *50*, 243–246. [CrossRef]
39. Sefc, K.M.; Regner, F.; Turetschek, E.; Glössl, J.; Steinkellner, H. Identification of microsatellite sequences in *Vitis riparia* and their applicability for genotyping of different *Vitis* species. *Genome* **1999**, *42*, 367–373. [CrossRef] [PubMed]
40. Scott, K.D.; Eggler, P.; Seaton, G.; Rosseto, M.; Abblet, E.M.; Lee, L.S.; Henry, R.J. Analysis of SSRs derived from grape ESTs. *Theor. Appl. Genet.* **2000**, *100*, 723–726. [CrossRef]
41. Lefort, F.; Kyvelos, C.; Zervou, M.; Edwards, K.; Roubelakis-Angelakis, K. Characterization of new microsatellite loci from *Vitis vinifera* and their conservation in some *Vitis* species and hybrids. *Mol. Ecol. Resour.* **2002**, *2*, 20–21. [CrossRef]
42. Cipriani, G.; Spadotto, A.; Jurman, I.; Di Gaspero, G.; Crespan, M.; Meneghetti, S.; Frare, E.; Vignani, R.; Cresti, M.; Morgante, M.; et al. The SSR-based molecular profile of 1005 grapevine (*Vitis vinifera* L.) accessions uncovers new synonymy and parentages and reveals a large admixture amongst varieties of different geographic origins. *Theor. Appl. Genet.* **2010**, *121*, 1569–1585. [CrossRef]
43. Dalbó, M.A.; Ye, G.N.; Weeden, N.F.; Steinkellner, H.; Sefc, K.M.; Reisch, B.I. A gene-controlling sex in grapevines is placed on a molecular marker-based genetic map. *Genome* **2000**, *43*, 333–340. [CrossRef]
44. This, P.; Jung, A.; Boccacci, P.; Borrego, J.; Botta, R.; Costantini, L.; Crespan, M.; Dangl, G.S.; Eisenheld, C.; Ferreira-Monteiro, F.; et al. Development of a standard set of microsatellite reference alleles for the identification of grape cultivars. *Theor. Appl. Genet.* **2004**, *109*, 1448–1458. [CrossRef]
45. Marsal, G.; Mateo, J.M.; Canals, J.M.; Zamora, F.; Fort, F. SSR analysis of 338 accessions planted in Penedès (Spain) reveals 28 unreported molecular profiles of *Vitis vinifera* L. *Am. J. Enol. Vitic.* **2016**, *67*, 466–470. [CrossRef]
46. Marsal, G.; Bota, J.; Martorell, A.; Canals, J.M.; Zamora, F.; Fort, F. Local cultivars of *Vitis vinifera* L. in Spanish islands: Balearic Archipelago. *Sci. Hortic.* **2017**, *226*, 122–132. [CrossRef]

47. Fort, F.; Marsal, G.; Mateo-Sanz, J.M.; Pena, V.; Canals, J.M.; Zamora, F. Molecular characterisation of the current cultivars of *Vitis vinifera* L. in Lanzarote (Canary Islands, Spain) reveals nine individuals which correspond to eight new varieties and two new sports. *OENO One* **2022**, *56*, 281–295. [CrossRef]
48. Fort, F.; Lin-Yang, Q.; Suárez-Abreu, L.R.; Sancho-Galán, P.; Canals, J.M.; Zamora, F. Study of molecular biodiversity and population structure of *Vitis vinifera* L. ssp. vinifera on the volcanic island of El Hierro (Canary Islands, Spain) by using microsatellite markers. *Horticulturae* **2023**, *9*, 1297. [CrossRef]
49. Fort, F.; Lin-Yang, Q.; Valls, C.; Sancho-Galán, P.; Canals, J.M.; Zamora, F. Characterisation and identification of vines from Fuerteventura (Canary Volcanic Archipelago (Spain)) using simple sequence repeat markers. *Horticulturae* **2023**, *9*, 1301. [CrossRef]
50. Fort, F.; Lin-Yang, Q.; Valls, C.; Sancho-Galán, P.; Canals, J.M.; Zamora, F. Analysis of the diversity presented by *Vitis vinifera* L. in the volcanic island of La Gomera (Canary Archipelago, Spain) using simple sequence repeats (SSRs) as molecular markers. *Horticulturae* **2024**, *10*, 14. [CrossRef]
51. Vitis International Variety Catalogue (VIVC). Available online: https://vivc.de (accessed on 22 May 2025).
52. Paetkau, D.; Calvert, W.; Stirling, I.; Strobeck, C. Microsatellite analysis of population structure in Canadian polar bears. *Mol. Ecol.* **1995**, *4*, 347–354. [CrossRef]
53. Peakall, R.; Smouse, P.E. GenAlEx 6.5: Genetic analysis in Excel. Population genetic software for teaching and research-an update. *Bioinformatics* **2012**, *28*, 2537–2539. [CrossRef]
54. Pritchard, J.K.; Stephens, M.; Donnelly, P. Inference of population structure using multilocus genotype data. *Genetics* **2000**, *155*, 945–959. [CrossRef]
55. Falush, D.; Stephens, M.; Pritchard, J.K. Inference of population structure using multilocus genotype data: Linked loci and correlated allele frequencies. *Genetics* **2003**, *164*, 1567–1587. [CrossRef] [PubMed]
56. Evanno, G.; Regnaut, S.; Goudet, J. Detecting the number of clusters of individuals using the software structure: A simulation study. *Mol. Ecol.* **2005**, *14*, 2611–2620. [CrossRef] [PubMed]
57. Paetkau, D.; Slade, R.; Burden, M.; Estoup, A. Genetic assignment methods for the direct, real-time estimation of migration rate: A simulation-based exploration of accuracy and power. *Mol. Ecol.* **2004**, *13*, 55–65. [CrossRef]
58. Three-Dimensional Plotting in Matplotlib (Python Data Science Handbook). Available online: https://jakevdp.github.io/PythonDataScienceHandbook/04.12-three-dimensional-plotting.html (accessed on 20 May 2025).
59. Kumar, S.; Stecher, G.; Tamura, K. MEGA7: Molecular Evolutionary Genetics Analysis version 7.0 for bigger datasets. *Mol. Biol. Evol.* **2016**, *33*, 1870–1874. [CrossRef]
60. Saitou, N.; Nei, M. The neighbor-joining method: A new method for reconstructing phylogenetic threes. *Mol. Biol. Evol.* **1987**, *4*, 406–425. [CrossRef]
61. Fort, F.; Suárez-Abreu, L.R.; Lin-Yang, Q.; Deis, L.; Canals, J.M.; Zamora, F. Preliminary Clonal Characterization of Malvasia volcanica and Listan prieto by Simple Sequence Repeat (SSR) Markers in Free-Phylloxera Volcanic Vineyards (Lanzarote and Fuerteventura (Canary Island, Spain)). *Horticulturae* **2025**, *11*, 823. [CrossRef]
62. Rodriguez-Torres, I. *Variedades de vid Cultivadas en Canarias. Descriptores Morfológicos. Caracterización Morfológica, Molecular, Agronómica y Enológica*, 2nd ed.; Instituto Canario de Investigaciones Agrarias. Gobierno de Canarias: Santa Cruz de Tenerife, Spain, 2018. Available online: https://www.icia.es/icia/download/Publicaciones/Variedades_Vid_Canarias.pdf (accessed on 4 March 2025).
63. Zerolo, J.; Cabello, F.; Espino, A.; Borrego, J.; Ibañez, J.; Rodriguez-Torres, I.; Muñoz-Organero, G.; Rubio, C.; Hernández, M. *Variedades de Vid de Cultivo Tradicional en Canarias*, 1st ed.; Instituto Canario de Calidad Agroalimentaria. Gobierno de Canarias: Santa Cruz de Tenerife, Spain, 2006; ISBN 978-84-606-3977-0.
64. Bacilieri, R.; Lacombe, T.; Cunff, L.L.; Di Vecchi-Staraz, M.; Laucou, V.; Genna, B.; Perós, J.P.; This, P.; Boursiquot, J.M. Genetic structure in cultivated grapevines is linked to geography and human selection. *BMC Plant Biol.* **2013**, *13*, 25. [CrossRef]
65. Fort, F.; Suárez-Abreu, L.R.; Lin-Yang, Q.; Deis, L.; Canals, J.M.; Zamora, F. Origin and Possible Members of the 'Malvasia' Family: The New Fuencaliente de La Palma Hypothesis on the True 'Malvasia'. *Horticulturae* **2025**, *11*, 561. [CrossRef]
66. Ibañez, J.; De Andrés, M.T.; Molino, A.; Borrego, J. Genetic study of key Spanish grapevine varieties using microsatellite analysis. *Am. J. Enol. Vitic.* **2003**, *54*, 22–30. [CrossRef]
67. Vélez, M.D.; Ibáñez, J. Evaluation of the uniformity and stability of Microsatellite markers in grapevine. *Acta Hortic.* **2009**, *827*, 163–168. [CrossRef]
68. Cabezas, A.; Ibañez, J.; Lijavetzky, D.; Vélez, D.; Bravo, G.; Rodríguez, V.; Carreño, I.; Jermakow, A.M.; Carreño, J.; Ruiz-García, L.; et al. A 48 SNP set for grapevine cultivar identification. *BMC Plant Biol.* **2011**, *11*, 153. [CrossRef] [PubMed]
69. Galet, P. *Dictionnaire Encyclopédique des Cépages*, 1st ed.; Hachette: Paris, France, 2000.
70. Emanuelli, F.; Lorenzi, S.; Grzeskowiak, L.; Catalano, V.; Stefanini, M.; Troggio, M.; Myles, S.; Martinez-Zapater, J.M.; Zyprian, E.; Moreira, F.M.; et al. Genetic diversity and population structure assessed by SSR and SNP markers in a large germplasm collection of grape. *BMC Plant Biol.* **2013**, *13*, 39. [CrossRef] [PubMed]

71. Dong, Y.; Duan, S.; Xia, Q.; Liang, Z.; Dong, X.; Margaryan, K.; Musayev, M.; Goryslavets, S.; Zdunić, G.; Bert, P.-F.; et al. Dual domestications and origin of traits in grapevine evolution. *Science* **2023**, *379*, 892–901. [CrossRef] [PubMed]
72. Casanova, J.; Mozas, P.; Ortiz, J.M. Ampelography and microsatellite DNA analysis of autochthonous and endangered grapevine cultivars in the province of Huesca (Spain). *Span. J. Agric. Res.* **2011**, *9*, 790–800. [CrossRef]
73. Otamendi, J.J. La Malvasía Atlántica. Dudas, Errores y algunas Certezas. Available online: https://dolanzarote.com/wp-content/uploads/2011/06/Malvasia-atlantica.pdf (accessed on 6 March 2025).

MDPI AG
Grosspeteranlage 5
4052 Basel
Switzerland
Tel.: +41 61 683 77 34

Horticulturae Editorial Office
E-mail: horticulturae@mdpi.com
www.mdpi.com/journal/horticulturae

www.ingramcontent.com/pod-product-compliance
Lightning Source LLC
Chambersburg PA
CBHW061449210926
51850CB00008B/1655